Noel W. Murray · Georg Thierauf

Tables for the Design and Analysis of Stiffened Steel Plates

Entwurfs- und Berechnungstabellen für ausgesteifte Stahlplatten

Noel W. Murray · Georg · Thierauf

Tables for the Design and Analysis of Stiffened Steel Plates

Entwurfs- und Berechnungstabellen für ausgesteifte Stahlplatten

Friedr. Vieweg & Sohn   Braunschweig/Wiesbaden

CIP-Kurztitelaufnahme der Deutschen Bibliothek

Murray, Noel W.:
Tables for the design and analysis of stiffened steel
plates = Entwurfs- und Berechnungstabellen für
ausgesteifte Stahlplatten / Noel W. Murray; Georg
Thierauf. — Braunschweig; Wiesbaden: Vieweg, 1981.
  ISBN-13: 978-3-528-08673-2   e-ISBN-13: 978-3-322-86307-2
  DOI: 10.1007/978-3-322-86307-2
NE: Thierauf, Georg:

Noel W. Murray ist Professor of Civil Engeneering an der Monash University
in Clayton, Victoria / Australia

Georg Thierauf ist Professor an der Gesamthochschule Essen / Bundesrepublik
Deutschland

Umschlagentwurf: Peter Neitzke, Köln

Buchbinderische Verarbeitung: W. Langelüddecke, Braunschweig

ISBN-13: 978-3-528-08673-2

Contents · Inhalt

A Theory
General Background and Use of the Tables

Table of Contents · Part A

Preface

Stiffened steel plates have been used for many years especially in the fields of bridges, ships and towers. They are very efficient components of a structure, having both a high strength-to-weight ratio and a pleasing appearance. However, in some cases they have a tendency to buckle in one mode or even two simultaneous modes. In existing codes of practice designers have been supplied with many comprehensive rules so that they can calculate failure loads (collapse conditions) and the loads at which first yield occurs (serviceability conditions). A key calculation in each of these determinations is that which gives the critical stress. However, the critical stress depends upon many features such as the boundary conditions along the sides and at the ends and the geometry of the cross-section. A short stiffened plate may buckle locally whereas a longer panel with the same cross-section may buckle as a wide Euler column.

The labour involved in checking individual trial designs is often formidable. It is the aim of this book to try to ease this burden for designers. They will find that it is a simple matter to use this book as if it were a catalogue of sectional properties. Indeed the tables contained herein present the usual section properties but they also indicate the stresses at which local buckling occurs and the panel length beyond which global (i.e. Euler type) buckling governs.

The design data presented here has been derived from a finite element program which uses the finite strip method of analysis. Details of this method are presented here for the sake of completeness. It is not necessary for designers to understand all of these details but Sections 1, 5.2 and 7 to 8.4 should be read so that the tables can be more fully understood.

Although it is difficult to compare results from a computer which are based upon purely theoretical considerations and those from a code of practice, where of necessity there must be some empiricism, nevertheless this has been attempted in Examples 8.1 and 8.2. The first shows how the tables of this book may be used in conjunction with a code of practice and the second gives a comparison between a theoretical critical stress from the tables of this book and that obtained from a code of practice. Not only are the results of that comparison interesting but readers will see that the amount of calculation is many times less in the former case.

The authors would especially like to thank Frau Christel Hausmann for the meticulous manner in which she has carried out the work of data preparation and the presentation of results. Frau Monika Mehl is thanked for her work preparing the diagrams. Thanks are also due to Herr Artur Senftleben for checking the manuscript and to Frau Ulrike Riewe and Mrs. Pam Smith for preparing the manuscript. The authors and collaborators have carefully checked the information contained in this book but there may still be errors in it. The authors and the publisher cannot accept responsibility for any losses incurred by users of this book, either through errors or interpretation of the information. The work described has been sponsored in part by Deutsche Forschungsgemeinschaft. All numerical results have been obtained using the computers at Universität Düsseldorf, Germany and Monash University, Australia.

Noel W. Murray.
Georg Thierauf.

1 Introduction

The problem of the elastic stability of stiffened plates
has many special features which do not occur in the stability of struts and frameworks. The differences can be
summarized by referring to Fig. 1. In the latter cases
it is found from first order theory (second order theory
gives almost the same result) that there is a horizontal
asymptote at the buckling load and the behaviour of a strut
with some initial imperfection δ_o closely follows the
hyperbola shown in Fig.1(a).

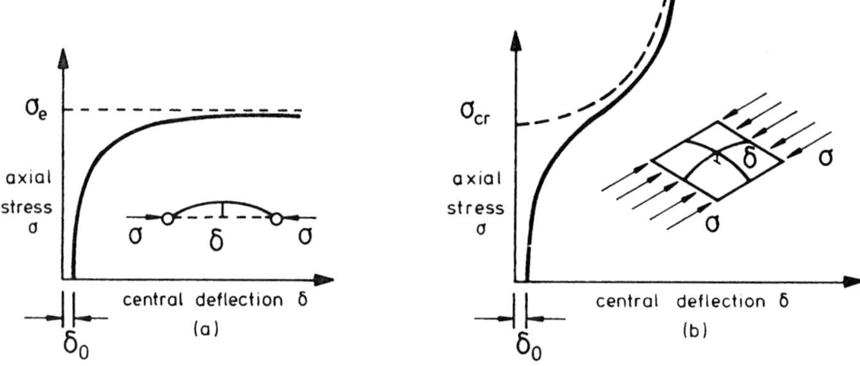

Fig. 1 Comparison of strut and plate behaviour

The value of the critical stress, in this case the Euler
stress, σ_e , and the mode of buckling can be found by
solving the eigenvalue problem

$$(\underline{K} + \sigma_e \underline{K}^G)\ \underline{r} = \underline{O} \tag{1}$$

where $\underline{K}$ is the usual stiffness matrix, $\underline{K}^G$ is the geometric
stiffness matrix [1] and $\underline{r}$ is a vector which defines the
geometry of the buckling mode. One must be careful to en-
sure that the smallest value of σ_e which satisfies equation
(1) is obtained. The buckling shape of the strut (or
structure) is obtained as part of the solution of equation
(1) but the magnitude of the buckles cannot be evaluated

from an eigenvalue problem. It is, of course, understood
that the boundary conditions, i.e., the conditions of
restraint at the ends of the strut, influence both
the value of the buckling stress and the buckling mode.

In the case of a simple plate the asymptote is often a
rising curve (Fig. 1(b)). The derivation of this asymptote
requires the application of large deflection theory[2] to a
plate with zero initial imperfection. Because of the
non-linear nature of this problem, which involves second-
order theory, solutions are obtained by iterative
procedures, such as the Newton-Raphson method. However, by
using linear (i.e., first order) theory the values of the
critical stress σ_{cr} and the buckling mode may be found.
This requires the solution of an equation of the same form
as equation(1). For design purposes the critical stress at
which an isolated and perfectly flat plate buckles is a
valuable piece of information and it assumes the same
importance for the designer as the Euler stress and the
yield stress.

The conditions along the boundaries strongly influence the
behaviour of the plate. Along the edges it is necessary to
specify both the in-plane and the out-of-plane conditions.
For example, an edge may be pinned and at the same time
prevented from moving inwards or it may be free-free, i.e.,
zero moment is applied and the edge may deflect freely out
of the original plane of the plate. Various combinations of
in-plane and out-of-plane restraint may occur in practice.

In the case of a stiffened plate which consists of an
assembly of flat plates the designer will need to know the
value of the stress at which so-called <u>local buckling</u>
occurs. At this stress the shape of the cross-section
changes. Flat plate elements become dished and usually the
longitudinal nodal lines, i.e., the join of two or more

plates which are not coplanar, remain straight. Two other important things happen when local buckling occurs. Firstly, the axial stress is redistributed, becoming more concentrated in the vicinity of the longitudinal nodal lines and reducing in magnitude away from them. Secondly, there is a sudden reduction in the axial stiffness due to the reduction in effective width of the plate. To illustrate the importance of this, two examples are quoted. A perfectly-flat square pin-sided plate with zero out-of-plane deflections and free in-plane deflections at the edges has a stiffness given by the value of the elastic modules E before buckling but its stiffness reduces by about 50% after buckling. If the same plate has one side released so that it can deflect out-of-plane the stiffness after buckling is zero. This point has important implications for the design of thin-walled channel, Z, I, and angle sections and for stiffened plates, which often have stiffeners with unsupported edges.

Not only can a stiffened plate buckle locally, but it may undergo global buckling. For example, a thin-walled and short strut of square box section and with pin ends will undergo local buckling. If its length is increased far enough it will behave as an Euler column undergoing no distortion of the cross-section during buckling. For the designer the problem is to determine which type of buckling, local or global, will occur first in a given strut. He may be tempted to try to "optimize" the length and cross-section so that the local and global buckling occur at the same axial stress. This approach should be avoided because there is an interaction between buckling modes , meaning that the actual buckling stress is lower than both the local and global buckling stresses calculated on the assumption that they do not interact.[3,4] Furthermore, initial imperfections have a greater influence in increasing the deflections when these two characteristic stresses are roughly equal. Such structures are said to be imperfection-sensitive.

In this book the buckling problems are treated as linear eigenvalue problems, i.e., only the minimum critical stress and buckling mode are calculated. In doing this it is necessary to evaluate the <u>critical length</u> of a given cross-section, i.e., that length of strut beyond which global buckling occurs first and below which local buckling occurs first. This concept is described more fully in Section 5.3. No attempt is made here to define the post-buckling behaviour of the structures.

A comprehensive set of results of various stability analyses to the year 1970 has been given by Bulson[3]. It is seen that the problem of elastic stability of stiffened plates can be solved in a number of ways. Although many methods are available for accurately solving particular problems, many difficulties arise when there is a great number of problems to solve in a systematic way. Such difficulties become acute when design tables are being prepared because of the enormous amount of numerical work involved. The best example of such a systematic analysis are the tables of Klöppel and Scheer[5] which are frequently used in German and other design offices. They cover the whole range of plates with stiffeners which are attached symmetrically to the plate with respect to its middle plane. The Ritz energy method was used to obtain explicit solutions to the eigenvalue problem.

A comparable handbook is not available for stiffeners which are attached eccentrically with respect to the middle plane of the plate. It does appear at first sight that the greater number of possible geometrical configurations would prohibit a similar approach to that used by Klöppel and Scheer[5]. Furthermore the complete solution of a buckling

problem should involve both plastic theory and the recently
developed notions of the catastrophe theory probably using
finite element analysis. However, in the near future it
seems unlikely that huge finite element calculations can be
performed for every practical design of a stiffened plate.
Designers have a great need for preliminary information so
that they can make their initial decisions about the
geometry of the structure. In the case of a stiffened plate
this information will be all of the normal sectional
properties, the yield stress and the stresses at which local and global buckling occur. It is the purpose of this
volume to provide designers with this information for a
wide variety of stiffened plates. Designers of stiffened
plates will find methods for the evaluation of critical
stresses in the Merrison Design Rules[6] and in DIN 4114[7].
The relevant formulae have had to be simplified for use in
design offices   and   they   often   give   approximate
(conservative) values of the critical stress.

In the preparation of the tables in this book the only
computational problems of any consequence arise from the
evaluation of the stress at which local buckling occurs.
The authors have sought the most efficient method of solution which can be generalized easily to include high order effects without much increase in complexitiy. The
finite strip method , a special variant of the well-known
finite element method, appears to offer the best possibility of fulfilling these requirements. As a first step a
pilot problem (PLATE) was developed and it is described
herein. The theoretical background has been described
earlier by Cheung and his co-workers[8]. For numerical solution some improvements have been made here and they have
lead to the possibility of using the finite strip method
for large-scale problems with comparatively small computing
times. Eventually three other plate programs (PLATE 1,PLATE
2 and PLATE 3 - see Sections 6 and 7 below) were written in
order to produce the tables contained herein.

2 Theoretical Background

2.1   Geometry

"The finite strip method is an extension of the well-known finite element method. This method is, however, semi-analytical in nature ...." (Cheung[8])

The reader familiar with the finite element method will recognize that the finite strip is only a special finite element with assumed displacement functions satisfying certain boundary conditions of a rectangular plate. A typical strip element is shown in Fig. 2 in Cartesian coordinates.

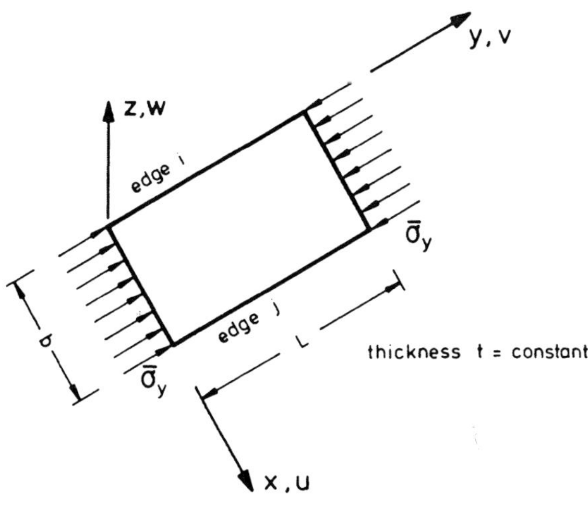

Fig 2   Finite strip element

It is in equilibrium with the load $\bar{\sigma}_y bt$ acting along the boundaries y = constant. The coordinates are local element coordinates. The strip is considered as part of a more general, longitudinally stiffened structure shown in Fig. 3.

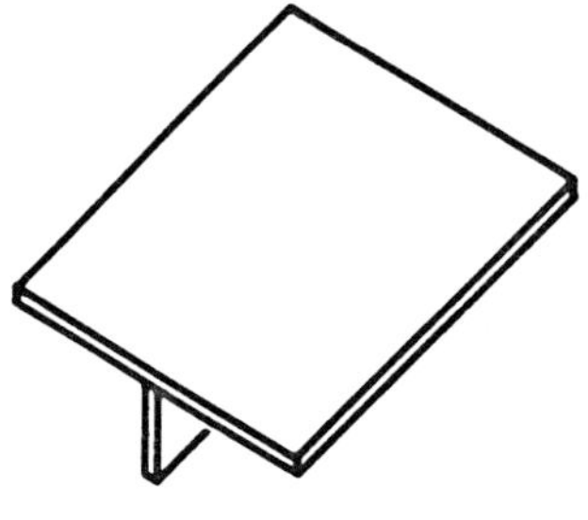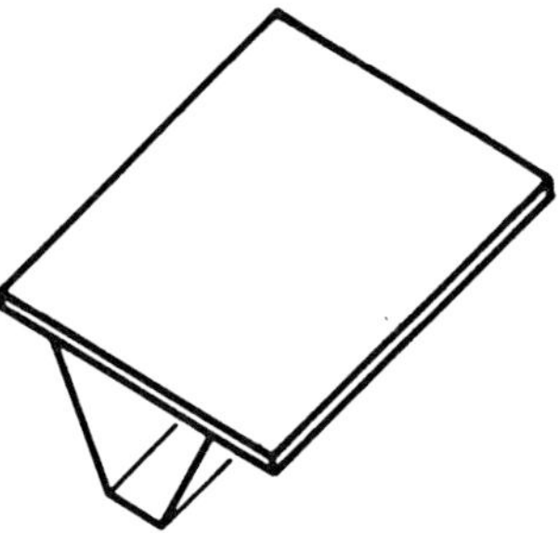

Fig. 3   Typical finite strip structures

2.2   Potential Energy

Let us consider a strip (Fig. 2) which is flat but at the
point of   buckling when   the   uniformly   applied stress
is $\bar{\sigma}_y$. We say that in this condition the total potential
energy of the strip is zero, this being an arbitrary datum
from which we can measure the changes in total potential
energy. Now let us displace the strip into its buckled
form, i.e., the displacement of an arbitrary point (x, y) is
(u, v, w). The stress components there change as follows:

in the x-direction from 0 to σ_x
in the y-direction from $\bar{\sigma}_y$ to $\bar{\sigma}_y + \sigma_y$
the shear stress   from 0 to τ_{xy}

The local strain around the point changes as follows

in the x-direction the strain increases by ε_x
in the y-direction the strain increases by ε_y
the shear strain increases by γ_{xy}

In a small element with dimensions dx × dy the increase in
total potential energy during buckling is the sum of,
firstly, the work done by the increases in the stresses

acting through corresponding strains, viz.,

$$t\left[\frac{1}{2}\,\sigma_x\,\varepsilon_x + \frac{1}{2}\,\sigma_y\,\varepsilon_y + \frac{1}{2}\,\tau_{xy}\,\gamma_{xy}\right] dx\ dy\ ,$$

secondly, the work done by the bending and twisting moments acting on the boundaries of the element, viz.,

$$\left[-\frac{1}{2}\,M_x\,\frac{\partial^2 w}{\partial x^2} - \frac{1}{2}\,M_y\,\frac{\partial^2 w}{\partial y^2} + M_{xy}\,\frac{\partial^2 w}{\partial x\,\partial y}\right] dx\ dy\ ,$$

and finally of the work done by $\bar{\sigma}_y$ acting through an apparent strain $\bar{\varepsilon}_y$ which makes allowance for the rotations of the element about axes parallel to the x- and z- directions (see Section 2.3).

The total potential energy of a strip element is therefore given by

$$U = \frac{t}{2}\int_0^L\int_0^b (\sigma_x\,\varepsilon_x + \sigma_y\varepsilon_y + \tau_{xy}\,\gamma_{xy})\ dx\ dy$$

$$+ \frac{1}{2}\int_0^L\int_0^b \left(-M_x\,\frac{\partial^2 w}{\partial x^2} - M_y\,\frac{\partial^2 w}{\partial y^2} + 2M_{xy}\,\frac{\partial^2 w}{\partial x\,\partial y}\right)\ dx\ dy \qquad (2)$$

$$+ t\int_0^L\int_0^b \bar{\sigma}_y\,\bar{\varepsilon}_y\ dx\ dy$$

The positive directions of M_x, M_y, M_{xy} are shown in Fig.4.

The first term in (2) is the strain energy of the plane stress problem, the second is the strain energy of the plate bending problem and the third is the potential energy of the external forces.

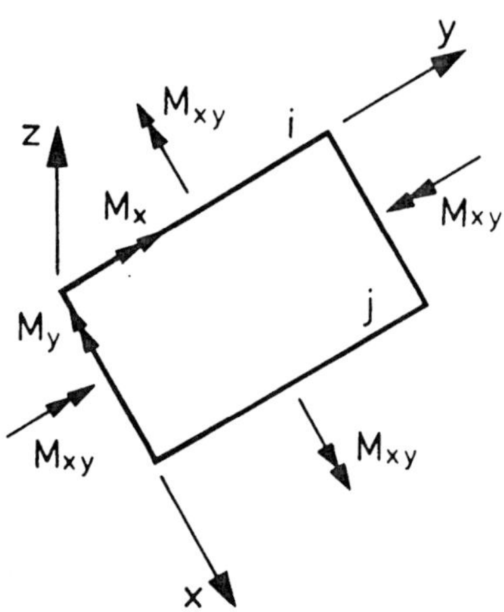

Fig. 4 Positive plate bending and twisting moments.
(Right hand screw rule is used)

The stress-strain relation for ideal elastic, isotropic material under plane stress is given by

$$
\begin{bmatrix} \sigma_x \\ \sigma_y \\ \tau_{xy} \end{bmatrix}
= \frac{E}{1-\mu^2}
\begin{bmatrix} 1 & \mu & 0 \\ \mu & 1 & 0 \\ 0 & 0 & \frac{1-\mu}{2} \end{bmatrix}
\begin{bmatrix} \varepsilon_x \\ \varepsilon_y \\ \gamma_{xy} \end{bmatrix}
\tag{3}
$$

where E is the modulus of elasticity and μ is Poisson's ratio.

The bending moments are obtained from the second derivatives of the displacements w (Fig.2) :

$$
\begin{bmatrix} M_x \\ M_y \\ M_{xy} \end{bmatrix}
= D
\begin{bmatrix} 1 & \mu & 0 \\ \mu & 1 & 0 \\ 0 & 0 & 1-\mu \end{bmatrix}
\begin{bmatrix} -\dfrac{\partial^2 w}{\partial x^2} \\ -\dfrac{\partial^2 w}{\partial y^2} \\ \dfrac{\partial^2 w}{\partial x\, \partial y} \end{bmatrix}
\tag{4}
$$

where $D = \dfrac{Et^3}{12(1-\mu^2)}$ is the plate stiffness.

It is assumed now that the strain field in the plane of the element can be obtained from a displacement field (u, v, w) by

$$\varepsilon_x = \frac{\partial u}{\partial x}$$

$$\varepsilon_y = \frac{\partial v}{\partial y} \tag{5}$$

$$\gamma_{xy} = \frac{\partial u}{\partial y} + \frac{\partial v}{\partial x}$$

where the nonlinear terms for ε_x and γ_{xy} of the Lagrangian strain tensor have been neglected.

2.3   Orthogonality of the Assumed Displacements

The essential feature of the finite strip formulation is the assumption of an orthogonal displacement field which satisfies all boundary conditions at y = 0 and y = L. The following Fourier series satisfy the homogenous boundary conditions for a strip which is simply supported at each end,

$$u = \sum_{m=1}^{\infty} \left[\left(1 - \frac{x}{b}\right) u_{im} + \left(\frac{x}{b}\right) u_{jm} \right] \sin \beta_m y \ ,$$

$$v = \sum_{m=1}^{\infty} \left[\left(1 - \frac{x}{b}\right) v_{im} + \left(\frac{x}{b}\right) v_{jm} \right] \cos \beta_m y \ , \tag{6}$$

$$w = \sum_{m=1}^{\infty} \left[\left(1 - \frac{3x^2}{b^2} + \frac{2x^3}{b^3}\right) w_{im} + \left(x - \frac{2x^2}{b} + \frac{x^3}{b^2}\right) \theta_{im} \right.$$

$$\left. + \left(\frac{3x^2}{b^2} - \frac{2x^3}{b^3}\right) w_{jm} + \left(\frac{x^3}{b^2} - \frac{x^2}{b}\right) \theta_{jm} \right] \sin \beta_m y$$

with $\beta_m = \dfrac{m\pi}{L}$, m = 1,2, ... and the suffices i and j referring to the edges i and j of the strip, respectively (Fig. 2).

The assumed displacement field is linear in x for the in-plane displacements u,v and cubic in x for the out-of-plane displacements w. All displacement functions are simple trigonometric functions in y.

For y = 0 and y = L we obtain u = 0, w = 0 and

$$\sigma_y = \frac{E}{1-\mu^2}(\mu\varepsilon_x + \varepsilon_y) = \frac{E}{1-\mu^2}\left(\mu\,\frac{\partial u}{\partial x} + \frac{\partial v}{\partial y}\right) = 0$$

The displacements v are linear functions in x at each end. For x = 0 and x = b the boundary displacements are Fourier series in y

$$x = 0: \qquad u = \sum_m u_{im}\,\sin\,\beta_m y$$

$$v = \sum_m v_{im}\,\cos\,\beta_m y \qquad\qquad (7)$$

$$w = \sum_m w_{im}\,\sin\,\beta_m y$$

$$x = b: \qquad u = \sum_m u_{jm}\,\sin\,\beta_m y$$

$$v = \sum_m v_{jm}\,\cos\,\beta_m y \qquad\qquad (8)$$

$$w = \sum_m w_{jm}\,\sin\,\beta_m y$$

The rotation $\dfrac{\partial w}{\partial x}$ is obtained from (6) as

$$\frac{\partial w}{\partial x} = \sum_m \left[\left(-\frac{6x}{b^2} + \frac{6x^2}{b^3}\right) w_{im} + \left(1 - \frac{4x}{b} + \frac{3x^2}{b^2}\right)\Theta_{im}\right.$$
$$\left. + \left(\frac{6x}{b^2} - \frac{6x^2}{b^3}\right) w_{jm} + \left(\frac{3x^2}{b^2} - \frac{2x}{b}\right)\Theta_{jm}\right] \sin\,\beta_m y$$

At $x = 0$ we have

$$\frac{\partial w}{\partial x} = \sum_m \theta_{im} \; \sin \beta_m y \tag{9}$$

and at $x = b$:

$$\frac{\partial w}{\partial x} = \sum_m \theta_{jm} \; \sin \beta_m y \tag{10}$$

The displacement functions define the stress and strain fields according to (3) and (5) and the distribution of bending moment by (4). Some of the corresponding terms in the potential energy (the first two integrals in equation (2)) also lead to integrals of the form

$$\frac{2}{L} \int_0^L \sin \beta_i y \; \sin \beta_j y \; dy = \delta_{ij} \tag{11}$$

where δ_{ij} is the Kronecker delta. The orthogonality of the assumed displacement functions therefore facilitates the calculation of the strain energy considerably. The third term in the total potential energy is due to the applied stress $\bar{\sigma}_y$. The corresponding strain is an approximation for the Lagrangian strain tensor

$$\bar{\varepsilon}_y = \frac{\partial v}{\partial y} + \frac{1}{2}\left[\frac{\partial u}{\partial y}\right]^2 + \frac{1}{2}\left[\frac{\partial w}{\partial y}\right]^2 \tag{12}$$

The last two terms represent the axial shortening due to small rotations of a fibre of length dy. The orthogonality in equation (2) also leads to integrals of the form

$$\frac{2}{L} \int_0^L \cos \beta_i y \; \cos \beta_j y \; dy = \delta_{ij} \tag{13}$$

with the same advantages as mentioned above.

It is therefore possible to calculate the potential energy separately for each mode m:

$$U_m = U_m^{(1)} \qquad + U_m^{(2)} \qquad + U_m^{(3)} \qquad + U_m^{(4)} \qquad (14)$$

in-plane   out-of-plane   in-plane   out-of-plane
first order   second order

3 Element Matrices

The potential energy is a quadratic form in the displacement components and can be calculated for each element separately in local coordinates. Separating the in-plane and out-of-plane displacement components by

$$\underline{r}_m^T = (u_i \ v_i \ u_j \ v_j)_m \tag{15}$$

and

$$\underline{\delta}_m^T = (w_i \ \Theta_i \ w_j \ \Theta_j)_m \tag{16}$$

where the superscript T denotes the transpose of the vector $\underline{r}_m$.

The following form of the potential energy of one element for one particular mode $m^{+)}$ can be obtained:

$$U_m = \frac{1}{2} \underline{r}_m^T \underline{k}_m^{(1)} \ \underline{r}_m + \frac{1}{2} \underline{\delta}_m^T \underline{k}_m^{(2)} \underline{\delta}_m$$

$$+ \frac{1}{2} \underline{r}_m^T \underline{k}_m^{(3)} \underline{r}_m + \frac{1}{2} \underline{\delta}_m^T \underline{k}_m^{(4)} \underline{\delta}_m \tag{17}$$

The four element matrices $\underline{k}^{(1)}$ to $\underline{k}^{(4)}$ follow from lengthy but elementary integrations.

+) Subscripts referring to the element number are partly omitted for simplicity.

3.1 First Order In-Plane Stiffness Matrix

$$\underline{k}^{(1)} = \begin{bmatrix} U_1+U_2 & U_3-U_4 & -U_1+\dfrac{U_2}{2} & U_3+U_4 \\ & U_5+U_6 & -U_3-U_4 & \dfrac{U_5}{2}-U_6 \\ \text{symmetric} & & U_1+U_2 & -U_3+U_4 \\ & & & U_5+U_6 \end{bmatrix} \tag{18}$$

$$U_1 = \frac{LEt}{2(1-\mu^2)b} \qquad U_4 = \frac{L\beta_m Gt}{4}$$

$$U_2 = \frac{Lb\beta_m^2\,Gt}{6} \qquad U_5 = \frac{Lb\beta_m^2\,Et}{6(1-\mu^2)} \tag{19}$$

$$U_3 = \frac{L\beta_m\mu Et}{4(1-\mu^2)} \qquad U_6 = \frac{LGt}{2b}$$

$$\left(G = \frac{E}{2(1+\mu)}\ ,\ \text{shear modulus}\right)$$

3.2 First Order Out-of-Plane Stiffness Matrix

$$\underline{k}^{(2)} = \begin{bmatrix} W_1 & W_2 & W_3 & W_4 \\ & W_5 & -W_4 & W_6 \\ & & W_1 & -W_2 \\ \text{symmetric} & & & W_5 \end{bmatrix} \tag{20}$$

$$W_1 = DL\left(\frac{13}{70}b\beta_m^4 + \frac{6}{5b}\beta_m^2 + \frac{6}{b^3}\right)$$

$$W_2 = DL\left(\frac{11}{420}b^2\beta_m^4 + \left(\frac{\mu}{2}+\frac{1}{10}\right)\beta_m^2 + \frac{3}{b^2}\right)$$

$$W_3 = DL\left(\frac{9}{140}b\beta_m^4 - \frac{6}{5b}\beta_m^2 - \frac{6}{b^3}\right)$$

$$W_4 = DL\left(-\frac{13}{840}b^2\beta_m^4 + \frac{1}{10}\beta_m^2 + \frac{3}{b^2}\right) \tag{21}$$

$$W_5 = DL\left(\frac{b^3\beta_m^4}{210} + \frac{2}{15}b\beta_m^2 + \frac{2}{b}\right)$$

$$W_6 = DL\left(-\frac{b^3\beta_m^4}{280} - \frac{1}{30}b\beta_m^2 + \frac{1}{b}\right)$$

3.3   Second Order In-Plane Stiffness Matrix

$$\underline{k}^{(3)} = \frac{tL\beta_m^2}{2}
\begin{bmatrix}
\frac{b}{3} & 0 & \frac{b}{6} & 0 \\
0 & 0 & 0 & 0 \\
\frac{b}{6} & 0 & \frac{b}{3} & 0 \\
0 & 0 & 0 & 0
\end{bmatrix} \tag{22}$$

3.4   Second Order Out-of-Plane Stiffness Matrix

$$\underline{k}^{(4)} = \frac{tL\beta_m^2}{2}
\begin{bmatrix}
\frac{13}{35}b & \frac{11}{210}b^2 & \frac{9}{70}b & -\frac{13}{420}b^2 \\
 & \frac{1}{105}b^3 & \frac{13}{420}b^2 & -\frac{1}{140}b^3 \\
\text{symmetric} & & \frac{13}{35}b & -\frac{11}{210}b^2 \\
 & & & \frac{1}{105}b^3
\end{bmatrix} \tag{23}$$

3.5   Complete Element Matrices in Local Coordinates

For computional purposes it is convenient to reassemble the
element matrices in a conventional, first order stiffness
matrix $\underline{k}^{(o)}$ and in a geometric, second order stiffness
matrix $\underline{k}^{(g)}$. The nodal displacements are rearranged in the
following form.

$$\underline{\delta}_m^T = (\underset{1}{u_i} \quad \underset{2}{v_i} \quad \underset{3}{w_i} \quad \underset{4}{\theta_i} \quad \underset{5}{u_j} \quad \underset{6}{v_j} \quad \underset{7}{w_j} \quad \underset{8}{\theta_j}) \tag{24}$$

This results in the following matrices:

$$
\underline{k}^{(0)} =
\begin{bmatrix}
U_1+U_2 & U_3-U_4 & & & -U_1+\dfrac{U_2}{2} & U_3+U_4 & & \\[2mm]
 & U_5+U_6 & & & -U_3-U_4 & \dfrac{U_5}{2}-U_6 & & \\[2mm]
 & & W_1 & W_2 & & & W_3 & W_4 \\[2mm]
 & & & W_5 & & & -W_4 & W_6 \\[2mm]
 & & & & U_1+U_2 & -U_3+U_4 & & \\[2mm]
 \text{symmetric} & & & & & U_5+U_6 & & \\[2mm]
 & & & & & & W_1 & -W_2 \\[2mm]
 & & & & & & & W_5
\end{bmatrix}
\tag{25}
$$

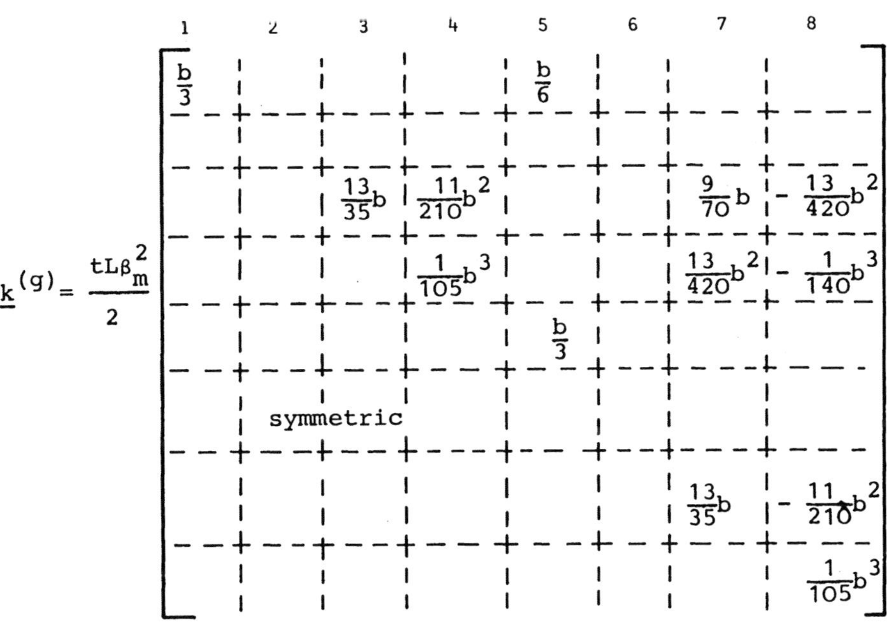

$$
\underline{k}^{(g)} = \frac{tL\beta_m^2}{2}
\begin{bmatrix}
\dfrac{b}{3} & & & & \dfrac{b}{6} & & & \\[2mm]
 & & & & & & & \\[2mm]
 & & \dfrac{13}{35}b & \dfrac{11}{210}b^2 & & & \dfrac{9}{70}b & -\dfrac{13}{420}b^2 \\[2mm]
 & & & \dfrac{1}{105}b^3 & & & \dfrac{13}{420}b^2 & -\dfrac{1}{140}b^3 \\[2mm]
 & & & & \dfrac{b}{3} & & & \\[2mm]
 & \text{symmetric} & & & & & & \\[2mm]
 & & & & & & \dfrac{13}{35}b & -\dfrac{11}{210}b^2 \\[2mm]
 & & & & & & & \dfrac{1}{105}b^3
\end{bmatrix}
$$

3.6 Coordinate Transformation to Global Coordinates

The potential energy of an element is now given by

$$U_m = \tfrac{1}{2}\underline{\delta}_m^T\left(\underline{k}^{(o)} + \bar{\sigma}_y\underline{k}^{(g)}\right)\underline{\delta}_m \qquad (26)$$

The coordinate transformation (rotation α)

$$\underline{\delta} = \underline{C}\,\bar{\underline{\delta}} \qquad (27)$$

from local (δ) to global coordinatates ($\bar{\delta}$) results in

$$U_m = \tfrac{1}{2}\bar{\underline{\delta}}^T\,\underline{C}^T\,(\underline{k}^{(o)} + \bar{\sigma}_y\underline{k}^{(g)})\,\underline{C}\,\bar{\underline{\delta}} \qquad (28)$$

where $\underline{C}=(\underline{C}^T)^{-1}$ is the rotation matrix (Fig. 5):

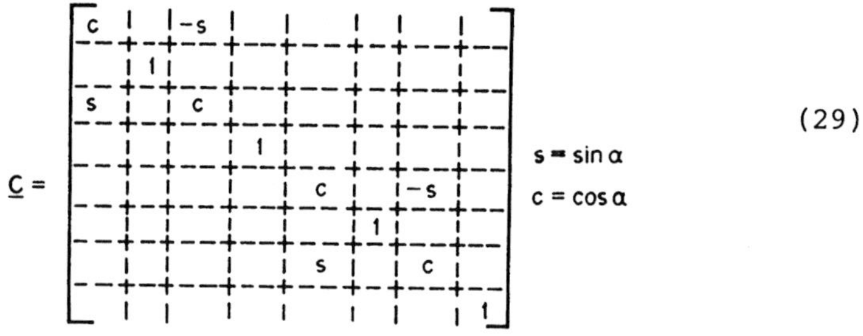

$$ (29) $$

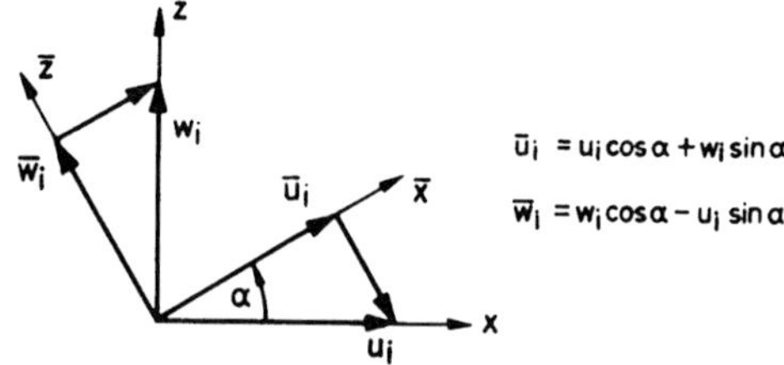

Fig. 5 Transformation to global coordinates

To avoid zero multiplications the matrix products in (28) are carried out explicitly for a general symmetric matrix

$$\underline{A} = (a_{ij}), \qquad i, j = 1,\ldots,8$$

taking into account the special structure, e.g.,

$$a_{13} = a_{14} = a_{17} = a_{18} = 0$$

This leads to the general form of the coordinate transformation given in equation (30) on the next page.

4 Assembly of Elements

The assembly of the elements is carried out using the concept of the direct stiffness method. The result is the total potential energy, in general terms for the first r modes:

$$U_r = \frac{1}{2} \underline{\delta}_1^T (\underline{K}_1^O + \bar{\sigma}_y \underline{K}_1^G) \underline{\delta}_1 + \ldots + \frac{1}{2} \underline{\delta}_r^T (\underline{K}_r^O + \bar{\sigma}_y \underline{K}_r^G) \underline{\delta}_r \qquad (31)$$

$\underline{K}_s^O$ is the complete stiffness matrix resulting from the conventional element stiffness matrices $\underline{K}^O$ in the s-th mode $(m - s)$ and $\underline{K}_s^G$ is the corresponding geometric stiffness matrix.

For a structure consisting of N elements, zero matrices $\underline{K}_s^O$ and $\underline{K}_s^G$, which are each of order $(4N \times 4N)$, are generated first. For each element $(1,\ldots,N)$ the submatrices $\underline{A}_{ii}, \underline{A}_{ij}, \underline{A}_{jj}$ from (cf. equ. 30)

$$\underline{C}^T \underline{A} \underline{C} = \begin{bmatrix} \underline{A}_{ii} & \underline{A}_{ij} \\ \\ \underline{A}_{ij} & \underline{A}_{jj} \end{bmatrix}$$

are added to these zero matrices. The numbers i and j are the node numbers of the element under consideration (Fig.6)

All matrices $\underline{K}^O$ and $\underline{K}^G$ are symmetric. For chain-type structures, such as stiffened panels, the maximum difference of node numbers in one element is small compared to 4N and the stiffness matrices are banded matrices (Fig.7)

$$\underline{C}^T\underline{A}\,\underline{C} =
\begin{bmatrix}
c^2a_{11}+s^2a_{33} & ca_{12} & -csa_{11}+csa_{33} & sa_{34} & c^2a_{15}+s^2a_{37} & ca_{16} & -csa_{15}+csa_{37} & sa_{38} \\
 & a_{22} & -sa_{21} & 0 & ca_{25} & a_{26} & -sa_{25} & 0 \\
 & & s^2a_{11}+c^2a_{33} & ca_{34} & -sca_{15}+sca_{37} & -sa_{16} & s^2a_{15}+c^2a_{37} & ca_{38} \\
 & & & a_{44} & sa_{47} & 0 & ca_{47} & a_{48} \\
 & & & & c^2a_{55}+s^2a_{77} & ca_{56} & -sca_{55}+sca_{77} & sa_{78} \\
 & \text{symmetric} & & & & a_{66} & -sa_{56} & 0 \\
 & & & & & & s^2a_{55}+c^2a_{77} & ca_{78} \\
 & & & & & & & a_{88}
\end{bmatrix}
\quad (30)$$

where blocks $\textcircled{i}$ and $\textcircled{j}$ label the row and column partitions.

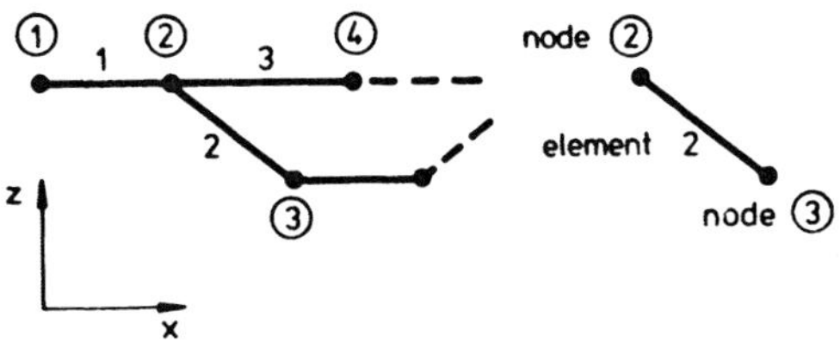

Fig. 6 Assembly of Elements

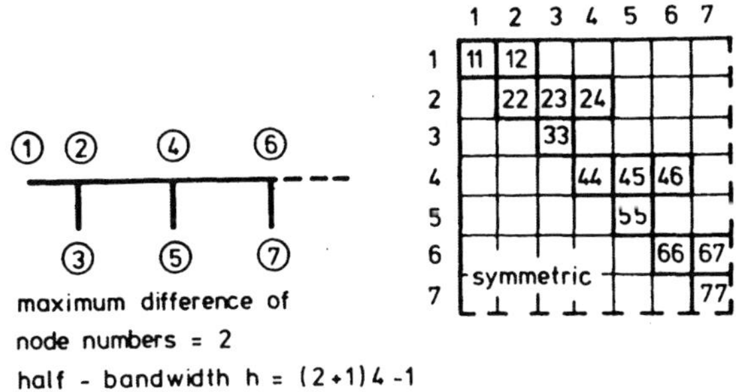

Fig. 7 Structure of stiffness matrices

Special boundary conditions can be introduced along the boundaries x = const. Deleting rows and columns of $\underline{K}^O$ and $\underline{K}^G$ is identical to restricting the corresponding degrees of freedom to zero along the boundary x = const.

5 The Eigenvalue Problem

5.1 General Form

The principle of minimum potential energy applied to (31) results in a system of eigenvalue problems

$$m = 1: \quad (\underline{K}_1^O + \bar{\sigma}_y \underline{K}_1^G) \, \underline{\delta}_1 = \underline{0}$$

$$m = 2: \quad (\underline{K}_2^O + \bar{\sigma}_y \underline{K}_2^G) \, \underline{\delta}_2 = \underline{0} \tag{32}$$

$$\vdots$$

$$m = r: \quad (\underline{K}_r^O + \bar{\sigma}_y \underline{K}_r^G) \, \underline{\delta}_r = \underline{0}$$

The lowest eigenvalue $\bar{\sigma}_y = \sigma_{cr}$ is the critical stress of the structure; the corresponding eigenvector determines the deflected shape of the structure by means of the assumed displacement functions (6). The index of $\underline{\delta}$ relates to the wavelength of the sine and cosine functions with

$$\beta_m = \frac{m\pi}{L} \tag{33}$$

It is not known a priori which mode is critical and theoretically the eigenvalue problems (32) for all r must be solved. This is certainly a drawback to the otherwise elegant formulation of the finite strip buckling analysis. In practice, however, it is sufficient to solve equation (32) for the first 10 or 30 modes.

5.2 Critical Stress and First Euler Length

The designer of a stiffened plate is faced with the following problem. Firstly, he must decide on the cross-section using some rational basis and secondly,he has to select the spacing of the transverse stiffeners. The tables presented in this book allow him to make both of these decisions. It is the purpose of this section to explain the logic and the meaning of the terms critical stress and first Euler length used in the tables because these are the only values the designer will require when he wishes to evaluate the buckling stress.

The solution of equation (32) could be found for a given cross-section by choosing a value of L and then allowing m to take the values 1,2,3, and so on, but this way is tedious and may not give the minimum stress. A more systematic way is to let m = 1 and then to increase L until, for practical purposes, all of the minima of $\bar{\sigma}_y$ have been obtained. In this way the variable β is changed more continuously and, for every value of β, the stress $\bar{\sigma}_y$ required to develop a buckle with wavelength $\frac{\pi}{\beta}$ (see equation (33)) is derived.

It is found that when this method is applied to stiffened
plate structures at least four types of buckling plots are
obtained. They are summarised in Fig. 8 . In Fig. 8(a) it
is seen that there is no minimum (critical) stress. The
longer L is made the lower is the critical stress. A cross-
section in this class is buckling like an Euler column. An
example of such a cross-section would be a flat plate which
is free along each of its longitudinal edges. Another
example would be a strut which buckles by torsional insta-
bility.

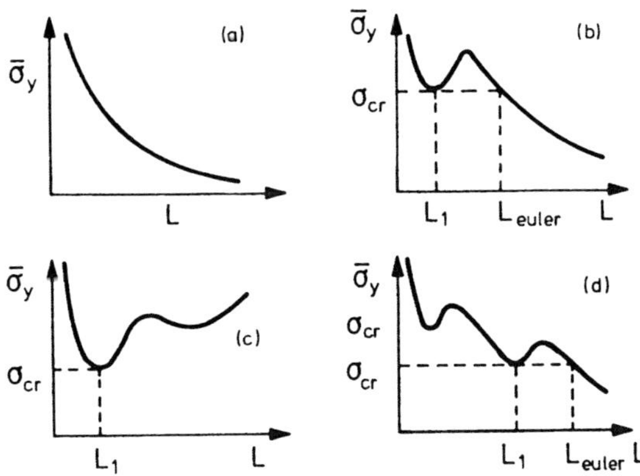

Fig. 8   Types of buckling plots

In Fig. 8(b) it is seen that a local buckle involving
distortion of the cross-section occurs at σ_{cr} with a comparatively small wavelength. As the assumed length is
increased, eventually the stiffened plate buckles as an
Euler column without distortion of the cross-section.
Obviously if the spacing between the transverse stiffeners
in a stiffened plate is made large enough it prefers to
buckle as an Euler column at a stress lower than σ_{cr}. The
length of a column for which the two buckling stresses are
equal is called the <u>first Euler length</u> to indicate that as

the length is increased it is the first length at which the column will begin to buckle as an Euler column. An example of a stiffened plate in this class is one which has stiffeners of almost any form but has edges which are free of both in-plane and out-of-plane restraint. A very wide stiffened plate will exhibit a similar behaviour pattern because the effect of any edge restraint is only felt locally.

In Fig. 8(c) the Euler buckling curve is not developed and only local buckling is significant. An example of a stiffened plate which is in this class is a long narrow strip of plate with one or two longitudinal stiffeners and which has its edges prevented from moving out of the plane of the plate. In Section 7 (Examples 7.3 and 7.4) two cross-sections, one free and the other pinned along each edge, are compared. They exhibit buckling curves which correspond to Fig. 8(b) and 8(c), respectively.

In Fig.8(d) the situation is similar to that illustrated in Fig.8(b) except that there are two modes of local buckling. An example of this type of behaviour is a channel section which, for certain geometries, exhibits both a symmetrical and an antisymmetrical local buckling mode or possibly a torsional buckling mode.

In preparing the tables it has been recognized that a designer requires only the lowest critical stress and the length, denoted here as L_{euler}, beyond which the Euler buckling stress is less than σ_{cr}. In the case of Fig. 8(a) Euler buckling governs for all L so the tables merely state the word "EULER", meaning that

$$\sigma_{cr} = \frac{\pi^2 E}{(L/r)^2} \tag{34}$$

where r is the radius of gyration of the cross section. In

these cases it is left to the designer to evaluate the buckling stress from equation (34) or, in rare cases, to check for torsional buckling [4].

When the tables of this book were being prepared the computer program was stopped automatically when a preset value of L, called L_{max}, was reached. The choice of L_{max} at either 500, 1000 or 1500 times the plate thickness was determined by practical considerations because it is usually necessary to incorporate transverse stiffeners at about these spacings. Thus, for some panels whose behaviour is similar to the curves of Fig. 8(b) and (c), the tables in this book will merely show under the heading L_{euler} the entry " >500" or " >2000". For these cases it is recommended that transverse stiffening within these spacings should be used.

The buckling plots shown in Fig.8 are the results of certain convenient mathematical manipulations already described, i.e., we let m = 1 and increase the length of the buckle L in steps. In Figs. 8(b) and (d) it is seen that a local buckle with wavelength L develops when the applied stress is σ_{cr}. If the length of the strut is made 2 L_1 (and assuming that $2L_1 < L_{euler}$) there is just space for two local buckles each of wavelength L_1 form and the buckling stress remains at σ_{cr}. Similarly for a strut of length 3 L_1 (provided 3 $L_1 < L_{euler}$) and so on until n $L_1 > L_{euler}$. Thus in such cases the buckling plots between L_1 and L_{euler} are not significant and should be drawn as shown in Fig.9. However for the purposes of the tables it is only neccessary to quote the values of σ_{cr} and L_{euler} as was stated above.

To reduce the size of the tables is has been necessary to non-dimensionalize the problem. This is done by letting the plate have unit thickness. For convenience a unit of length (thickness, depth, spacing etc.) is assumed to be 1 mm. This enables the results to be scaled easily as is explained in Section 8.

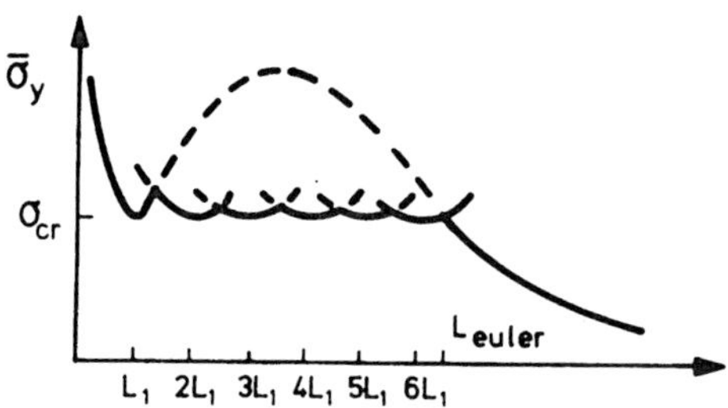

Fig. 9   Several local buckles of wavelength L_1
can develop in a strut whose length is $< L_{euler}$

6 Some Remarks on the Numerical Solution

The banded structure of the matrices is considered to be a significant feature of the problem. Only methods which take advantage of this feature are taken into account.

6.1   Iterative Solution (Program PLATE 1)

First the triangular decomposition of the band matrix $\lambda \underline{K}^O + \underline{K}^G$ is performed. The eigenvector $\underline{X}$ associated with the eigenvalue nearest to λ (inverse of σ_{cr}) is calculated by inverse iteration[9]. A better estimate of the minimum value of λ is calculated by the Rayleigh-Quotient

$$(\underline{X}^T \underline{K}^G \underline{X}) / (\underline{X}^T \underline{K}^O \underline{X})$$

The convergence to the lowest critical stress follows from the convergence of the Rayleigh-Quotient to the greatest eigenvalue $\lambda_{max} = 1/\sigma_{min}$ where $\sigma_{min} = \sigma_{cr}$. The full band of the matrices is used as storage.

6.2   Solution by Bisection (PLATE 2)

The general eigenvalue problem (32) is first reduced to the

specific one$(\underline{A} - \lambda\underline{E})$ $\underline{X} = \underline{O}$. In contrast to the usual transformation with the inverse of $\underline{K}^O$ which destroys the banded
structure, a special reduction is carried out which retains
the banded matrices[9].The resulting matrices are then reduced to symmetric tridiagonal form by Givens-Reduction[9].
The greatest eigenvalue, corresponding to the lowest
(critical) stress is found by Given's method of bisection
from the resulting tridiagonal matrices[9]. Only the upper
triangle of the symmetric band matrices including the
diagonals are used.

7 The Programs PLATE 1, PLATE 2 and PLATE 3

The programs are almost identical in structure, the
difference being due to the storage allocation and the
eigenvalue solution.

PLATE 1 and 2 calculate the critical stress, PLATE 3 is a
plot-program derived from PLATE 2 for the critical length
solution.

7.1 Sample Problems

To show the kind of output which is obtained from the
computer three sample cross-sections are considered here
(Fig. 10).

Example 7.1

In this case the structure is a flat plate (Fig.10(a)) with
free edges. Two elements each of width 10 are used. One
expects this type of structure to buckle as an Euler column
at the stress given by equation (34). This is indeed what
happens as is shown in Fig. 11. In the tables presented in
this book the designer is simply told to treat this
structure as an Euler column ("EULER").

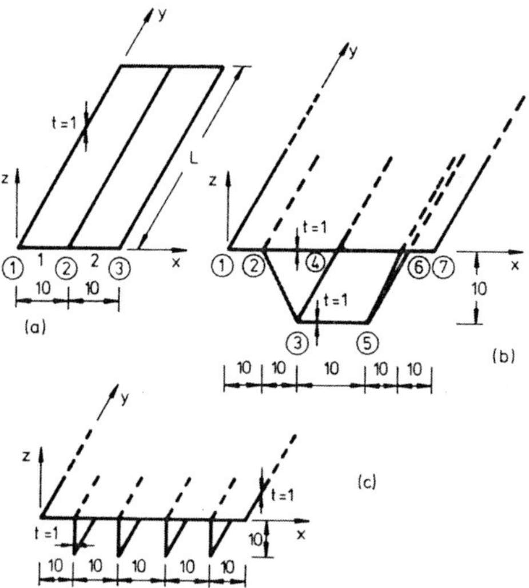

Fig. 10   Dimensions of stiffened plates used in Examples 7.1 to 7.7

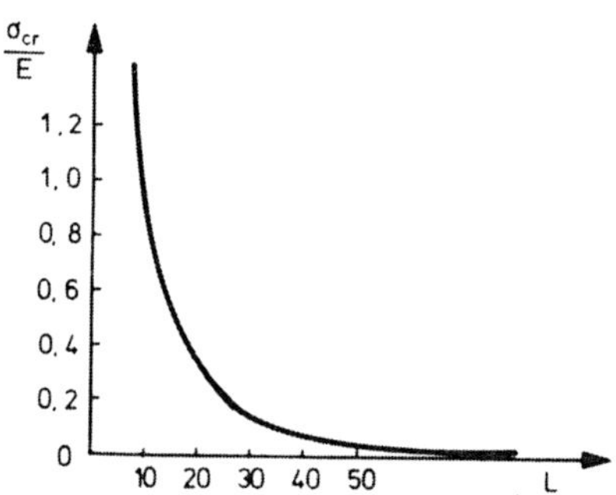

Fig. 11   Critical stress plot for Example 7.1

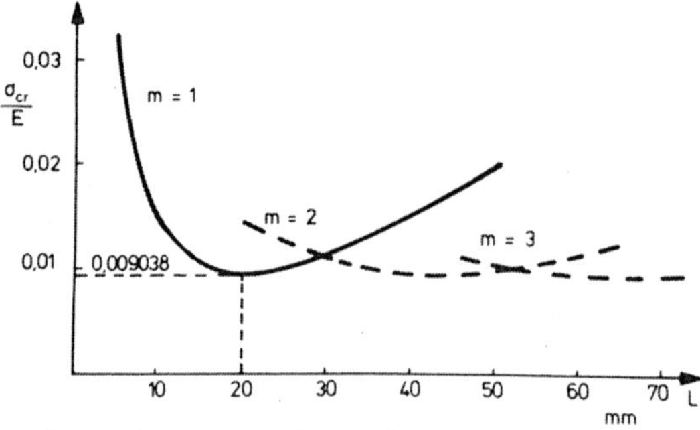

Fig. 12   Critical stress plot for Example 7.2

Example 7.2

In this case the same flat plate (Fig. 10(a)) is used but the longitudinal edges are pinned and free to move in the plane of the plate. Fig. 12 shows how this form of constraint affects the buckling stress. There is a minimum value of σ_{cr} when the plate is square. If the length L of the plate is doubled two buckles of opposite sign can develop so strictly speaking the first dotted curve should be included. Similarly other curves for three or more buckles should be drawn. However, the designer needs only to know the minimum value of σ_{cr} and that this structure does not buckle as an Euler column. This information is given in the tables (see Section 8).

Example 7.3

In this example a single trough stiffener is used (Fig. 10(b)) with the longitudinal edges free. As the length L is varied the plot shown in Fig. 13 is obtained. It is seen that this structure is one of the type shown in Fig. 8(b). For L < 156 mm local buckles of wavelength 24.39 mm occur at a critical stress of $5.543 \times 10^{-3}E$. Obviously a whole family of curves similar to those shown as dotted lines in Fig. 9 could be drawn in Fig. 13. These would indicate (correctly) that when the length of the panel was, say, 48.78 mm (= 2 x 24.39 mm) the buckling stress would still be $5.543 \times 10^{-3}E$ because two buckles could form in that length. These curves have been omitted for clarity. For L > 156 mm the structures should be treated as an Euler column. In the tables the values of both σ_{cr} and L_{euler} would be given for a case such as this.

Example 7.4

This example is the same as Example 7.3 except that the

edges are pinned (w = 0) but as before they are free to move in the plane of the plate. This stucture is one of the type shown in Fig. 8(c). A long structure of this type develops local buckles with L_1 = 22.29 mm over its entire length at a critical stress of 6.029 x 10^{-3}E. As seen from Fig. 14 no Euler buckling occurs. The tables would indicate this information.

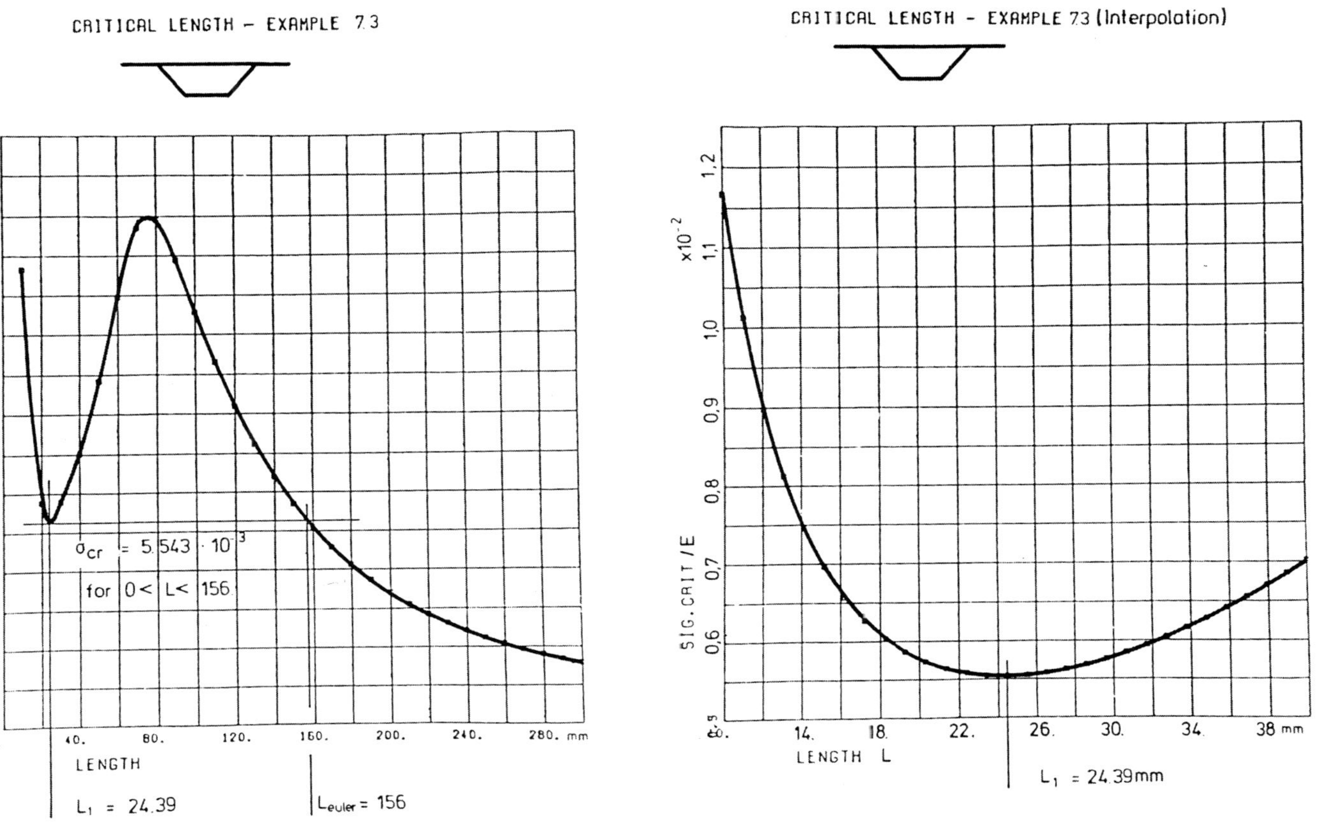

Fig. 13a : Example 73 Boundaries unconstrained

Fig. 13b : Enlargement of figure 13a

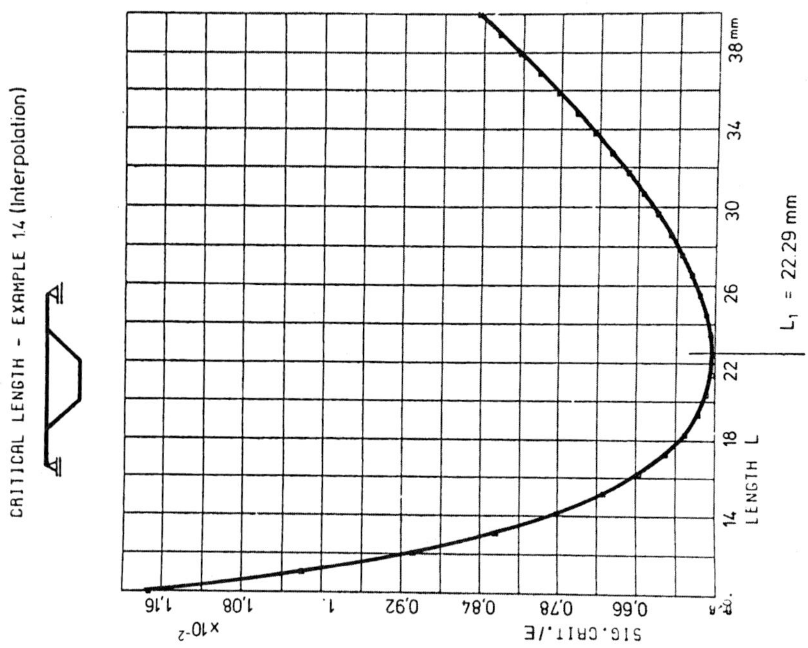

Fig. 14b : Enlargement of figure 14a

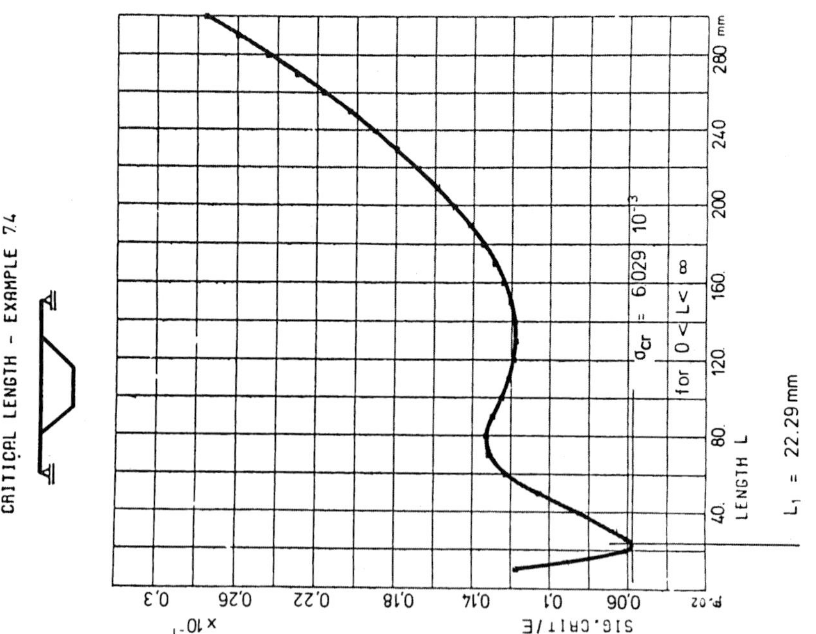

Fig. 14a : Example 74 Boundaries constrained

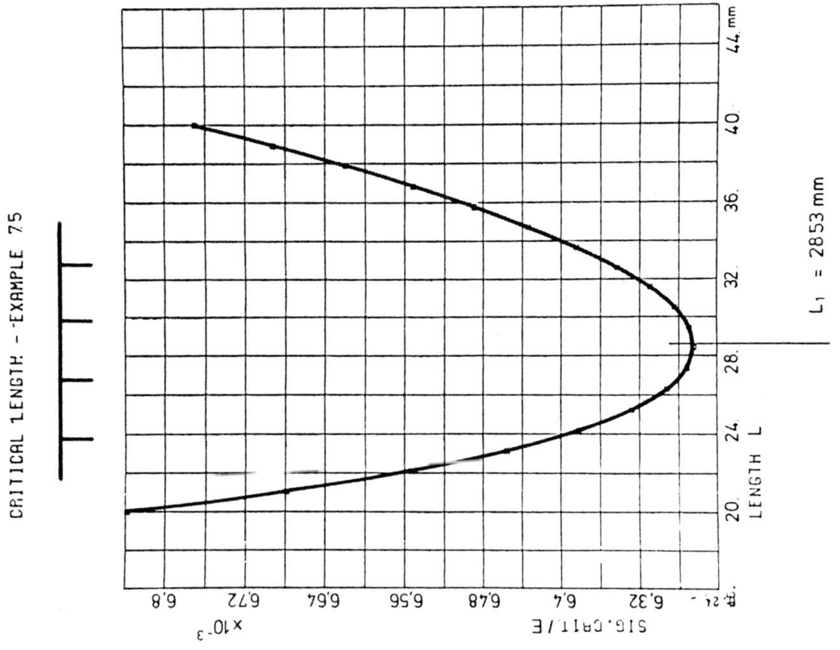

Fig. 15b : Enlargement of figure 15 a

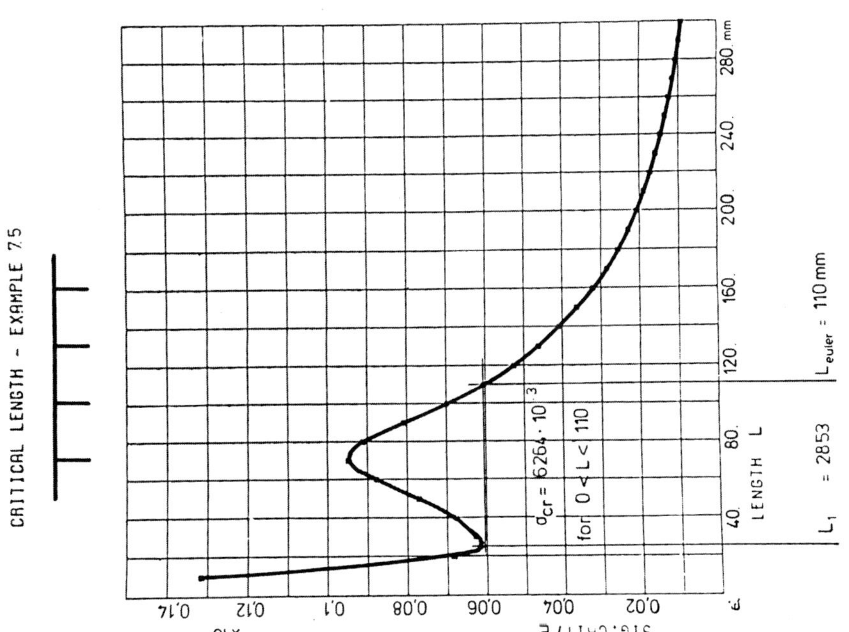

Fig. 15a : Example 75

Example 7.5

The cross-section is given in Fig. 10(c). It is seen from the results plotted in Fig. 15 that this panel behaves like that analysed as Example 7.3.

Example 7.6

In this example two plates with thin rectangular stiffeners are compared. In the first case the neutral axis of the stiffener lies in the central plane of the plate and in the other it is eccentric (Fig.16). The former case has been treated by Klöppel and Scheer[5].

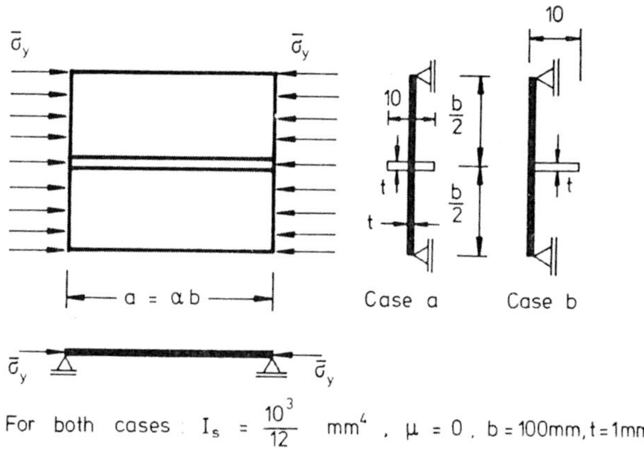

Fig. 16   Test cases to study effect of using eccentric stiffener instead of symmetrical stiffener

The two sets of results are plotted in the one graph in Fig 17. It is seen that in this case the buckling stress is almost doubled by using an eccentric stiffener instead of one which is symmetrical. It should not be concluded that this is always so. If a very deep and thin stiffener is used the panel could have a higher buckling stress when it is joined symmetrically. This is because very deep stiffeners are weak (see Section 8.3).

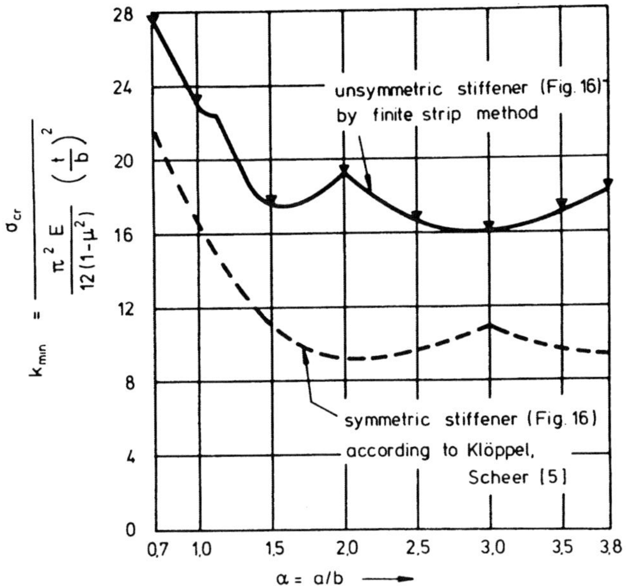

Fig. 17   Buckling factor k_{min} for Example 7.6

8 The Use of the Tables

8.1 Information Listed in the Tables

It will be seen from the tables that the information presented can be divided into three types.

(a)   The first set of figures present the geometry of the
cross-section. The   dimensions are taken relative to
the   thickness t_1 of the plate which is assumed to be
unity by the computer in the preparation of the tables.
This is merely a convenient device to reduce the size
of the tables. Other plate thicknesses can be considered as explained below. Each stiffened plate is given
a code which reflects the geometry of its cross-section. The first letter refers to the shape of the
stiffener and so on as indicated in Fig.18. A few examples of the coding are given in Fig.19.

(b)   The   second   set of figures are the usual section pro-
      perties   used   in design, e.g., area of cross-section,
      distance   from upper surface of plate to neutral axis,
      second   moment   of area about the neutral axis, and so
      on.

(c)   The third   set   of   figures   are   the   results   of the
      buckling analysis   described in the previous sections,
      i.e., the critical stress and the first Euler length.

8.2   Scaling Results for Stiffened Plates with Plate Thickness $t_1 \neq 1$.

The results   presented in the tables   can be used   directly
when the plate thickness t_1 = 1. When $t_1 \neq 1$ the figures in
the table   are simply multiplied by t_1 raised to   the power
indicated   in the heading entitled "to scale multiply   by".
For example, when t_1 is 8 mm   and the value   of the   search
moment of area 1,   in the table is given as 1200, the value
of 1 this stiffened   plate will be 1200 x 8^4 = 4915200 mm^4.
All of   the   linear   dimensions   e.g. S, a, t_2, $\bar{y}$, L,   and
L_{euler}, are scaled   by 8, the area of cross-section. A must
be scaled by 8^2 = 64, and so on.   The value of σ_{cr}/E is not
scaled because   it is a dimensionless   quantity.   In   other
words, and exactly   scaled   model of a   larger   prototype
stiffened plate   has the   same buckling   stress on the prototype.

8.3   Standardization of Cross-Sections

The strength   of   a   thin-walled   cross-section can be very
sensitive   to   small   changes   of geometry. It is often the
case   that   by   adding   material   to   a stiffened plate its
strength   is   reduced.   For example, if the depth of a thin
rectangular   stiffener   is increased its buckling stress is

decreased because deep thin plates are very weak in compression when they have an unsupported (free) edge. Another way in which a stiffened plate can be weakened is for its overhang "a" (Fig. 20) to be increased, especially if the longitudinal edges are free. Fig. 21 shows the results of a study of the effect of increasing "a" and it is seen that for free edges it is unreasonable to allow "a" to increase beyond about 15 plate thicknesses. When the longitudinal edges are pinned there is no appreciable reduction in the critical stress until "a" is equal to the stiffener spacing. However, it has been found from other studies that when the number of stiffeners is small "a" has a strong influence upon σ_{cr}. For example, Cases 3 and 5 in Fig. 25 should be compared. So that designers can gauge what influence "a" will have upon σ_{cr} for a pin-sided panel two values are given in the tables, viz. a = S and a = 0.2 S, the former being designated by P and the latter by Q (see Fig. 18).

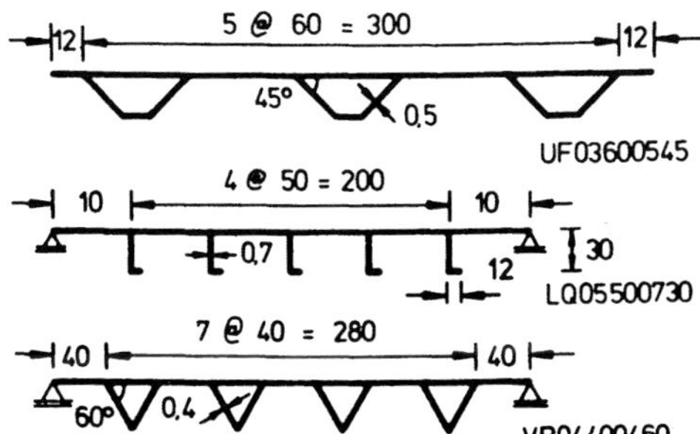

Fig. 19 Example of coding of cross sections (see also Fig. 18)

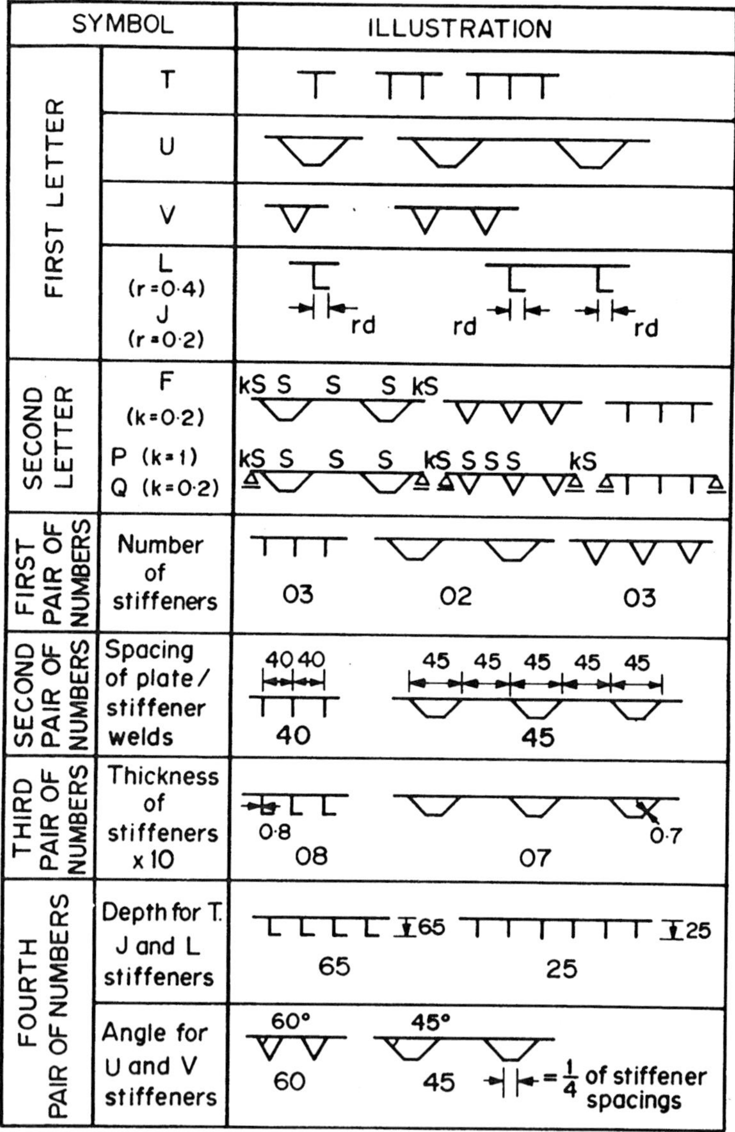

FIG. 18   Coding of Cross Section (see also FIG. 19)

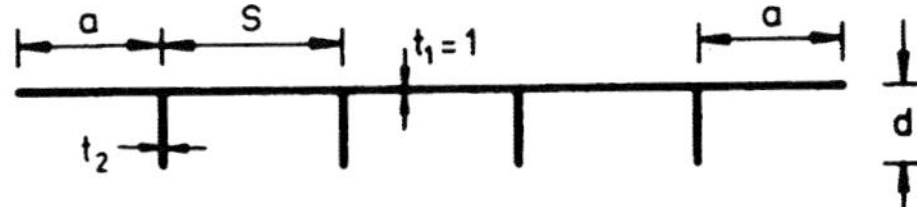

Fig. 20   Dimensions used in parametric studies

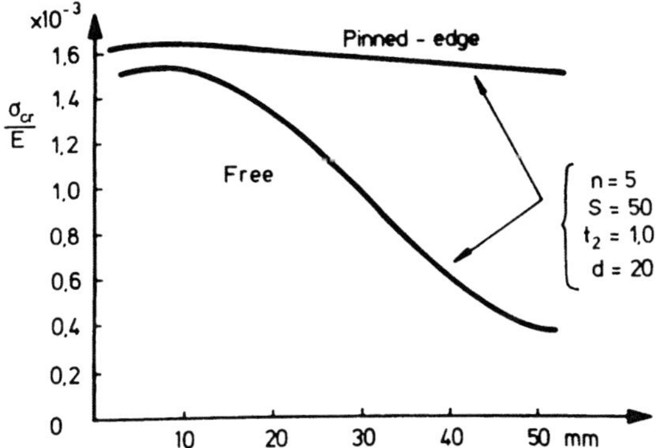

Fig. 21   Graphs showing effect on σ_{cr} of increasing
dimension a (Fig. 20)

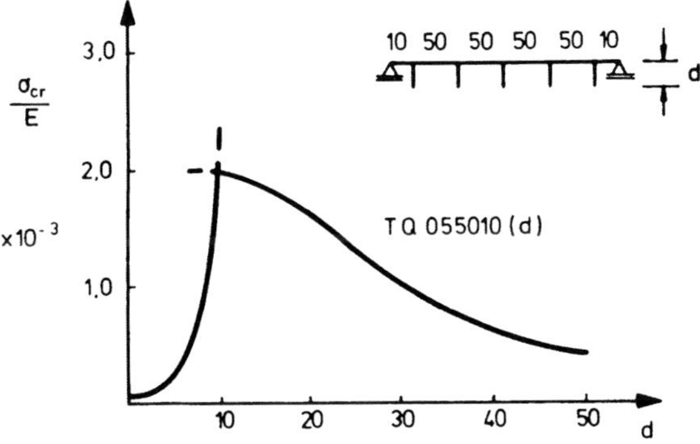

Fig. 22   Influence of depth of stiffener on σ_{cr}

Fig. 22 shows the result of a study of the effect of varying the depth of the stiffener. For small d the panel, which in this case has pinned sides, behaves like an unstiffened panel of width 220 (When d = 5 the buckling length is 389). As d is increased by a small amount, local buckling governs (When d = 10 the buckling length is 44.9). The buckling stress attains its maximum value at about d = 10. Beyond this value the stiffener becomes weaker and weaker as d is increased and the buckling stress decreases.

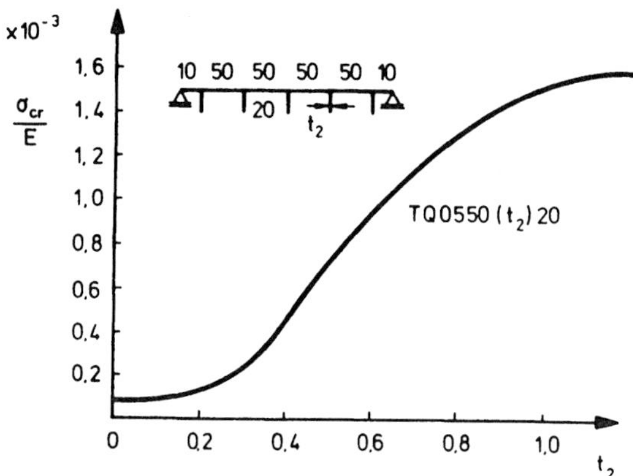

Fig. 23 Influence of thickness t_2 of stiffener on σ_{cr}

Fig.23 shows the effect of changing the thickness t_2 of the stiffeners in a similar panel. For very small t_2 the panel behaves like an unstiffened plate of width 220 and as t_2 increases local buckling governs. When t_2 exceeds about 0.8 the critical stress increases more and more slowly with increases in t_2 because the plate is the weaker component and governs the behaviour of the panel.

Another study was carried out to see how the critical stress of a plate with a given size of stiffener and a given stiffener spacing was affected by increasing the number of stiffeners; at the same time the overall width of the panel would, of course, be increasing. One can expect that the buckling pattern of two such stiffened panels

with, say, 5 and 6 stiffeners would be very similar so the
critical stresses would be nearly the same. If this were
so it would not be necessary to include in the tables
results for stiffened plates with a large number of
stiffeners. This was indeed found to be the case. Fig. 24
shows that for four panels with different geometry the
local buckling stress is essentially constant once the
number of stiffeners exceeds four. Therefore a designer
may use the results for n = 5 when the number of stiffeners
is 5 or more.

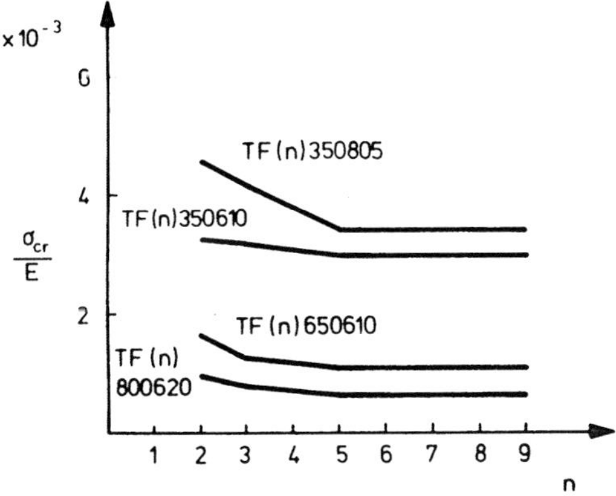

Fig. 24 Influence of number of stiffeners on σ_{cr}

Finally a study was made of the effects of the boundary
conditions to see whether some cases could be eliminated
from the tables. If it is assumed that for every stiffened
panel the longitudinal edges are always free to rotate and
to displace longitudinally there are four possible combinations of boundary conditions in the x- and z-directions,
viz., free or fixed in both directions (Cases 1 to 4,
Fig. 25). When such a plate is compressed in the y-
direction the Poisson's expansion in the x-direction can be
suppressed (Cases 2 and 4) but this requires enormous re-
straining forces supplied by a very large structure. These
two boundary conditions are therefore considered to be of

academic interest only and have been eliminated from the tables. Another study of the influence of the width a of the edge plate on σ_{cr} and L_1 showed that it had a strong effect (compare Cases 3 and 5). Therefore in the tables the boundary conditions illustrated in Cases 1, 3, and 5 have been chosen.

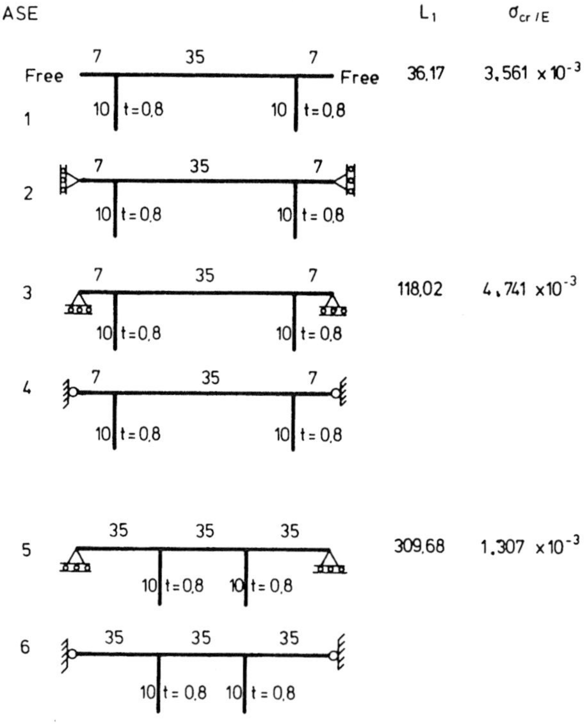

Fig. 25 Influence of boundary conditions on σ_{cr}

From these and other studies the following conclusions are drawn.

1. For a panel with free longitudinal edges the edge distance "a" should not exceed 0.2 S and the values of σ_{cr} and L in the tables are for this geometry. When a < 0.2 S the values of σ_{cr} in the tables may be used but they may be slightly unsafe.

2. For a panel with pinned longitudinal edges the edge distance "a" has little effect upon σ_{cr} when the number of stiffeners is large but it has been necessary to include two values of "a" (viz. a = S and a = 0.2 S) to cover those cases when the influence is stronger. For values of "a" between a = S and a = 0.2 S linear interpolation may be used.

3. The depth of the stiffener has a strong influence upon σ_{cr}. For very small and for very large depths the value of σ_{cr} can be very small, in the former case the panel behaves like an unstiffened plate and in the latter case the stiffeners are weak. In between these two extremes the stiffeners have an optimum depth. The tables cover a comprehensive range of depths to enable designers to evaluate σ_{cr} for practical situations when it is not possible to design for the optimum geometry.

4. The thickness t_2 of the stiffeners has a strong influence at first but as t_2 becomes larger σ_{cr} increases more slowly. For small t_2 the panel behaves like an unstiffened plate but when t_2 is large the plate between the stiffeners governs and buckles as a plate with built-in longitudinal edges. The tables give a comprehensive range of t_2 values.

5. The results presented in the tables are for the boundary conditions illustrated in Fig. 25 as Cases 1, 3 and 5. In the tables these are given symbols F, Q and P respectively. They may not be used in panels with boundary conditions illustrated as Cases 2, 4 and 6 (Fig. 25) as the results are then unsafe.

6. It is not "safe" to use a spacing of the transverse stiffeners greater than L_{euler} unless a thorough investigation is undertaken because in this region buckling stresses are suppressed by the interaction of the two modes of buckling[4]. Interaction is important in those cases where σ_{cr}/E is significantly less than the Euler buckling

$$\frac{\sigma_E}{E} = \frac{\pi^2 \, I}{(L_{euler})^2 \, A} \quad .$$

8.4 Use of Tables with Design Rules

In the analysis and design of stiffened plates it is
necessary to calculate the elastic critical load so that
the failure load can be estimated. The tables contained in
this book may be used in this part of the calculation in
place of the methods used in the design rules. In this
section a box-girder beam whose upper flange is a stiffened
plate is considered. In the first example the tables are
used in conjunction with the rules of the German DASt-
Richtlinie 012[10]. In the second example the critical
stress only is calculated by using the Merrison Rules[6] so
that a comparison can be made. It will be seen that the
tables of this book are easy to use, they give accurate
estimates of the critical stress without empiricism and
they indicate in a direct manner the spacing of the transverse stiffeners.

Example 8.1

The upper flange of the box-girder illustrated in Fig.26 is
to be checked over the length AB by using the tables in
this book and by relevant clauses of DASt-Richtlinie
012[10]. In this example the clauses and tables referred to
are those in Richtlinie 012 unless stated otherwise.

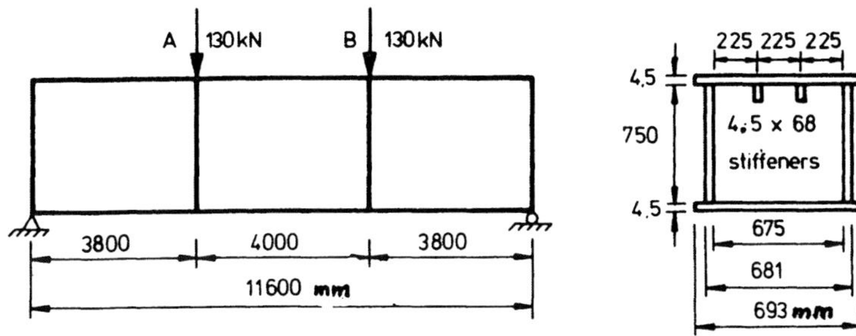

Fig. 26 Box girder analysed in Example 8.1

Section Properties

Area of cross-section $= 11349 \text{ mm}^2$

Distance from upper surface to neutral axis $=$ 362 mm

I_{NA} $= 1164.958 \times 10^6 \text{ mm}^4$

$Z_{upper\ flange}$ $=$ $3.218 \times 10^6 \text{ mm}^3$

Stress in upper flange

$$\text{axial stress} = \frac{\text{bending mt.}}{Z}$$

$$\sigma_1 = \frac{130000 \times 3800}{3.218 \times 10^6} = 153.51 \text{ MPa}$$

Local buckling analysis of upper flange

Steel St 52 , $\sigma_F = 360$ MPa (=yield stress)

In the next step it is necessary to find the stiffened
plate in the tables in this book, representing a scaled
version of the upper flange. The plate thickness is 4.5 mm
so the model in the tables in this book has the following
dimensions

Plate thickness $= \dfrac{4.5}{4.5} = 1$ mm

Spacing of stiffeners $= \dfrac{225}{4.5} = 50$ mm

$$\text{Depth of stiffeners} \quad = \quad \frac{68}{4.5} \quad = \quad 15 \text{ mm}$$

$$\text{Thickness of stiffeners} = \quad \frac{4.5}{4.5} \quad = \quad 1 \text{ mm.}$$

Thus the tabulated properties of TP 02501015 apply, i.e.,

$$\frac{\sigma_{cr}}{E} = 1.6095 \times 10^{-3} \;,\; L_1 = 49.1 \text{ mm}, \; L_{euler} = 326 \text{ mm}$$

For the box girder flange the following properties apply

$$\sigma_{cr} = 210000 \times 1.6095 \times 10^{-3} \qquad\qquad = \quad 338 \text{ MPa}$$
$$L_1 = 49.1 \times 4.5 \qquad\qquad\qquad\qquad = \quad 221 \text{ mm}$$
$$L_{euler} = 326 \times 4.5 \qquad\qquad\qquad\qquad = \quad 1467 \text{ mm}$$

These last figures mean that at an applied stress of 338 MPa local buckles of wavelength 221 mm will appear. At the same stress longer Euler buckles could also exist unless they are suppressed by transverse stiffeners with a spacing of less than 1467 mm. This can be achieved by using 2 transverse stiffeners with a spacing of 4000/3 = 1333 mm.

In Richtlinie 012 Clause 6.1.3

$$\sigma_{vki} = \sigma_{ki} = 338$$

$$\frac{\sigma_1}{\sigma_{ki}} = \frac{153.51}{338} = 0.454 \quad (= \frac{1}{\nu_B} = \frac{1}{\text{safety factor}})$$

$$\frac{\sigma_F}{\sigma_{ki}} = \frac{360}{338} = 1.06$$

From Clause 6.2 the reduced stress is

$$\sigma_{vk} = \sigma_F \; (1.474 - 0.677\sqrt{\frac{\sigma_F}{\sigma_{ki}}})$$
$$= 360 \; (1.474 - 0.677 \sqrt{1.06}) \qquad\qquad = 279.71 \text{ MPa}$$

Relative to the yield stress this is

$$\bar{\sigma}_{vk} = \frac{279.71}{360} = 0.777$$

From Clause 6.3.3.2 the relative limiting stress $\sigma_G = \sigma_{vk}$
The available calculated buckling safety factor (Cl.6.4)

$$\nu_B = \frac{\sigma_G}{\sigma_1} = \frac{279.71}{153.51} = 1.82$$

The required buckling safety factor (Clause 8 Table 7) is

$$\nu_B = 1.32 + 0.19(1 + \psi_B) = 1.32 + 0.19(1+1.0) = 1.70 < 1.82$$
$$\text{(satisfactory)}$$

Example 8.2

Use of the Merrison Rules to obtain the critical stress of the flange analysed in Example 8.1.

In the following analysis the numbers of the clause and the tables and the symbols refer to those in the Merrison Rules

Initial imperfection (Table 23.1)

For $t < 25$ mm :
$$\Delta_x = \frac{G}{30t}\left(1 + \frac{b}{5000}\right)$$

$$= \frac{225}{30 \times 4.5}\left(1 + \frac{225}{5000}\right) = 1.74 \text{ mm}$$

Residual stress (Cl.7.3.2) $= \dfrac{10\,A}{bt}$ (A = area of weld)

$$= \frac{10 \times 10.1}{225 \times 4.5} = 0.1 \text{ MPa}$$

Ignore residual stress so $\Sigma\delta = \delta_o$. From Clause 18.1.2

$$\delta_o = \frac{1.2\,b\,\Delta_x^3}{G}\cdot\frac{1}{1+N}$$

$$= \frac{1.2 \times 225 \times 1.74^3}{225}\cdot\frac{1}{1+2} = 1.45 \text{ mm}$$

$$\frac{\Sigma \delta}{t} \quad = \frac{1.45}{4.5} \qquad\qquad = 0.32$$

For an unrestrained panel use Fig. 18.4 with

$$\sigma_{ecr} = 4 \times \frac{\pi^2 E}{12(1-\mu^2)} \frac{t}{b}^2 = 4 \frac{\pi^2 \times 206000}{10.92} \cdot \frac{4.5}{225}^2 = 298 \text{ MPa}$$

$$\frac{\sigma_e}{\sigma_{ecr}} = \frac{153.51}{298} \qquad\qquad = 0.52$$

Hence K_{bt} = 0.77 and the effective width of plating associated with a stiffener is 0.77 x 225 = 173 mm. I_{ox} of this cross section is 408314 mm^4 (Cl.20.1.4)

$$\mu_x = \mu_y = 0.3$$

$$D_x = \frac{E\,I_{ox}}{b} + (1-K_{bt}) \frac{Et^3}{12(1-\mu_x\mu_y)}$$

$$= 206000 \; \frac{408314}{225} + (1-0.77)\frac{4.5^3}{12(1-0.3^2)} = 374 \; \times 10^6 \text{Nmm.}$$

For stiffeners in one direction only (Cl.20.14)

$$H = \frac{G\,t^3}{6} + \frac{G\,J_x}{2b}$$

$$= 80\;000 \; \frac{4.5^3}{6} + \frac{68 \times 4.5^3}{3} \times \frac{1}{2\times225} = 1.582 \times 10^6 \text{Nmm.}$$

To find the next buckling stress (Cl.20.1.5.1 a) first assume L = 4400 mm, whence $\bar{\phi} = \frac{4400}{675} = 6.518$

$$t_{eff} = t \; (1 + \frac{N\,A_s}{b_s t})$$

$$= 4.5 \; (1 + \frac{2 \times 68 \times 4.5}{675 \times 4.5}) \qquad = 5.41 \text{ mm}$$

$$D_y = D = \frac{Et^3}{12(1-\mu^2)} = \frac{206000 \times 4.5^3}{12 \times 0.91} = 1.7206 \times 10^6 \text{Nmm}$$

$$\sigma_{cr1} = \frac{\pi^2}{b_s^2\, t_{eff}}\, \frac{D_x}{\bar{\phi}^2} + 2H + D_y\bar{\phi}^2$$

$$= \frac{\pi^2}{675^2 \times 5.41}\, \frac{374 \times 10^6}{6.518^2} + 2 \times 1.582 \times 10^6 + 1.7206 \times 10^6 \times 6.518^2$$

$$= 340.6 \text{ MPa}$$

Now assume $L = 2200$ mm (i.e., two buckling lengths)

$$\bar{\phi} = \frac{2200}{675} \qquad\qquad = \quad 3.26$$

$$\sigma_{cr1} = \frac{\pi^2}{675^2 \times 5.41}\, \frac{374 \times 10^6}{3.26^2} + 2 \times 1.582 \times 10^6 + 1.7206 \times 10^6 \times 3.26^2$$

$$= 226.7 \text{ MPa} \qquad\qquad < \ 340.6 \text{ MPa}$$

Now try $L = 1467$ mm (i.e., three buckling lengths)

$$\bar{\phi} = \frac{1467}{675} \qquad\qquad = 2.173$$

$$\sigma_{cr1} = \frac{\pi^2}{675^2 \times 5.41}\, \frac{374 \times 10^6}{2.173^2} + 2 \times 1.582 \times 10^6 + 1.7206 \times 10^6 \times 2.173^2$$

$$= 362.3 \text{ MPa} \qquad\qquad 226.7 \text{ MPa}$$

Hence the estimated critical stress by the Merrison Rules
is 226.7 MPa with a buckling length of 2200 mm (compare 338
MPa by using the tables of this book). However, by using
two transverse stiffeners as is required in Example 8.1 the
critical stress can be raised to 362.3 MPa which is in good
agreement with the buckling analysis the tables of this
book.

9 List of Symbols

$\underline{A}$ =   General symmetric matrix

a =   Width of edge strip (Fig. 20)

a_{ij} =   Element of general symmetric matrix A

b =   Width of panel

$\underline{C}$ =   Transformation (rotation) matrix for deflections (equn. (27))

D = $\dfrac{Et^3}{12(1-\mu^2)}$ = plate stiffness

d =   Depth of stiffener

E =   Young's modulus

G =   Shear modulus

h =   Half band width of stiffness matrix

I_s =   Moment of inertia of stiffener

$\underline{K}$ =   Stiffness Matrix

$\underline{K}^G$ =   Geometric stiffness matrix

$\underline{K}^O_m$ =   Complete stiffness matrix of element due to mth term of Fourier series

$\underline{K}^G_m$ =   Geometric stiffness matrix of element due to m-th term of Fourier series

k =   Buckling factor $= \dfrac{\sigma_{cr}}{E}\ \dfrac{12(1-\mu^2)}{\pi^2}\ \left(\dfrac{b}{t}\right)^2$

k   = S/a = Factor   for   relative   width   of   edge   strip
         (Fig.20)

$\underline{k}^{(o)}$, $\underline{k}^{(g)}$ = First   and   second   order   element   stiffness
         matrix, resp.

$\underline{k}^{(1)}$, $\underline{k}^{(2)}$, $\underline{k}^{(3)}$, $\underline{k}^{(4)}$ =   Element stiffness   matrices (equns.
         (18), (20), (22) and (23))

L   = Length of panel

L_1 = Length of local buckles

L_{euler} = First Euler length (Section 5.2 and Fig.8)

M   = Bending and twisting moments per unit length (Fig.4)

m   = Number of term in Fourier series

N   = Number of elements

n   = Number of stiffeners

r   = Factor   for relative width of stiffener flange for L
         and J plates (Fig.18)

r   = Radius of gyration

r_m = Vector of in-plane edge displacements (equn.(15))

$\underline{r}$ = Eigenfunction (buckling mode)

S   = Spacing of stiffener (Fig.18)

t   = Thickness of element (Fig.2)

t_1 = Thickness of deck plate (Fig.20)

t_2 = Thickness of stiffener (Fig.18)

U = Total potential energy of strip element (equn.(2))

U_m = Total potential energy of element from mth term in Fourier series

$U_m^{(1)}$, $U_m^{(2)}$ = In-plane and out-of-plane total potential energy from first order analysis of mth term in Fourier series

$U_m^{(3)}$, $U_m^{(4)}$ = In-plane and out-of-plane total potential energy from second order analysis of mth term in Fourier series

U_1 to U_6 = Functions defined in equation (19)

$u,v,w,$ = Displacements in the x,y and z directions, resp.

u_i, v_i, w_i, θ_i = Displacements (Fig.2) and rotation of ith edge

u_j, v_j, w_j, θ_j = Displacements (Fig.2) and rotation of jth edge

$\bar{u}_i$, $\bar{v}_i$, $\bar{w}_i$ = Components of u_i, v_i, w_i, resp. after transformation from local to global coordinates

u_{im}, v_{im}, w_{im} = Maximum values of mth deflection term in Fourier series associated with ith edge (Fig.2)

u_{jm}, v_{jm}, w_{jm} = Maximum values of mth deflection term in Fourier series associated with jth edge (Fig.2)

W_1 to W_6 = Functions defined in equation (21)

$\underline{X}$ = Eigenvector (Section 6.1)

x, y, z = Element coordinate axes (Fig.2)

Z = Section modulus of beam (Example 8.1)

α = Angle of inclination of global axes to element axes (Fig.5)

α = Inclination of stiffener web to deck plate(Fig.18)

β = $\dfrac{m\,\pi}{L}$

γ_{xy} = Additional in-plane shear strain after buckling

δ = Total central deflection in a strut or plate (Fig.1)

δ_o = Initial central deflection in a strut or plate (Fig.1)

$\underline{\delta}_m$ = Vector of out-of-plane displacements (equn(16)or vector of all element edge displacements (equn.(24))

δ_{ij} = Kronecker delta (= 1 if i = j; = o if i $\neq$ j)

ε_x = Additional in-plane strain in x-direction after buckling

ε_y = Additional in-plane strain in y-direction after buckling

$\bar{\epsilon}_y$ = Apparent strain in y-direction (equn(2))

Θ_i, Θ_j = Maximum values of mth rotation term in Fourier series associated with ith and jth edges, resp. (Fig.2)

λ = $\dfrac{1}{\sigma_{cr}}$

μ = Poisson's ratio

Σ = Summation

σ = Axial stress (Fig.1)

σ_e = Euler stress

σ_F = Yield stress

σ_{cr} = Critical value of axial stress

σ_x = In-plane stress in x-direction

σ_y = Additional in-plane stress in y-direction

$\bar{\sigma}_y$ = Axial stress in y-direction at point of buckling (Fig.2)

In Example 8.1 the notation of DASt-Richtlinie 012 is generally used and in Example8.2 that of the Merrison Rules applies.

B Theorie

Allgemeine Grundlagen und Gebrauch der Tabellen

Deutsche Übersetzung von A. M. Thierauf

Inhaltsverzeichnis · Teil B

Vorwort

Ausgesteifte Platten sind seit langem im Brückenbau, im Schiffsbau und im Mastbau üblich. Sie stellen bewährte Konstruktionselemente dar mit sehr hoher Materialausnutzung bei geringem Gewicht. In gewissen Fällen zeigen sich jedoch Ausbeulungen in einer oder in mehreren gleichzeitig auftretenden Beulformen. Die bestehenden Bemessungsvorschriften enthalten Regeln für die Berechnung der Lasten, bei denen Versagen durch Beulen oder durch Erreichen der Fliessgrenze eintritt (Grenzlast, Gebrauchslast).

Die Berechnung einer kritischen Spannung nimmt einen wesentlichen Platz bei allen Nachweisen ein. Die kritische Spannung hängt von vielen Parametern ab, so z.B. von den Randbedingungen und der Geometrie des Plattenquerschnitts. Eine kurze ausgesteifte Platte zeigt im allgemeinen lokale Beulformen, eine längere Platte mit demselben Querschnitt kann hingegen wie ein Eulerstab durch Ausknicken versagen.

Für den Entwurf einer ausgesteiften Platte sind häufig mehrere aufwendige Vorberechnungen erforderlich. Es ist das Ziel des vorliegenden Buches, diese Arbeit zu erleichtern und so zu reduzieren, dass ein Vergleich mit einer Querschnittstafel möglich ist. Die hier vorliegenden Tafeln enthalten neben den üblichen Querschnittswerten die Spannung, bei der lokales Ausbeulen auftritt und die Länge, bei der sich die Platte wie ein Euler'scher Druckstab zu verhalten beginnt.

Die Bemessungsgrössen wurden mit einem Rechenprogramm auf der Grundlage der Methode der Finiten Elemente (Finite Strips, "endliche Streifen") berechnet. Einzelheiten der Berechnung werden der Vollständigkeit halber dargestellt. Fur den Gebrauch der Tafeln ist das Verständnis dieser theoretischen Grundlagen nicht erforderlich, wichtig hierfür sind jedoch die Abschnitte 1, 5.2 und 7 bis 8.4.

Obwohl bekanntlich der Vergleich von theoretisch begründeten Ergebnissen mit Bemessungsvorschriften, die ein gewisses Mass von Erfahrungswerten beinhalten, schwierig ist, wurde dies hier anhand der Beispiele 8.1 und 8.2 versucht.

Das erste Beispiel zeigt die Anwendungsmöglichkeit der Tafeln im Zusammenhang mit Bemessungrichtlinien; das zweite ergibt einen Vergleich für die kritische Spannung. Dieser Vergleich führt zu interessanten Ergebnissen, es zeigt aber auch den relative geringen Rechenaufwand, der sich durch die Benutzung der Tafeln ergibt.

Die Autoren danken Frau Chr. Hausmann fur die sorgfältige Aufbereitung der Daten und Darstellung der Rechenergebnisse, Frau M. Mehl für die Erstellung der Diagramme und Frau U. Riewe für die Anfertigung des Manuskripts. Herr Artur Senftleben hat dankenswerter Weise das Manuskript und die Berechnung der Beispiele auf Vollständigkeit und Richtigkeit überprüft.

Die Verfasser und ihre Mitarbeiter haben alle Ergebnisse dieses Buches sorgfältigst überprüft; es könnten jedoch trotzdem Fehler enhalten sein. Die Verfasser und der Verlag übernehmen keine Verantwortung fur Schaden, die durch die Benutzung dieses Buches entstehen könnten, sei es durch etwaige Fehler oder durch eine fehlerhafte Anwendung durch den Benutzer. Sollte der Leser Fehler oder Unstimmigkeiten auffinden, so wären die Verfasser für eine Mitteilung dankbar. Die vorliegende Arbeit wurde teilweise von der Deutschen Forschungsgemeinschaft unterstützt. Die numerischen Ergebnisse wurden mit den Rechenanlagen der Universitäten Düsseldorf und Monash (Melbourne, Australien) erhalten.

Noel W. Murray.
Georg Thierauf.

1 Einleitung

Die elastische Stabilität ausgesteifter Platten weist viele
Besonderheiten auf, wie sie bei der Stabilität von Stützen
und Rahmentragwerken nicht auftreten. Die Unterschiede sind
zusammenfassend in Fig. 1 dargestellt. In den dort gezeig-
ten Fällen erhält man nach Theorie 1. Ordnung (Theorie 2.
Ordnung ergibt fast dasselbe Resultat) eine horizontale
Asymptote bei Erreichen der Beullast; das Verhalten
einer Stütze mit Vorverformung δ_o folgt eng einer Hyperbel,
die in Fig. 1(a) dargestellt ist. Den Wert der kritischen
Spannung - in diesem Fall ist es die Eulerspannung σ_e - und
die Beulform erhält man durch Lösung des folgenden
Eigenwertproblems:

$$(\underline{K} + \sigma_e \underline{K}^G)\,\underline{r} = \underline{O} \tag{1}$$

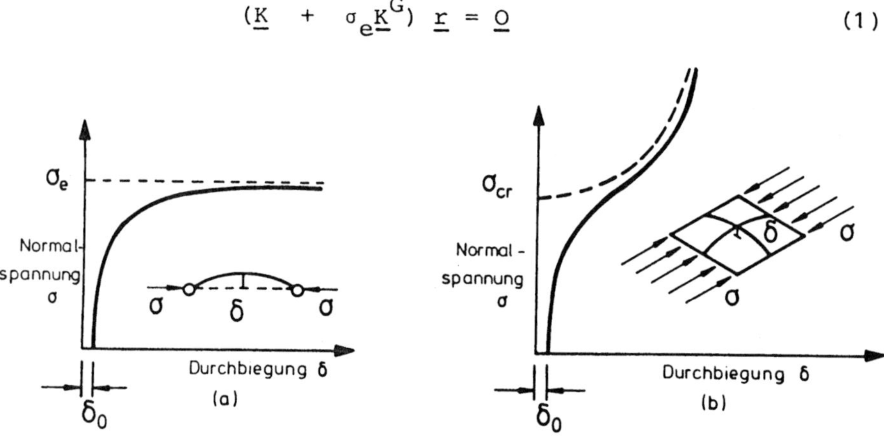

Fig. 1 Vergleich des Stab-und Plattenverhaltens

Hierbei ist $\underline{K}$ die bekannte Steifigkeitsmatrix, $\underline{K}^G$ ist die
geometrische Steifigkeitsmatrix [1] und $\underline{r}$ ist der Vektor,
der die Geometrie der Beulform bestimmt (Eigenvektor). σ_e
ist der kleinste Eigenwert von Gleichung (1). Die Beulform
der Platte (oder des Tragwerks) erhält man durch Lösung der
Gleichung (1), die Größe der Ausbeulung kann jedoch
nicht durch ein Eigenwertproblem berechnet werden.
Selbstverständlich beeinflussen die Randbedingungen, d.h.

die Verformungsbehinderungen an den Plattenenden sowohl die Beulspannung als auch die Beulform.

Im Fall einer einfachen Platte ist die Asymptote häufig eine ansteigende Kurve (Fig.1 (b)). Die Ableitung dieser Asymptote erfordert die Anwendung der Theorie großer Verformungen [2] auf eine Platte ohne Vorverformungen. Aufgrund der Nichtlinearität dieses Problems, das sich nach Theorie 2. Ordnung ergibt, erhält man die Lösungen nur durch iterative Verfahren wie mit dem Newton-Raphson-Verfahren. Es ist jedoch möglich, mit der linearen Stabilitätstheorie die Werte der kritischen Spannung σ_{cr} und die qualitative Beulform zu berechnen. Dies erfordert die Lösung eines Eigenwertproblemes wie in Gleichung (1). Für die Bemessung ist die kritische Spannung, bei der eine einzelne, ideal ebene Platte ausbeult, eine wertvolle Information; für den Ingenieur ist sie so wichtig wie die Euler'sche Knickbelastung oder die Fließspannung.

Die Randbedingungen beeinflussen das Verhalten der Platte wesentlich und es ist nötig, sowohl die Scheiben- als auch die Platten-Randbedingungen festzulegen. So kann z.B. ein Rand frei drehbar gelagert sein, bei gleichzeitiger Behinderung der Verschiebungen in der Ebene, oder er kann frei drehbar und frei verschieblich gelagert sein. In der Praxis können die verschiedensten Kombinationen von Scheiben- und Platten-Randbedingungen auftreten.

Im Fall einer ausgesteiften Platte, bestehend aus ebenen Plattenteilen, muß der Ingenieur die Spannung kennen, bei der lokales Ausbeulen auftritt. Bei dieser Spannung ändert sich die Form des Querschnitts. Ebene Plattenelemente beulen aus während die Längsknotenachsen, d.h. die Verbindung von zwei oder mehreren nicht-koplanaren Platten meist gerade bleiben. Zwei weitere Gesichtspunkte sind bei lokaler Ausbeulung zu beachten: Erstens wird die Normalspannung um-

gelagert; sie konzentriert sich auf die Nähe der Längsknotenachsen und wird mit zunehmendem Abstand kleiner. Zwei-
tens nimmt die Normalsteifigkeit plötzlich ab aufgrund der
Abnahme der mittragenden Plattenbreite. Zwei Beispiele sollen diese Aussage verdeutlichen. Eine idealebene, quadratische und gelenkig gelagerte Platte ohne Vorverformung und
mit frei verschieblichen Rändern in der Plattenebene
besitzt eine durch den Elastizitätsmodul E bestimmte
Steifigkeit vor dem Ausbeulen; nach dem Ausbeulen reduziert
sich die Steifigkeit jedoch um ca. 50 %. Wird nun ein Rand
frei verschieblich, so daß die Platte sich senkrecht zur
Plattenebene verformen kann, so ist die Steifigkeit nach
dem Ausbeulen Null. Diese Erscheinung hat wichtige
Konsequenzen für die Bemessung dünnwandiger Hohlprofile,
Z-, I- und Winkelquerschnitte und für ausgesteifte Platten,
deren Steifen meist freie Ränder haben.

Eine ausgesteifte Platte kann nicht nur lokal, sondern auch
__global ausbeulen__ . Eine dünnwandige, kurze Platte mit qua-
dratischem Hohlquerschnitt und mit gelenkigen Endlagerungen
beult z.B. lokal aus. Wird diese Platte jedoch hinreichend
verlängert, so verhält sie sich wie ein Eulerstab und die
Querschnittsform bleibt beim Verlust der Stabilität
erhalten. Für den Ingenieur ergibt sich das Problem
abzuschätzen, ob eine gegebene Platte zuerst lokal oder
global ausbeult. Er wird vielleicht versuchen, Länge und
Querschnitt zu "optimieren", so daß lokales und globales
Ausbeulen bei der gleichen Normalspannung auftreten. Diesen
Ansatz sollte man jedoch vermeiden, da eine
__Interaktion zwischen den Beulformen__ besteht. Dies bedeutet,
daß die tatsächlich vorhandene Beulspannung kleiner ist als
die Beulspannung, die sich unter der Annahme ergibt, daß
sich die Beulformen gegenseitig nicht beeinflussen [3,4] .
Außerdem haben Vorverformungen einen größeren Einfluß, wenn
diese beiden charakteristischen Spannungen nahezu gleich
sind. Derartige Tragwerke bezeichnet man als empfindlich
für Vorverformungen.

In diesem Buch werden die Beulprobleme als lineare Eigenwertprobleme behandelt, d.h. es werden nur minimale kritische Spannungen und Beulformen berechnet. Hierzu ist es notwendig, die <u>kritische Länge</u> für eine gegebene Querschnittsform zu errechnen. Bei Stützlängen, die größer sind als diese kritische Länge, tritt Knicken auf, bei Stützlängen, die kleiner sind als diese kritische Länge, tritt lokales Ausbeulen auf. Dieses Konzept ist ausführlicher in Abschnitt 5.3 beschrieben. Es wurde hier nicht versucht, das Nachbeulverhalten der Tragwerke zu bestimmen.

Eine Übersicht über verschiedene Stabilitätsuntersuchungen bis zum Jahr 1970 gibt Bulson [3]. Man erkennt, daß das Problem der elastischen Stabilität ausgesteifter Platten auf verschiedene Weise gelöst werden kann. Obwohl viele Verfahren zur genauen Lösung spezieller Probleme zur Verfügung stehen, tauchen doch Schwierigkeiten auf, wenn eine große Anzahl von Problemen systematisch bearbeitet werden soll. Diese Schwierigkeiten treten insbesondere bei der Erstellung von Bemessungstabellen wegen des enormen Aufwandes an numerischer Arbeit auf. Das beste Beispiel einer systematischen Berechnung dieser Art sind die Tabellen von Klöppel und Scheer [5], die in Deutschland und im Ausland häufig benutzt werden. Sie erfassen ein breites Feld von Platten mit symmetrischen Aussteifungen zur Plattenmittelebene. Dabei wurde das Ritz'sche Verfahren zur Berechnung expliziter Lösungen des Eigenwertproblems angewendet.

Ein vergleichbares Handbuch für Platten mit einseitig ange- ordneten Aussteifungen existiert nicht. Auf den ersten Blick scheint es, daß die Vielfalt der möglichen geometrischen Formen einen Ansatz wie den von Klöppel/Scheer [5] verbiete. Weiterhin sollte eine vollständige Lösung des Beulproblems das plastische

Werkstoffverhalten, möglicherweise auf der Grundlage der Finiten-Element-Methode, erfassen. Für die nächste Zukunft scheint es jedoch unwahrscheinlich, daß größere Berechnungen mit Finiten Elementen für jede einzelne ausgesteifte Platte in der Praxis durchgeführt werden können. Beim Entwurf eines Tragwerks müssen eine Reihe von Größen festgelegt werden. Im Fall einer ausgesteiften Platte sind dies alle üblichen Querschnittswerte, die Fließspannung und die Spannungen, bei denen lokales und globales Ausbeulen auftritt. Es ist das Ziel dieses Handbuches, dem Ingenieur diese Größen für eine Vielzahl ausgesteifter Platten bereitzustellen.

Für den Entwurf ausgesteifter Platten sind Berechnungsverfahren für die kritische Spannung in den "Merrison Design Rules" [6] und in DIN 4114 [7] festgelegt. Diese Berechnungsverfahren sind Näherungsverfahren; sie ergeben trotztdem i.A. Werte für die kritische Spannung, die auf der sicheren Seite liegen.

Bei der Vorbereitung der Tabellen für dieses Buch war das schwierigste Problem die Berechnung der Spannung, bei der lokales Ausbeulen auftritt. Die Autoren waren bemüht das leistungsfähigste Verfahren zu wählen, welches außerdem leicht erweitert werden kann zur Berücksichtigung von Einflüssen höherer Ordnung ohne dabei zu kompliziert zu werden. Die <u>Methode der Finiten Streifen</u> , eine Variante der bekannten Methode der Finiten Elemente, scheint diesen Anforderungen am besten gerecht zu werden. Als erster Schritt wurde ein Testprogramm (PLATE) entwickelt, daß nachfolgend beschrieben ist. Die theoretischen Grundlagen dafür finden sich bei Cheung [8] . Für die numerische Lösung wurden hier spezielle Eigenschaften ausgenutzt, die es ermöglichen, das Verfahren der Finiten Streifen für Probleme größeren Umfangs bei relativ geringem Aufwand an Rechenzeit einzusetzen. Schließlich wurden drei weitere

Rechenprogramme (PLATE 1, PLATE 2 und PLATE 3 - siehe
Abschn. 6 und 7) zur Berechnung der vorliegenden Tabellen
erstellt.

2 Theoretische Grundlagen

2.1   Geometrie

"The finite strip method is an extension of the well-known
finite element method. This method is, however, semi-analytical in nature ...." (Cheung [8]) - (Das Verfahren
der Finiten Streifen ist eine Erweiterung der bekannten Methode der Finiten Elemente. Diese Methode ist jedoch semi-
analytisch.......)

Der Leser, der mit der Methode der Finiten Elemente
vertraut ist, wird erkennen, daß der Finite Streifen nur
ein besonderes Finites Element mit einem angenommenen
Verschiebungsansatz ist, der gewisse Randbedingungen einer
rechteckigen Platte erfüllt. Ein typischer Elementstreifen
ist in Fig.2 in kartesischen Koordinaten dargestellt.

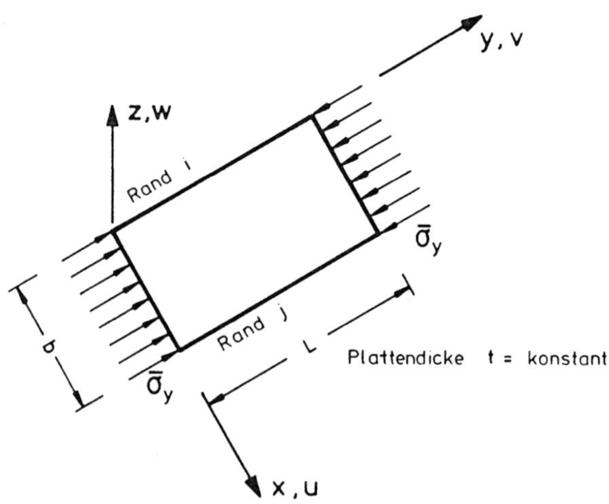

Fig. 2 Finites Streifenelement

Das Element befindet sich im Gleichgewicht mit der
Last $\bar{\sigma}_y$bt, die entlang der Ränder y = konstant angreift.
Die Koordinaten sind lokale Element-Koordinaten. Der

Streifen wird als Teil eines allgemeineren, längsversteif-
ten Tragwerks (Fig.3) betrachtet.

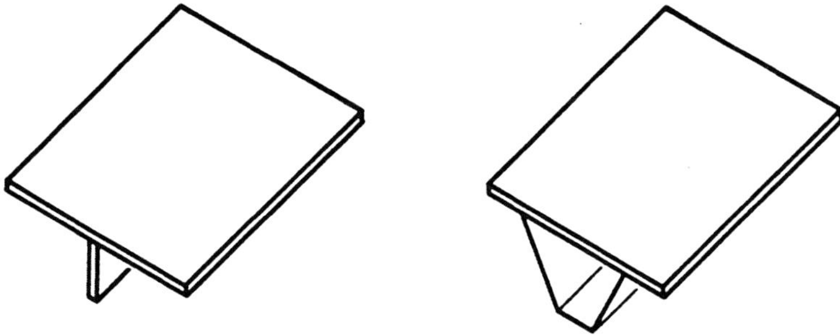

Fig. 3 Typische finite Streifentragwerke

2.2   Potentielle Energie

Man stelle sich einen Streifen (Fig.2) vor, der bis zum Be-
ginn des Ausbeulens eben ist und unter der konstanten Rand-
spannung $\bar{\sigma}_y$ steht. Wir nehmen an, daß in diesem Zustand die
gesamte potentielle Energie des Streifens Null ist; dies
stellt einen beliebigen Referenzpunkt dar, von dem aus wir
die Änderungen des Gesamtpotentials messen können. Wir den-
ken uns nun den Streifen in seine ausgebeulte Form verscho-
ben, so daß die Verschiebungen eines beliebigen Punktes
(x,y) in (u,v,w) übergehen. Die Spannungskomponenten ändern
sich dann wie folgt:

$$\text{in x-Richtung von} \quad 0 \text{ nach } \sigma_x$$
$$\text{in y-Richtung von} \quad \bar{\sigma}_y \text{ nach } \bar{\sigma}_y + \sigma_y$$
$$\text{die Schubspannung von} \quad 0 \text{ nach } \tau_{xy}$$

Die Dehnungen ändern sich in diesem Punkt wie folgt:

$$\text{in x-Richtung vergrößert sich die Dehnung um } \varepsilon_x$$
$$\text{in y-Richtung vergrößert sich die Dehnung um } \varepsilon_y$$
$$\text{die Schubverzerrung vergrößert sich um } \gamma_{xy}$$

Bei einem Element mit den Abmessungen dx x dy   setzt sich die Vergrößerung der gesamten potentiellen Energie während des Ausbeulens wie folgt zusammen:

Aus der Arbeit von Sekundärspannungen mit den zugehörigen Dehnungen, nämlich

$$t\left[\frac{1}{2}\,\sigma_x\,\varepsilon_x + \frac{1}{2}\,\sigma_y\,\varepsilon_y + \frac{1}{2}\,\tau_{xy}\,\gamma_{xy}\right]dx\,dy\ ,$$

aus der Arbeit der Biege- und Torsionsmomente, die auf die Ränder des Elementes einwirken, nämlich

$$\left[-\frac{1}{2}\,M_x\,\frac{\partial^2 w}{\partial x^2} - \frac{1}{2}\,M_y\,\frac{\partial^2 w}{\partial y^2} + M_{xy}\,\frac{\partial^2 w}{\partial x\,\partial y}\right]dx\,dy\ ,$$

und schließlich aus der Arbeit von $\bar{\sigma}_y$ an virtuellen Dehnungen $\bar{\varepsilon}_y$, durch welche Verdrehungen des Elementes um Achsen parallel zur x- und zur z-Achse ermöglicht werden (s.Abschn. 2.3).

Die gesamte potentielle Energie eines Elementstreifens ergibt sich deshalb zu

$$U = \frac{t}{2}\int_0^L\int_0^b (\sigma_x\,\varepsilon_x + \sigma_y\,\varepsilon_y + \tau_{xy}\,\gamma_{xy})\ dx\,dy$$

$$+ \frac{1}{2}\int_0^L\int_0^b (-M_x\,\frac{\partial^2 w}{\partial x^2} - M_y\,\frac{\partial^2 w}{\partial y^2} + 2M_{xy}\,\frac{\partial^2 w}{\partial x\,\partial y}\ )\ dx\,dy \qquad (2)$$

$$+ t\int_0^L\int_0^b \bar{\sigma}_y\,\bar{\varepsilon}_y\ dx\,dy$$

Die positiven Richtungen von M_x, M_y, M_{xy} sind in Fig.4 dargestellt.

Das erste Glied in (2) ist die Verzerrungsenergie eines

ebenen   Spannungsproblems, das zweite ist die Verzerrungsenergie eines Plattenbiegeproblems und das dritte Glied ist
die potentielle Energie der äußeren Kräfte.

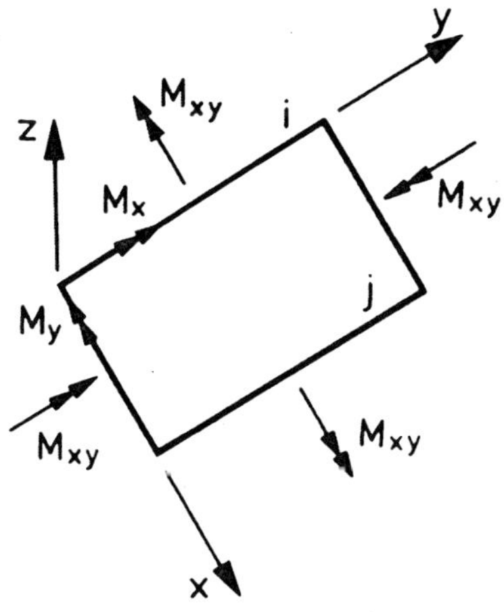

Fig. 4   Positive Plattenbiege - und Drillmomente
(Schraubenregel für pos. Richtung)

Die Spannungs-Dehnungs-Beziehung des ebenen Spannungszu-
standes für ideal-elastische, isotrope Werkstoffe ist
gegeben durch

$$\begin{bmatrix} \sigma_x \\ \sigma_y \\ \tau_{xy} \end{bmatrix} = \frac{E}{1-\mu^2} \begin{bmatrix} 1 & \mu & 0 \\ \mu & 1 & 0 \\ 0 & 0 & \frac{1-\mu}{2} \end{bmatrix} \begin{bmatrix} \varepsilon_x \\ \varepsilon_y \\ \gamma_{xy} \end{bmatrix} \tag{3}$$

wobei E der Elasitizitätsmodul und μ die
Poisson'sche Zahl ist.

Die Biegemomente erhält man als zweite Ableitungen der Verschiebungen w (Fig.2):

$$
\begin{bmatrix} M_x \\ M_y \\ M_{xy} \end{bmatrix} = D
\begin{bmatrix} 1 & \mu & 0 \\ \mu & 1 & 0 \\ 0 & 0 & 1-\mu \end{bmatrix}
\begin{bmatrix} -\dfrac{\partial^2 w}{\partial x^2} \\ -\dfrac{\partial^2 w}{\partial y^2} \\ \dfrac{\partial^2 w}{\partial x\, \partial y} \end{bmatrix}
\tag{4}
$$

Hierbei ist $D = \dfrac{Et^3}{12\,(1-\mu^2)}$ die Plattensteifigkeit.

Man nimmt nun an, daß das Verzerrungsfeld in der Ebene des Elementes aus einem Verschiebungsfeld (u, v, w) mit

$$
\begin{aligned}
\varepsilon_x &= \frac{\partial u}{\partial x} \\[2ex]
\varepsilon_y &= \frac{\partial v}{\partial y} \\[2ex]
\gamma_{xy} &= \frac{\partial u}{\partial y} + \frac{\partial v}{\partial x}
\end{aligned}
\tag{5}
$$

abgeleitet werden kann. Hierbei sind die nichtlinearen Glieder für ε_x und γ_{xy} des Lagrange'schen Verzerrungstensors vernachlässigt.

2.3   Orthogonalität der Verschiebungsansätze

Ein wesentliches Merkmal der Finiten Streifen ist die Annahme eines orthogonalen Verschiebungsfeldes, das alle Randbedingungen bei $y = 0$ und $y = L$ erfüllt. Die folgenden Fourierreihen erfüllen die homogenen Randbedingungen für einen Streifen, der an beiden Enden gelenkig gelagert ist.

$$u = \sum_{m=1}^{\infty} \left[\left(1 - \frac{x}{b}\right) u_{im} + \left(\frac{x}{b}\right) u_{jm} \right] \sin \beta_m y \quad ,$$

$$v = \sum_{m=1}^{\infty} \left[\left(1 - \frac{x}{b}\right) v_{im} + \left(\frac{x}{b}\right) v_{jm} \right] \cos \beta_m y \quad , \tag{6}$$

$$w = \sum_{m=1}^{\infty} \left[\left(1 - \frac{3x^2}{b^2} + \frac{2x^3}{b^3}\right) w_{im} + \left(x - \frac{2x^2}{b} + \frac{x^3}{b^2}\right) \Theta_{im} \right.$$
$$\left. + \left(\frac{3x^2}{b^2} - \frac{2x^3}{b^3}\right) w_{jm} + \left(\frac{x^3}{b^2} - \frac{x^2}{b}\right) \Theta_{jm} \right] \sin \beta_m y$$

Hierbei ist $\beta_m = \frac{m\pi}{L}$, $m = 1,2, \ldots$; die Indizes i und j verweisen auf die jeweiligen Ränder i und j des Streifens (Fig.2).

Das angenommene Verschiebungsfeld ist linear in x für die Verschiebungen u, v in der Ebene; es ist kubisch in x für die Verschiebungen w senkrecht zur Ebene. Alle Verschiebungsfunktionen sind trigonometrische Funktionen in y. Für y = O und y = L erhalten wir u = O, w = O und

$$\sigma_y = \frac{E}{1-\mu^2}(\mu\varepsilon_x + \varepsilon_y) = \frac{E}{1-\mu^2}\left(\mu\,\frac{\partial u}{\partial x} + \frac{\partial v}{\partial y}\right) = 0$$

Die Verschiebungen v sind lineare Funktionen in x an jedem Ende.

Für x = O und x = b sind die Randverschiebungen Fourierreihen in y

$$x = O: \quad u = \sum_m u_{im} \sin \beta_m y$$

$$v = \sum_m v_{im} \cos \beta_m y \tag{7}$$

$$w = \sum_m w_{im} \sin \beta_m y$$

$$x = b: \quad u = \sum_m u_{jm} \sin \beta_m y$$

$$v = \sum_m v_{jm} \cos \beta_m y \tag{8}$$

$$w = \sum_m w_{jm} \sin \beta_m y$$

Die Verdrehung $\frac{\partial w}{\partial x}$ erhält man aus (6) wie folgt:

$$\frac{\partial w}{\partial x} = \sum_m \left[\left(-\frac{6x}{b^2} + \frac{6x^2}{b^3} \right) w_{im} + \left(1 - \frac{4x}{b} + \frac{3x^2}{b^2} \right) \Theta_{im} \right.$$

$$\left. + \left(\frac{6x}{b^2} - \frac{6x^2}{b^3} \right) w_{jm} + \left(\frac{3x^2}{b^2} - \frac{2x}{b} \right) \Theta_{jm} \right] \sin \beta_m y$$

Bei $x = 0$ ist

$$\frac{\partial w}{\partial x} = \sum_m \Theta_{im} \sin \beta_m y \tag{9}$$

und bei $x = b$:

$$\frac{\partial w}{\partial x} = \sum_m \Theta_{jm} \sin \beta_m y \tag{10}$$

Die Verschiebungsfunktionen bestimmen die Spannungs- und Dehnungsfelder entsprechend (3) und (5) und die Verteilung der Biegemomente durch (4). Die zugehörigen Anteile der potentiellen Energie (die ersten Integrale in Gleichung (2)) führen zu Integralen der Form:

$$\frac{2}{L} \int_0^L \sin \beta_i y \, \sin \beta_j y \, dy = \delta_{ij} \tag{11}$$

Dabei ist δ_{ij} das Kronecker-Delta. Die Orthogonalität der angenommenen Verschiebungsfunktionen erleichtert daher die Berechnung der Verzerrungsenergie erheblich. Das dritte Glied des Gesamtpotentials ist bedingt durch die äußere Spannung $\bar{\sigma}_y$. Die entsprechende Dehnung

$$\bar{\varepsilon}_y = \frac{\partial v}{\partial y} + \frac{1}{2} \left[\frac{\partial u}{\partial y} \right]^2 + \frac{1}{2} \left[\frac{\partial w}{\partial y} \right]^2 \tag{12}$$

ist eine Näherung für den Lagrange'schen Verzerrungstensor. Die letzten beiden Glieder ergeben sich durch die axiale Verkürzung aufgrund der kleinen Verdrehungen eines Strei-

fens der Länge dy. Die Orthogonalität in Gleichung (2)
führt auch zu Integralen der Form

$$\frac{2}{L} \int_0^L \cos \beta_i y \, \cos \beta_j y \, dy = \delta_{ij} \tag{13}$$

mit den oben erwähnten Vorteilen.

Deshalb ist es möglich, die potentielle Energie für jedes
Reihenglied m getrennt zu berechnen:

$$U_m = U_m^{(1)} + U_m^{(2)} + U_m^{(3)} + U_m^{(4)} \tag{14}$$

Scheibenanteil Plattenanteil   Scheibenanteil Plattenanteil
    Theorie 1. Ordnung   Theorie 2. Ordnung

3 Elementmatrizen

Die potentielle Energie ist eine quadratische Form in den
Verschiebungskomponenten und kann für jedes Element
getrennt in lokalen Koordinaten berechnet werden. Wir
trennen die Scheiben- und Plattenanteile der Verschiebungskomponenten durch

$$\underline{r}_m^T = (u_i \ v_i \ u_j \ v_j)_m \tag{15}$$

und

$$\underline{\delta}_m^T = (w_i \ \Theta_i \ w_j \ \Theta_j)_m \,. \tag{16}$$

Hierbei kennzeichnet T die Transponierte des Vektors $\underline{r}_m$.
Man erhält die folgende Form der potentiellen Energie eines
Elements für ein beliebiges Reihenglied m[+) :

$$U_m = \frac{1}{2} \underline{r}_m^T \underline{k}_m^{(1)} \underline{r}_m + \frac{1}{2} \underline{\delta}_m^T \underline{k}_m^{(2)} \underline{\delta}_m$$
$$+ \frac{1}{2} \underline{r}_m^T \underline{k}_m^{(3)} \underline{r}_m + \frac{1}{2} \underline{\delta}_m^T \underline{k}_m^{(4)} \underline{\delta}_m \tag{17}$$

[+)] Indizes zur Elementnumerierung wurden zur Vereinfachung
weggelassen.

Die vier Elementmatrizen $\underline{k}^{(1)}$ bis $\underline{k}^{(4)}$ ergeben sich durch elementare aber langwierige Integration.

3.1 Scheibenanteil der Steifigkeitsmatrix nach Theorie 1. Ordnung

$$\underline{k}^{(1)} = \begin{bmatrix} U_1+U_2 & U_3-U_4 & -U_1+\dfrac{U_2}{2} & U_3+U_4 \\ & U_5+U_6 & -U_3-U_4 & \dfrac{U_5}{2}-U_6 \\ \text{symmetrisch} & & U_1+U_2 & -U_3+U_4 \\ & & & U_5+U_6 \end{bmatrix} \tag{18}$$

$$U_1 = \frac{LEt}{2(1-\mu^2)b} \qquad\qquad U_4 = \frac{L\beta_m Gt}{4}$$

$$U_2 = \frac{Lb\beta_m^2 Gt}{6} \qquad\qquad U_5 = \frac{Lb\beta_m^2 Et}{6(1-\mu^2)} \tag{19}$$

$$U_3 = \frac{L\beta_m \mu Et}{4(1-\mu^2)} \qquad\qquad U_6 = \frac{LGt}{2b}$$

$$\left(G = \frac{E}{2(1+\mu)} \ , \ \text{Schubmodul}\right)$$

3.2 Plattenanteil der Steifigkeitsmatrix nach Theorie 1. Ordnung

$$\underline{k}^{(2)} = \begin{bmatrix} W_1 & W_2 & W_3 & W_4 \\ & W_5 & -W_4 & W_6 \\ & & W_1 & -W_2 \\ \text{symmetrisch} & & & W_5 \end{bmatrix} \tag{20}$$

$$W_1 = DL\left(\frac{13}{70}b\beta_m^4 + \frac{6}{5b}\beta_m^2 + \frac{6}{b^3}\right)$$

$$W_2 = DL\left(\frac{11}{420}b^2\beta_m^4 + \left(\frac{\mu}{2} + \frac{1}{10}\right)\beta_m^2 + \frac{3}{b^2}\right)$$

$$W_3 = DL\left(\frac{9}{140}b\beta_m^4 - \frac{6}{5b}\beta_m^2 - \frac{6}{b^3}\right)$$

$$W_4 = DL\left(- \frac{13}{840}b^2\beta_m^4 + \frac{1}{10}\beta_m^2 + \frac{3}{b^2}\right) \qquad (21)$$

$$W_5 = DL\left(\frac{b^3\beta_m^4}{210} + \frac{2}{15}b\beta_m^2 + \frac{2}{b}\right)$$

$$W_6 = DL\left(- \frac{b^3\beta_m^4}{280} - \frac{1}{30}b\beta_m^2 + \frac{1}{b}\right)$$

3.3 Scheibenanteil der Steifigkeitsmatrix nach Theorie 2. Ordnung

$$\underline{k}^{(3)} = \frac{tL\beta_m^2}{2}
\begin{bmatrix}
\frac{b}{3} & 0 & \frac{b}{6} & 0 \\
0 & 0 & 0 & 0 \\
\frac{b}{6} & 0 & \frac{b}{3} & 0 \\
0 & 0 & 0 & 0
\end{bmatrix} \qquad (22)$$

3.4 Plattenanteil der Steifigkeitsmatrix nach Theorie 2. Ordnung

$$\underline{k}^{(4)} = \frac{tL\beta_m^2}{2}
\begin{bmatrix}
\frac{13}{35}b & \frac{11}{210}b^2 & \frac{9}{70}b & - \frac{13}{420}b^2 \\
 & \frac{1}{105}b^3 & \frac{13}{420}b^2 & - \frac{1}{140}b^3 \\
\text{symmetrisch} & & \frac{13}{35}b & - \frac{11}{210}b^2 \\
 & & & \frac{1}{105}b^3
\end{bmatrix} \qquad (23)$$

3.5 Vollständige Elementmatrizen in lokalen Koordinaten

Aus numerischen Gründen ist es vorteihaft, die Elementmatrizen in der üblichen Form einer Elementsteifigkeitsmatrix $\underline{k}^{(0)}$ nach Theorie 1. Ordnung und in eine geometrische Steifigkeitsmatrix $\underline{k}^{(g)}$ nach Theorie 2. Ordnung darzustellen. Die Knotenverschiebungen werden deshalb in der folgenden Form zusammengefaßt:

$$\underline{\delta}^T = (\overset{1}{u_i} \quad \overset{2}{v_i} \quad \overset{3}{w_i} \quad \overset{4}{\theta_i} \quad \overset{5}{u_j} \quad \overset{6}{v_j} \quad \overset{7}{w_j} \quad \overset{8}{\theta_j}) \tag{24}$$

Daraus ergeben sich folgende Matrizen:

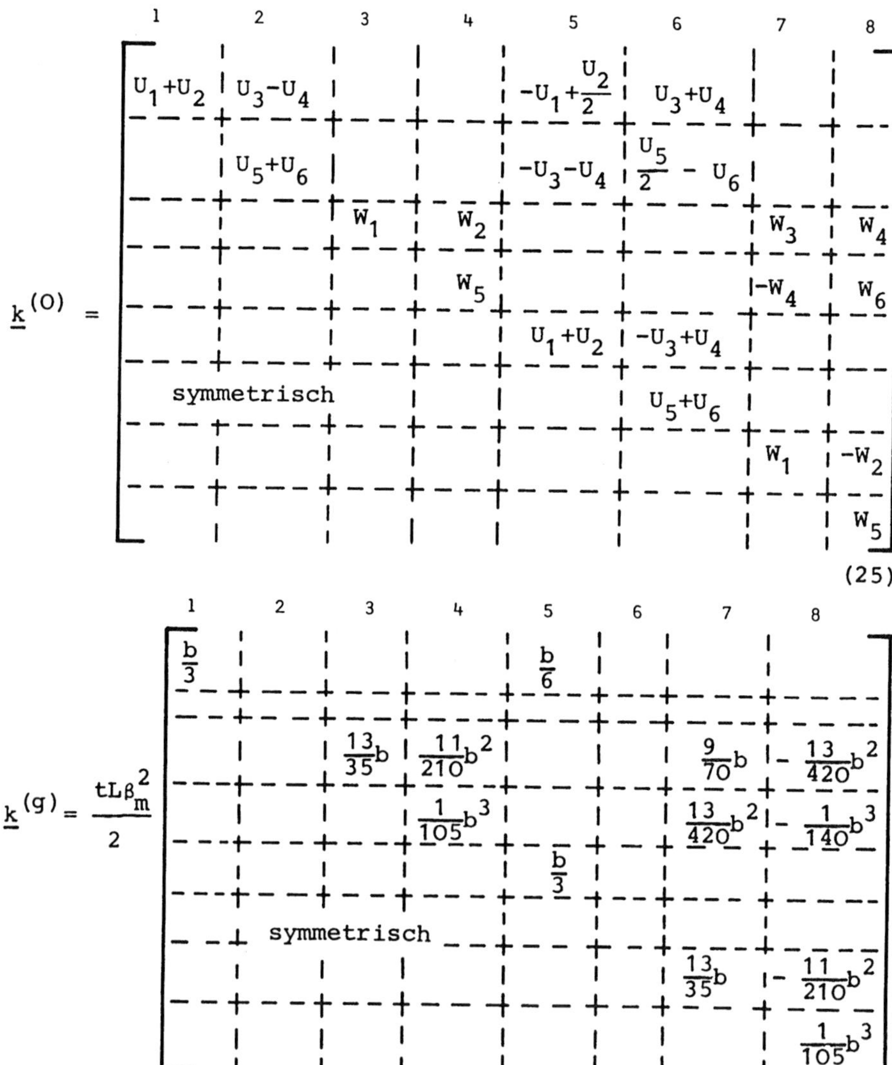

$$\underline{k}^{(0)} = \begin{bmatrix}
U_1+U_2 & U_3-U_4 & & & -U_1+\dfrac{U_2}{2} & U_3+U_4 & & \\
 & U_5+U_6 & & & -U_3-U_4 & \dfrac{U_5}{2}-U_6 & & \\
 & & \overline{W}_1 & \overline{W}_2 & & & \overline{W}_3 & \overline{W}_4 \\
 & & & W_5 & & & -W_4 & W_6 \\
 & & & & U_1+U_2 & -U_3+U_4 & & \\
\text{symmetrisch} & & & & & U_5+U_6 & & \\
 & & & & & & W_1 & -W_2 \\
 & & & & & & & W_5
\end{bmatrix} \tag{25}$$

$$\underline{k}^{(g)} = \frac{tL\beta_m^2}{2} \begin{bmatrix}
\dfrac{b}{3} & & & & \dfrac{b}{6} & & & \\
 & & & & & & & \\
 & & \dfrac{13}{35}b & \dfrac{11}{210}b^2 & & & \dfrac{9}{70}b & -\dfrac{13}{420}b^2 \\
 & & & \dfrac{1}{105}b^3 & & & \dfrac{13}{420}b^2 & -\dfrac{1}{140}b^3 \\
 & & & & \dfrac{b}{3} & & & \\
 & \text{symmetrisch} & & & & & & \\
 & & & & & & \dfrac{13}{35}b & -\dfrac{11}{210}b^2 \\
 & & & & & & & \dfrac{1}{105}b^3
\end{bmatrix}$$

3.6   Koordinatentransformation auf globale Koordinaten

Die potentielle Energie eines Elements stellt sich nun wie
folgt dar:

$$U_m = \tfrac{1}{2}\underline{\delta}_m^T\left(\underline{k}^{(o)} + \bar{\sigma}_y\underline{k}^{(g)}\right)\underline{\delta}_m \tag{26}$$

Die Koordinatentransformation (Verdrehung α)

$$\underline{\delta} = \underline{C}\,\bar{\underline{\delta}} \tag{27}$$

von lokalen ($\underline{\delta}$) auf globale Koordinaten ($\bar{\underline{\delta}}$) ergibt

$$U_m = \tfrac{1}{2}\bar{\underline{\delta}}^T\,\underline{C}^T\,(\underline{k}^{(o)} + \bar{\sigma}_y\underline{k}^{(g)})\,\underline{C}\,\bar{\underline{\delta}} \tag{28}$$

Hierbei ist $\underline{C}=(\underline{C}^T)^{-1}$ die Drehungsmatrix (Fig.5):

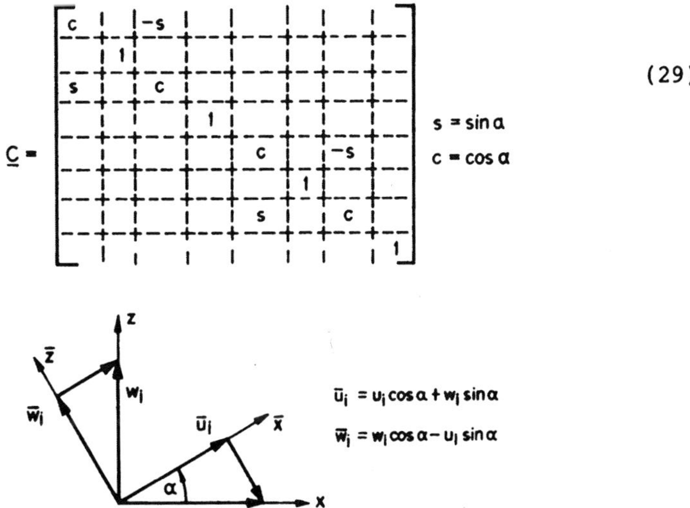

$$\tag{29}$$

Fig.5 Transformation auf globale Koordinaten

Um Nullmultiplikationen   zu   vermeiden,   werden die
Matrizenprodukte nach (28) für eine allgemeine symmetrische
Matrix ausgeführt.

$$\underline{A} = (a_{ij}), \quad i, j = 1, \ldots, 8$$

Hierbei wird die spezielle Struktur der Matrizen berücksichtigt, so z.B.,

$$a_{13} = a_{14} = a_{17} = a_{18} = 0$$

Dies führt zu der allgemeinen Form der Koordinatentransformation wie sie in Gleichung (30) dargestellt ist.

4 Verknüpfung der Elemente

Die Verknüpfung der Elemente wird nach der direkten Steifigkeitsmethode durchgeführt. Das Ergebnis ist die gesamte potentielle Energie, in allgemeiner Form für die ersten r Reihenglieder:

$$U_r = \frac{1}{2} \underline{\delta}_1^T (\underline{K}_1^O + \bar{\sigma}_y \underline{K}_1^G) \underline{\delta}_1 + \ldots + \frac{1}{2} \underline{\delta}_r^T (\underline{K}_r^O + \bar{\sigma}_y \underline{K}_r^G) \underline{\delta}_r \quad (31)$$

$\underline{K}_s^O$ ist die vollständige Steifigkeitsmatrix, die sich aus den üblichen Elementsteifigkeitsmatrizen $\underline{K}^O$ im s-ten Glied (m = s), und $\underline{K}_s^G$ ist die entsprechende geometrische Steifigkeitsmatrix.

Für ein Tragwerk, das aus n Elementen besteht, werden zuerst Nullmatrizen $\underline{K}_s^O$ und $\underline{K}_s^G$ mit der Dimension (4N x 4N) generiert. Für jedes Element (1, ...N) werden Untermatrizen $\underline{A}_{ii}$, $\underline{A}_{ij}$, $\underline{A}_{jj}$ aus (vgl. Gleichung (30))

$$\underline{C}^T \underline{AC} = \begin{bmatrix} \underline{A}_{ii} & \underline{A}_{ij} \\ & \\ \underline{A}_{ij} & \underline{A}_{jj} \end{bmatrix}$$

zu diesen Nullmatrizen addiert. Die Zahlen i und j sind die Knotennummern des Elementes (Fig.6).

$$
\underline{C}^T \underline{A} \underline{C} =
\begin{array}{c}
\\ (i) \\ \\ \\ \\ (j) \\ \\ \\
\end{array}
\left[
\begin{array}{cccc|cccc}
 & & (i) & & & & (j) & \\
c^2a_{11}+s^2a_{33} & ca_{12} & -csa_{11}+csa_{33} & sa_{34} & c^2a_{15}+s^2a_{37} & ca_{16} & -csa_{15}+csa_{37} & sa_{38} \\
 & a_{22} & -sa_{21} & 0 & ca_{25} & a_{26} & -sa_{25} & 0 \\
 & & s^2a_{11}+c^2a_{33} & ca_{34} & -sca_{15}+sca_{37} & -sa_{16} & s^2a_{15}+c^2a_{37} & ca_{38} \\
 & & & a_{44} & sa_{47} & 0 & ca_{47} & a_{48} \\
\hline
 & & & & c^2a_{55}+s^2a_{77} & ca_{56} & -sca_{55}+sca_{77} & sa_{78} \\
 & & & & & a_{66} & -sa_{56} & 0 \\
 & \text{symmetrisch} & & & & & s^2a_{55}+c^2a_{77} & ca_{78} \\
 & & & & & & & a_{88}
\end{array}
\right] \qquad (30)
$$

Die Matrizen $\underline{K}^O$ und $\underline{K}^G$ sind symmetrisch. Für Tragwerke mit Kettenstruktur, wie z.B. ausgesteifte Platten, ist der Maximalunterschied der Knotennumerierung in einem Element klein, verglichen mit 4N, und die Steifigkeitsmatrizen sind Bandmatrizen (Fig.7).

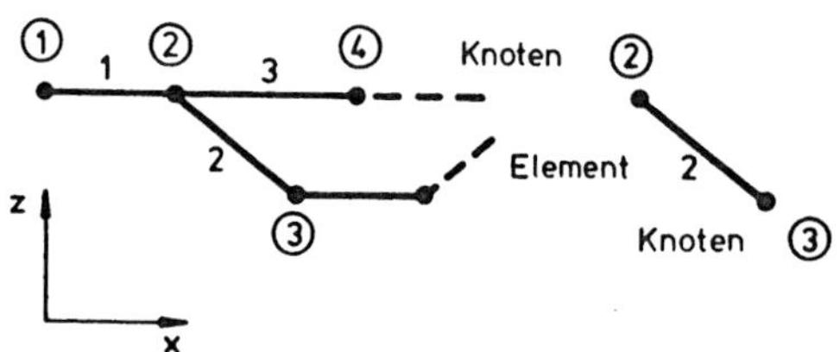

Fig. 6   Zusammenbau der Elemente

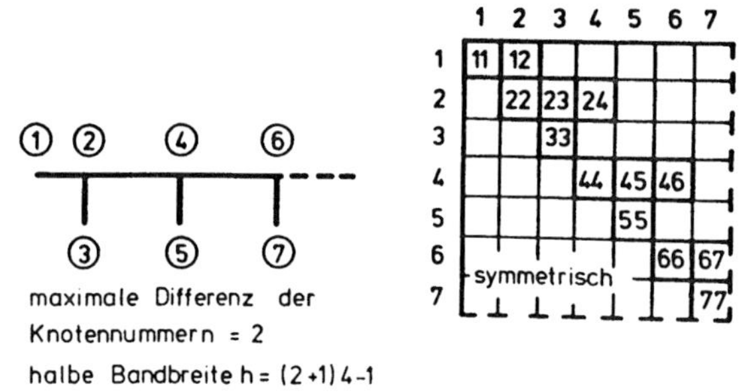

Fig. 7   Aufbau der Steifigkeitsmatrizen

An den Rändern x = konstant können spezielle Randbedingungen eingeführt werden. Das Streichen von Zeilen und Spalten in $\underline{K}^O$ und $\underline{K}^G$ ist identisch mit der Beschränkung der entsprechenden Freiheitsgrade der Verschiebungen auf Null an den Rändern x = konstant.

5 Das Eigenwertproblem

5.1 Allgemeine Form

Das Prinzip vom Minimum der potentiellen Energie, angewendet auf Gleichung (31), ergibt ein System von Eigenwertproblemen

$$m = 1: \quad (\underline{K}_1^O + \bar{\sigma}_y \underline{K}_1^G)\, \underline{\delta}_1 = \underline{O}$$

$$m = 2: \quad (\underline{K}_2^O + \bar{\sigma}_y \underline{K}_2^G)\, \underline{\delta}_2 = \underline{O} \tag{32}$$

$$\vdots$$

$$m = r: \quad (\underline{K}_r^O + \bar{\sigma}_y \underline{K}_r^G)\, \underline{\delta}_r = \underline{O}$$

Der kleinste Eigenwert $\bar{\sigma}_y = \sigma_{cr}$ ist die kritische Spannung des Tragwerks; der entsprechende Eigenvektor bestimmt die Verformung des Tragwerks durch die angenommene Verschiebungsfunktion (6). Der Index von $\underline{\delta}$ steht in Bezug zur Wellenlänge der Sinus- und Kosinusfunktionen

$$\beta_m = \frac{m\pi}{L} \tag{33}$$

Es ist nicht von vornherein bekannt, welche Beulform die kritische ist; theoretisch müssen alle Eigenwertprobleme für r in (32) gelöst werden. Dies ist sicherlich ein Nachteil der sonst sehr eleganten Formulierung der Beulberechnung mit Finiten Streifen. In der Praxis genügt es jedoch, Gleichung (32) für die ersten 10 bis 30 Eigenformen zu lösen.

5.2 Kritische Spannung und Erste Eulerlänge

Bei der Bemessung einer ausgesteiften Platte wird man mit dem Problem konfrontiert, zuerst die Querschnittsform festzulegen und dann den Abstand der Querstreifen wählen zu müssen. Die vorliegenden Tafeln erleichtern diese Aufgabe. In diesem Abschnitt sollen Sinn und Zweck der Bezeichnungen "Kritische Spannung" und "Erste Eulerlänge" erläutert werden. Es sind dies die wesentlichen Größen für eine Beuluntersuchung.

Für einen gegebenen Querschnitt erhält man die Lösung nach
Gleichung (32), indem man einen Wert für L vorgibt und dann
m die Werte 1, 2, 3 usw. durchlaufen läßt. Dieser Weg ist
jedoch langwierig und es ist nicht sicher, ob die minimale
Spannung erfaßt wird. Für eine systematische Berechnung ist
es zweckmäßig, m = 1 zu setzen und dann L anwachsen zu
lassen bis man im Rahmen einer praktischen Näherung alle
Minimalwerte von $\bar{\sigma}_y$ erhalten hat. Auf diese Weise wird die
Variable β stetig verändert und man findet für jeden Wert
von β die erforderliche Spannung $\bar{\sigma}_y$, bei der sich ein Ausbeulen mit Wellenlänge $\frac{\pi}{\beta}$ (s. Gleichung (33)) ergibt.

Wenn dieses Verfahren auf ausgesteifte Platten angewendet
wird, so ergeben sich im wesentlichen vier Typen von Beulkurven, die in Fig. 8 zusammengefaßt werden.

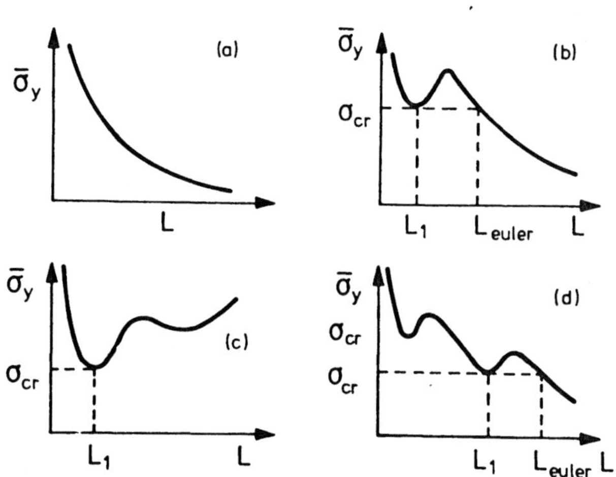

Fig. 8   Typische Beulkurven

Fig. 8 (a) zeigt, daß es keine minimale (kritische)
Spannung gibt; je länger L ist desto kleiner ist die
kritische Spannung. Ein Querschnitt mit diesem Verhalten
knickt wie ein Eulerstab aus. Ein Beispiel eines solchen
Querschnitts ist eine an den Längsrändern frei verschiebli-

che, ebene Platte. Ein weiteres Beispiel ist das Biegedrillknicken einer Stütze.

Fig. 8 (b) zeigt, daß lokales Ausbeulen mit Veränderung der Querschnittsform bei σ_{cr} mit relativ kleiner Wellenlänge eintritt. Wenn die angenommene Länge sich vergrößert, beult die ausgesteifte Platte letztlich wie ein Eulerstab mit konstantem Querschnitt aus. Wenn der Abstand zwischen den Querstreifen einer ausgesteiften Platte groß genug ist, tritt offensichtlich ein Euler'scher Stabilitätsfall mit einer Spannung kleiner als σ_{cr} ein. Die Länge des Stabes, bei der beide Beulspannungen gleich sind, wird als erste Eulerlänge bezeichnet. Damit soll darauf hingewiesen werden, daß bei wachsender Länge dies der erste Wert ist, bei dem der Stab wie ein Eulerstab ausknickt. Ein Beispiel für solches Verhalten ist eine ausgesteifte Platte mit nahezu beliebiger Querschnittsform und freien Längsrändern. Eine sehr breite ausgesteifte Platte mit unverschieblichen Längsrändern verhält sich ähnlich, da die Randbedingungen nur lokalen Einfluß haben.

In Fig. 8 (c) tritt kein Eulerknicken sondern nur lokales Ausbeulen ein. Ein Beispiel für dieses Verhalten einer ausgesteiften Platte ist ein langer schmaler Plattenstreifen mit ein oder zwei Längssteifen und mit gelenkig gelagerten Rändern. In Abschnitt 7 (Beispiel 7.3 und 7.4) werden zwei Querschnitte, einer mit freien und der andere mit gelenkig gelagerten Rändern, verglichen. Es ergeben sich Beulkurven, die jeweils mit Fig. 8(b) und 8(c) übereinstimmen.

In Fig. 8(d) handelt es sich um einen ähnlichen Fall wie in Fig. 8(b); es ergeben sich hier jedoch zwei verschiedene Möglichkeiten des lokalen Ausbeulens. Ein Beispiel für solches Verhalten ist ein Hohlquerschnitt mit bestimmten geometrischen Abmessungen, der symmetrisch wie auch antisymmetrisch lokal ausgebeult oder aber durch Drillknicken versagt.

Bei der Vorbereitung der Tafeln stellte sich heraus, daß
für die Bemessung nur die kleinste kritische Spannung und
die Länge, hier als L_{euler} bezeichnet, benötigt wird. Bei
Werten von $L < L_{euler}$ ergibt sich eine Beulspannung $\sigma < \sigma_{cr}$.
In Fig. 8(a) liegt der Euler-Fall für alle L vor, so daß in
den Tafeln vereinfachend die Bezeichnung "EULER" benutzt
wurde im Sinne von

$$\sigma_{cr} = \frac{\pi^2 E}{(L/r)^2} \tag{34}$$

wobei r der Trägheitsradius des Querschnittes ist. In diesen Fällen ist die Beulspannung aus Gleichung (34) zu berechnen oder, falls erforderlich, Drillknicken[4] nachzu-
weisen.

Bei der Erstellung der Tafeln wurde L zwischen 0 und L_{max}
variiert, d.h. bei Erreichen von L_{max} wurde der Programm-
lauf beendet. Die Festlegung von L_{max} auf den 500- oder
2000-fachen Wert der Einheitsplattendicke ergab sich aus
praktischen Überlegungen. Es ist üblicherweise notwendig,
innerhalb dieser Abstände Quersteifen vorzusehen. Deshalb
geben für einige Platten, deren Verhalten dem der Kurven
von Fig. 8(b) und (c) gleicht, die Tafeln nur die Eingangs-
werte " > 500", oder " > 2000" in der Spalte L_{euler} an. Für
diese Fälle wird empfohlen, Querstreifen zumindest in
diesen Abständen anzuordnen.

Die Beulkurven in Fig. 8 sind das Ergebnis der oben be-
schriebenen numerischen Berechnungen; für m = 1 ist über L
die Beulspannung aufgetragen.

In Fig. 8(b) und (d) entwickelt sich lokales Ausbeulen mit
Wellenlänge L bei der Spannung σ_{cr}. Beträgt die Länge 2 L
(unter der Annahme 2 $L_1 < L_{euler}$), so können sich gerade

zwei lokale Ausbeulungen der Wellenlänge L_1 ausbilden und die Beulspannung bleibt bei σ_{cr}. Ähnliches gilt für eine Stütze der Länge $3\,L_1$ (vorausgesetzt $3\,L_1 < L_{euler}$) und so weiter bis $n\,L_1 > L_{euler}$. Daher sind in diesen Fällen die Beulkurven zwischen L_1 und L_{euler} ohne Bedeutung und sollten wie in Fig. 9 dargestellt werden. Für die Erstellung der Tafeln war es deshalb notwendig, die Werte von σ_{cr} und L_{euler} aufzunehmen.

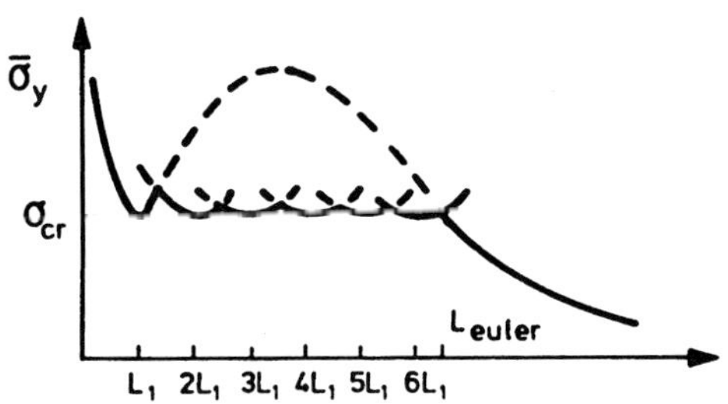

Fig. 9 Mehrere lokale Beulwellen der Wellenlänge L_1
(für $L < L_{euler}$)

Zur Beschränkung des Umfanges der Tafeln ist eine Darstellung in dimensionsloser Form nötig. Dies geschieht durch Festlegung einer Einheitsplattendicke: die Einheit der Länge (Dicke, Höhe, Abstand etc.) ist 1 mm. Damit sind die Ergebnisse leicht zu skalieren, wie in Abschnitt 8 erklärt.

6 Bemerkungen zur numerischen Lösung

Die Bandstruktur der Matrizen wird als wichtiges Merkmal des Problems angesehen. Es wurden nur Verfahren berücksichtigt, welche diese Eigenschaft ausnützen.

6.1 Iterative Lösung (Programm PLATE 1)

Zuerst wird eine Dreieckszerlegung der Bandmatrix $\lambda \underline{K}^O + \underline{K}^G$ durchgeführt. Der Eigenvektor $\underline{X}$, der zum größten Eigenwert λ gehört (Kehrwert von σ_{cr}), wird durch inverse Iteration[9] berechnet. Eine bessere Näherung für den Minimalwert von λ ergibt sich mit dem Rayleigh-Quotienten.

$$(\underline{x}^T \, \underline{K}^G \, \underline{x}) \; / \; (\underline{x}^T \, \underline{K}^O \, \underline{x})$$

Die Konvergenz zur kleinsten kritischen Spannung folgt aus der Konvergenz des Rayleigh-Quotienten zum größten Eigenwert $\lambda_{max} = 1/\sigma_{min}$ mit $\sigma_{min} = \sigma_{cr}$. Die volle Bandbreite der Matrizen wird als Speicher benützt.

6.2 Lösung durch Bisektion (PLATE 2)

Das allgemeine Eigenwertproblem (32) wird zuerst auf das spezielle Eigenwertproblem $(\underline{A} - \lambda\underline{E}) \, \underline{x} = \underline{0}$ reduziert. Im Gegensatz zur üblichen Transformation mit der Inversen von $\underline{K}^O$, welche die Bandstruktur zerstört, wird zur Erhaltung der Bandmatrizen eine spezielle Transformation durchgeführt [9]. Die sich daraus ergebenden Matrizen werden dann auf symmetrische Tridiagonalform mit Hilfe einer Given-Reduktion gebracht. Den größten Eigenwert, welcher der kleinsten (kritischen) Spannung entspricht, erhält man durch die Given'sche Methode der Bisektion aus den resultierenden Tridiagonalmatrizen [9]. Man benutzt nur das obere Dreieck der symmetrischen Bandmatrizen einschließlich der Diagonalen als Speicher.

7 Die Programme PLATE 1, PLATE 2 und PLATE 3

Die Programme sind in ihrem Aufbau fast identisch, sie unterscheiden sich nur durch Speichertechnik und Eigenwertberechnung. PLATE 1 und 2 berechnen die kritische Spannung, PLATE 3 ist ein Plotprogramm, das von PLATE 2 zur Untersuchung der kritischen Länge abgeleitet wurde.

7.1 Beispiele

Es werden im folgenden drei Querschnitte betrachtet, um die Ergebnisse einer Berechnung darzustellen (Fig. 10).

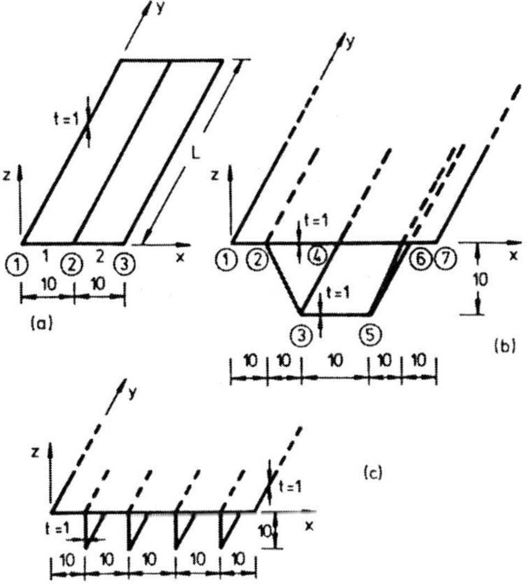

Fig. 10 Abmessungen der ausgesteiften Platten
in den Beispielen 71 bis 77

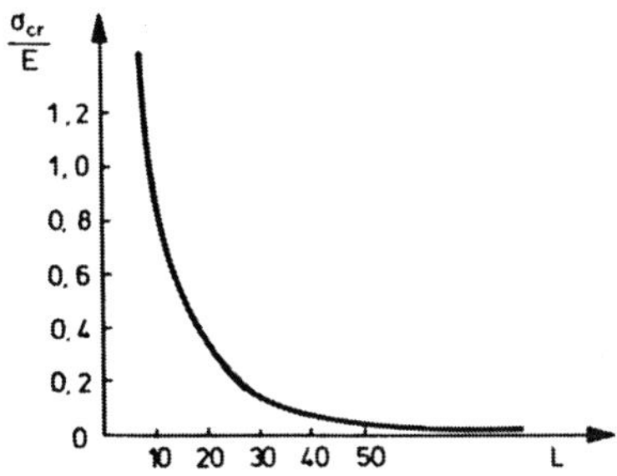

Fig. 11 Verlauf der kritischen Spannung für Beispiel 7

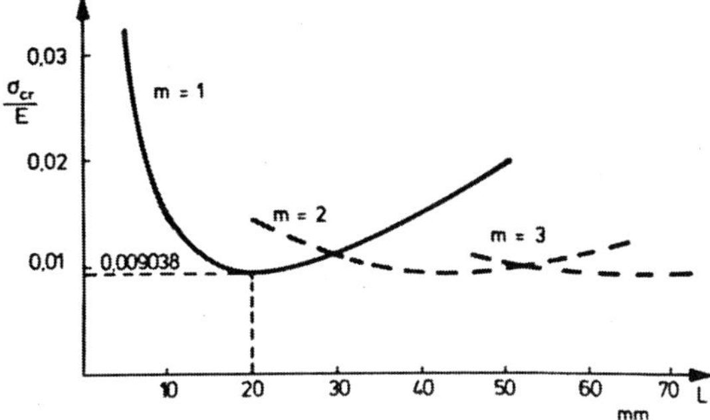

Fig. 12 Verlauf der kritischen Spannung für Beispiel 72

Beispiel 7.1

Das Tragwerk ist eine ebene Platte (Fig. 10(a)) mit freien
Rändern. Es werden zwei Elemente der Breite 10 benutzt. Wie
zu erwarten knickt dieses Tragwerk bei der Spannung die
sich nach Gleichung (34) ergibt wie ein Eulerstab aus
(Fig.11). Die vorliegenden Tabellen enthalten nur den
Hinweis, daß sich die Platte wie ein Eulerstab ("EULER")
verhält.

Beispiel 7.2

In diesem Fall wird die gleiche ebene Platte von Fig. 10(a)
betrachtet, jedoch sind die Längsränder gelenkig und frei
verschieblich in der Plattenebene gelagert. Fig. 12 zeigt,
wie diese Randbedingungen die Beulspannung beeinflussen.
Ein Minimalwert von σ_{cr} erhält man für eine quadratische
Platte. Wird die Länge L der Platte verdoppelt, so können
sich zwei Beulen mit entgegengesetzter Krümmung bilden, so
daß man strenggenommen die erste gepunktete Kurve mit ein-
beziehen müßte. Ebenso sollten weitere Kurven für drei oder
mehr Beulformen gezeichnet werden. Für die Bemessung ist
jedoch nur der Minimalwert von σ_{cr} erforderlich und die
Aussage, daß sich die Platte nicht wie ein Eulerstab
verhält. Dies geht aus den Tafeln (s.Abschn.8) hervor.

Beispiel 7.3

In diesem Beispiel wird eine einzelne Hohlsteife
(Fig.10(b)) mit freien Längsrändern betrachtet. Durch die
Veränderung der Länge L erhält man die in Fig. 13
dargestellte Kurve. Für L < 156 mm tritt lokales Ausbeulen
ein mit einer Wellenlänge von 24.39 mm bei einer kritischen
Spannung von 5.543×10^{-3} . Offensichtlich könnte in Fig.
13 eine ganze Kurvenschar gezeichnet werden, die jenen
Kurven gleicht, die in Fig. 9 als gepunktete Linien

dargestellt sind; zur besseren Übersicht wurden diese Kurven weggelassen. Für L >156 mm sollte das Tragwerk wie ein Eulerstab behandelt werden. Für solche Fälle sind die Werte σ_{cr} und L_{euler} in den Tafeln angegeben.

Beispiel 7.4

Es wird dasselbe System wie in Beispiel 7.3 benutzt, jedoch sind hier die Ränder gelenkig gelagert (w = o), aber wie oben in der Plattenebene frei verschieblich. Ein solches Tragwerk ist in Fig. 8 (c) dargestellt. Eine lange Platte dieser Art beult lokal mit 22,29 mm über die gesamte Länge aus, und zwar bei einer kritischen Spannung von 6.029 x 10^{-3}E. Wie aus Fig. 14 hervorgeht, handelt es sich hier nicht um einen Eulerfall; dies ist in den Tabellen dargestellt.

Beispiel 7.5

Der Querschnitt ist in Fig. 10(c) angegeben. Das Ergebnis in Fig. 15 zeigt, daß hier ähnliche Verhältnisse wie in Beispiel 7.3 vorliegen.

Beispiel 7.6

Es werden hier zwei Platten mit dünner Rechtecksteife verglichen. Im ersten Fall liegt die neutrale Achse in der Plattenmittelebene und im zweiten Fall ist sie exzentrisch (Fig. 16). Der erste Fall wurde von Klöppel und Scheer [5] behandelt.

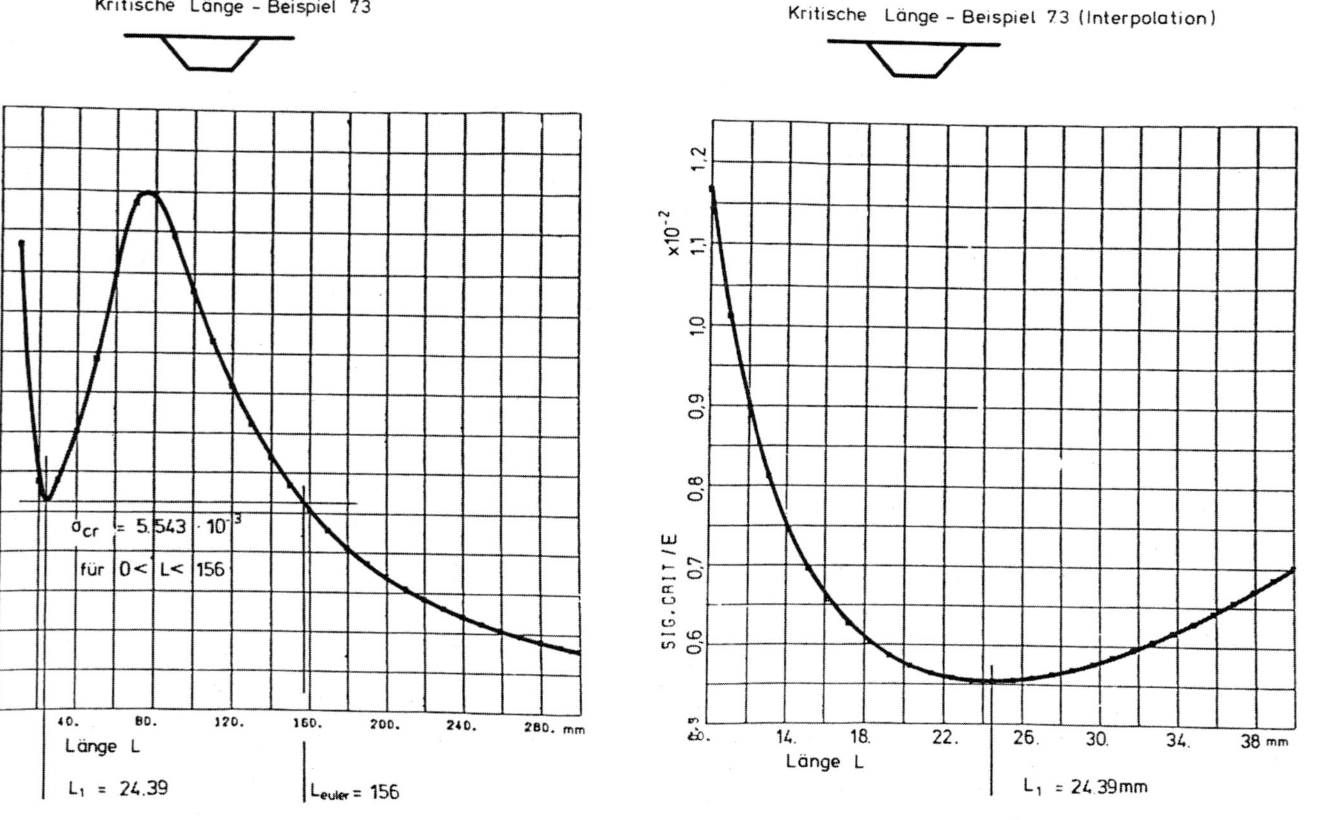

Fig.13a : Beispiel 73, Ränder frei

Fig. 13b : Ausschnitt aus Fig. 13a

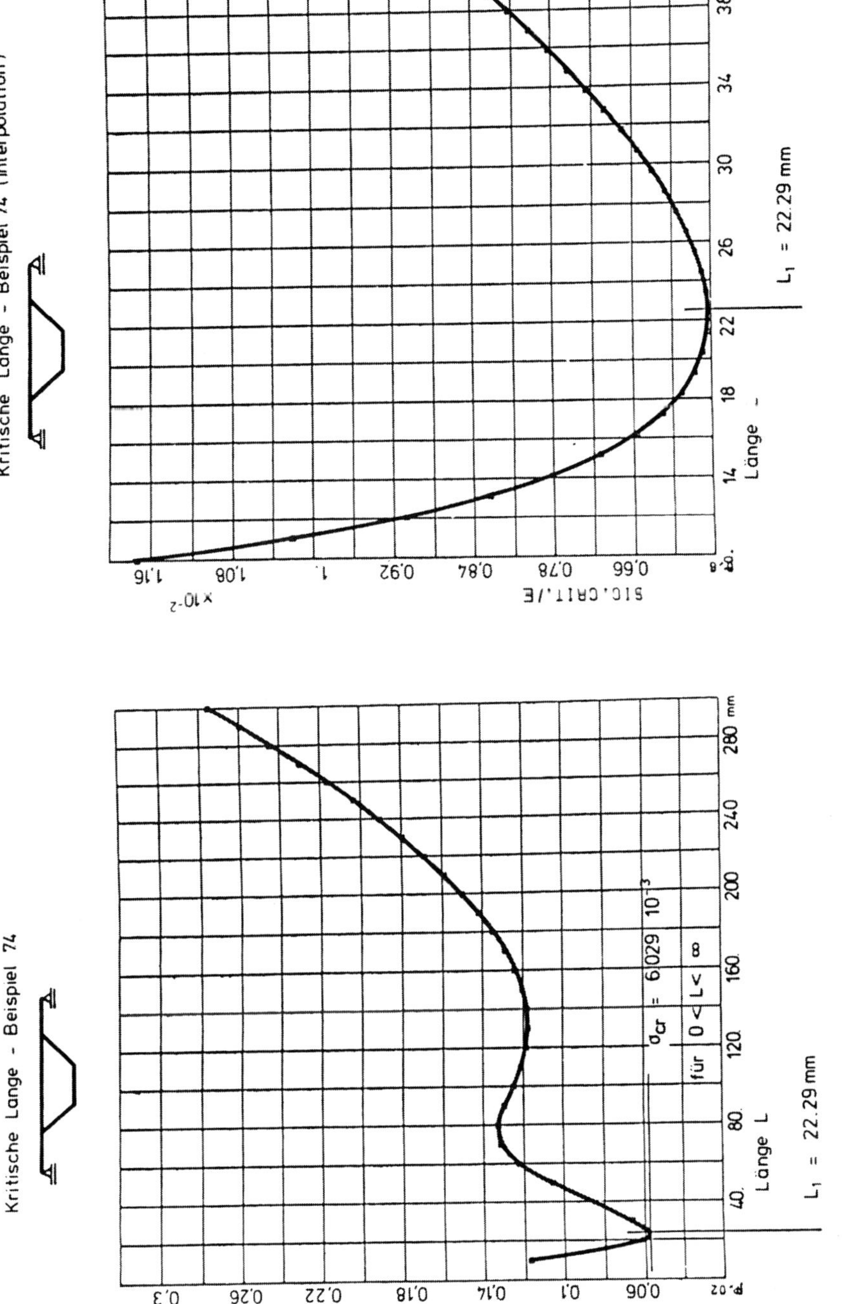

Fig 14b   Ausschnitt aus Fig. 14a

Fig. 14a : Beispiel 7.4, Ränder gelenkig gelagert

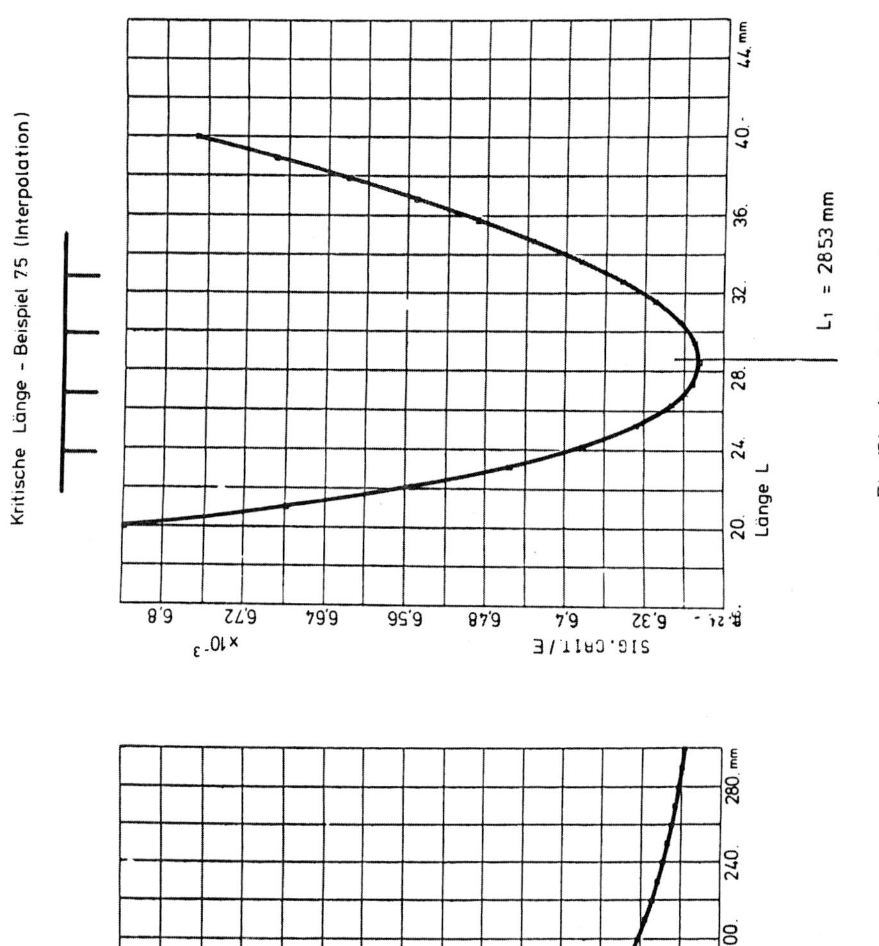

Fig. 15b: Ausschnitt aus Fig. 15a

Fig. 15a: Beispiel 7.5, Ränder frei

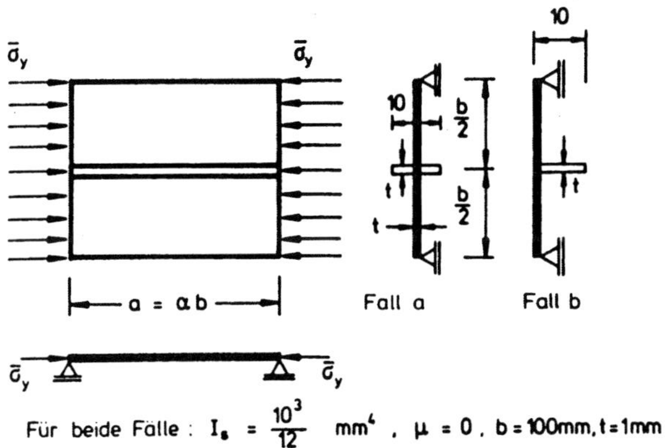

Fig. 16 : Testbeispiel für exentrische und symmetrische
Anordnung der Steifen

Die Ergebnisse beider Berechnungen sind in Fig. 17 dargestellt. Es zeigt sich, daß sich mit exzentrischer Steife
die doppelte Beulspannung ergibt. Dies kann jedoch nicht
verallgemeinert werden: Wird eine sehr schlanke Steife benutzt, so könnte die Beulspannung der Platte höher sein,
wenn die Steife symmetrisch angeordnet wäre. Dies erklärt
sich aus der geringen Tragfähigkeit von sehr schlanken
Steifen (siehe Abschnitt 8.3).

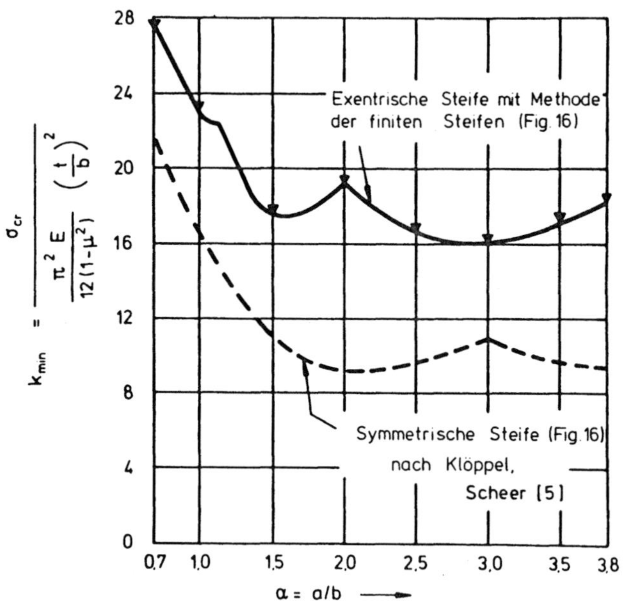

Fig. 17 : Beulfaktoren k_{min} für Beispiel 7.6

8 Anwendung der Tabellen

8.1 Erklärung der Tabellenwerte

a) Der erste Satz von Tabellenwerten beschreibt die Geometrie eines Querschnitts. Die Abmessungen sind relativ zur angenommenen Plattendicke von 1 mm. Dies dient zur Einschränkung des Tabellenumfangs. Für andere Plattendicken wird auf die nachfolgenden Erklärungen verwiesen. Jede ausgesteifte Platte erhält eine Kodierung,welche die Geometrie ihres Querschnitts beschreibt. Der erste Buchstabe bezieht sich auf die Form der Steife usw., wie in Fig.18 gezeigt. In Fig, 19 werden einige Beispiele dieser Kodierung angegeben.

b) Die zweite Gruppe von Tabellenwerten besteht aus den üblichen Querschnittswerten, die zur Bemessung nötig sind, z.B. Querschnittsfläche, Abstand von der Plattenoberkante zur neutralen Achse, Flächenträgheitsmoment bezogen auf die neutrale Achse usw.

c) Die dritte Gruppe von Tabellenwerten ist das Ergebnis der in den vorhergehenden Abschnitten dargestellten Beuluntersuchung, d.h. Kritische Spannung und Erste Eulerlänge.

8.2 Skalierung der Ergebnisse für ausgesteifte Platten mit Plattendicke $t \neq 1$

Die in den Tafeln vorliegenden Ergebnisse können bei einer Plattendicke $t = 1$ direkt benutzt werden. Für $t \neq 1$ sind die Dimensionen der zu skalierenden Größen maßgebend; der Tafelwert muß mit t in der Potenz der Dimension multipliziert werden. Wenn z.B. $t = 8$ mm und der Wert des Flächenträgheitsmoments I mit 1200 mm^4 aus der Tafel entnommen wird, dann ergibt sich der Wert von I für den größeren Querschnitt (alle anderen Dimensionen sind mit Faktor 8 skaliert) zu $1200 \times 8^4 = 4915200$ mm^4. Der Wert von σ_{cr}/E wird durch Vergrößerung des Querschnitts nicht verändert, so daß dieser Wert direkt aus der Tabelle entnommen werden kann. Der Wert der ersten Eulerlänge wird mit dem Faktor t multipliziert. In dem hier betrachteten Beispiel ergibt sich bei einem Tafelwert von 1200 mm die erste Eulerlänge für den größeren Querschnitt zu $1200 \times 8 = 9600$ mm.

8.3 Standardisierung der Querschnitte

Die Tragfähigkeit eines dünnwandigen Querschnitts wird stark beeinflußt durch kleine Veränderungen der Geometrie.

Durch Hinzufügen von Material wird die Tragfähigkeit einer ausgesteiften Platte oft reduziert. Wenn z.B. die Höhe einer schlanken Rechteckssteife vergrößert wird, vermindert sich die Beulspannung, weil schlanke Platten nur eine sehr geringe Normalbelastung tragen können, insbesondere, wenn ein Rand frei verschieblich ist.

Ein weiterer Fall, bei dem die Tragfähigkeit einer ausgesteiften Platte reduziert wird, ist das Anwachsen des Randfeldes "a" (Fig. 20), insbesondere wenn die Längsränder frei verschieblich sind. Fig. 21 zeigt den Einfluß einer Vergrößerung von "a". Es stellt sich heraus, daß es nicht sinnvoll ist, "a" größer als etwa das 15-fache der Plattendicke zu wählen. Wenn die Längsränder gelenkig gelagert sind, gibt es keine nennenswerte Minderung der kritischen Spannung bis "a" gleich dem Abstand der Steife ist. Aus anderen Untersuchungen zeigt sich jedoch, daß "a" großen Einfluß auf σ_{cr} ausübt, wenn die Zahl der Steifen gering ist. Es empfiehlt sich z.B., die Fälle 3 und 5 in Fig. 25 zu vergleichen. Für die Beurteilung des Einflusses von "a" auf σ_{cr} sind für eine gelenkig gelagerte Platte zwei Werte in den Tabellen angegeben, nämlich a = S und a = 0.2S: der erste Wert wird dabei mit P, der letzte Wert mit Q (s.Fig. 18) bezeichnet.

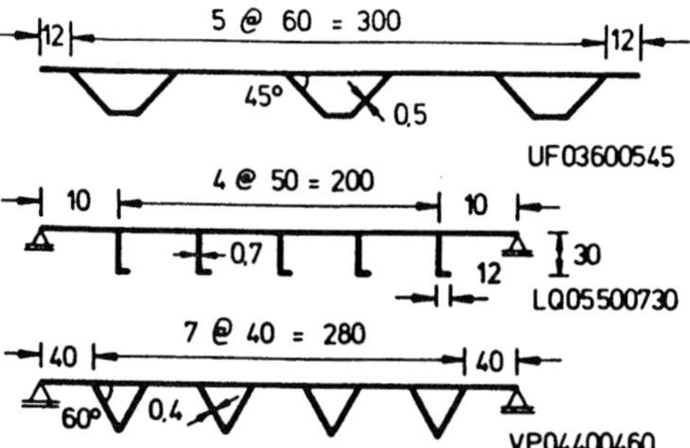

Fig. 19 : Beispiele zur Kodierung der Querschnitte (vgl. auch Fig. 18)

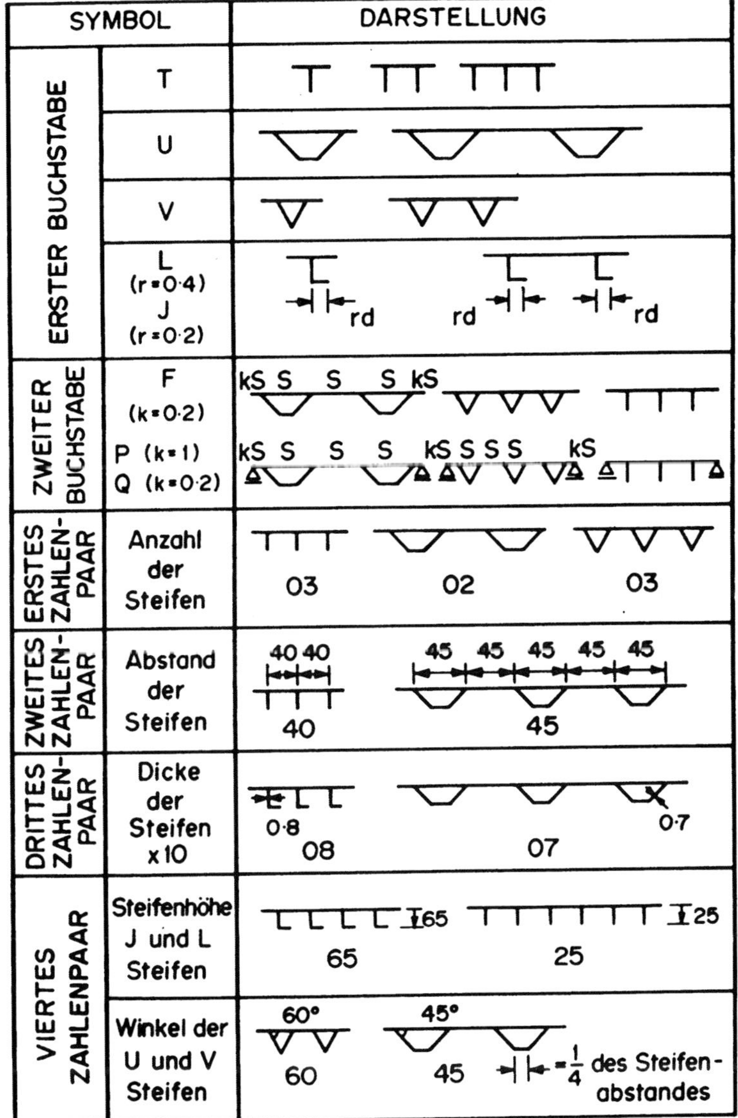

FIG. 18: Kodierung der Querschnitte (vgl. auch FIG. 19)

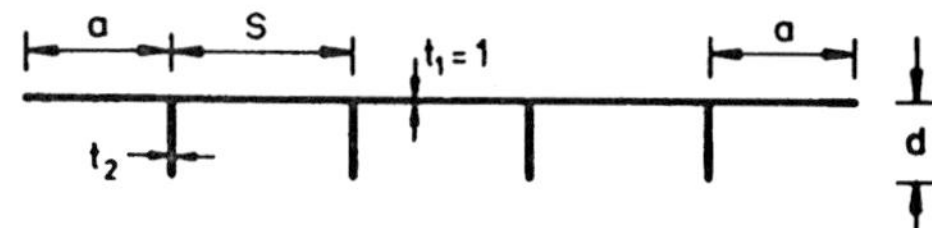

Fig. 20 : Dimensionen für die Parameterstudie

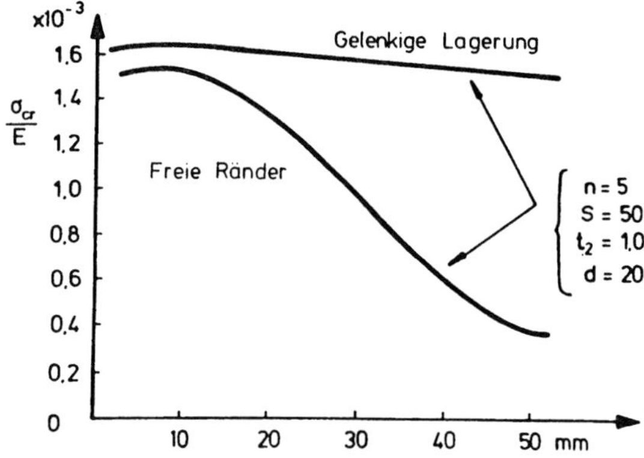

Fig. 21 : Darstellung des Einflußes von a auf σ_{cr}

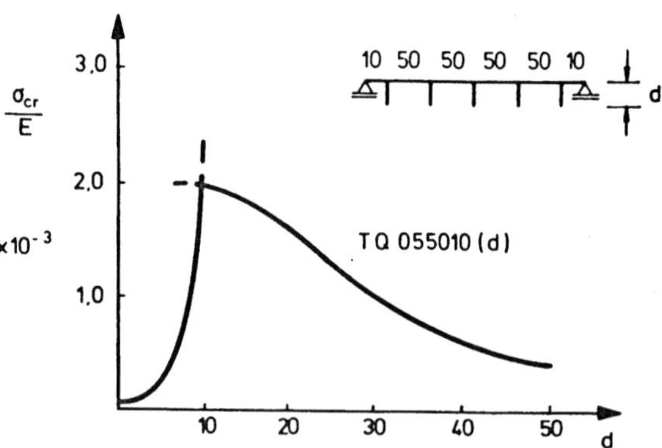

Fig. 22 : Einfluß der Steifenhöhe auf σ_{cr}

Fig. 22 zeigt das Ergebnis einer Untersuchung über die Aus-
wirkung einer Veränderung der Steifenhöhe. Für
kleinere Werte von d verhält sich die Platte, die in diesem
Fall gelenkig gelagerte Längsränder besitzt, wie eine
unversteifte Platte der Breite 220 (wenn d = 5, beträgt die
Beullänge 389). Wenn d etwas vergrößert wird, herrscht
lokales Ausbeulen vor (wenn d = 10, beträgt die Beullänge
44.9). Die Beulspannung erreicht ihren maximalen Wert bei
etwa d = 10; über diesen Wert hinaus verliert die Steife
mit wachsendem d mehr und mehr an Tragfähigkeit.

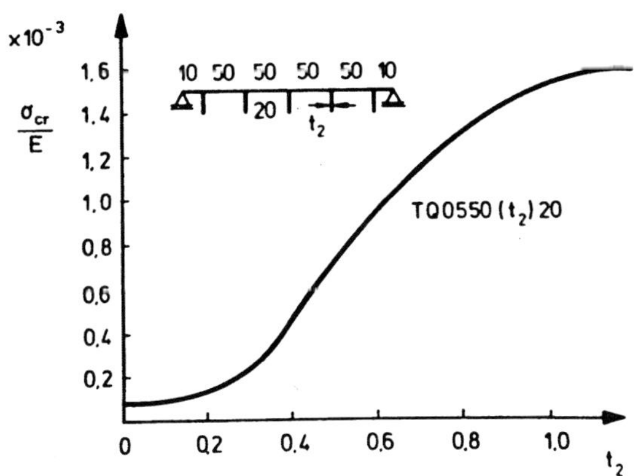

Fig. 23 : Einfluß der Dicke t_2 der Steifen auf σ_{cr}

Fig. 23 zeigt, welchen Einfluß die Veränderung der Dicke
t_2 der Steifen bei einer ähnlichen Platte hat. Bei sehr
geringem Wert t_2 verhält sich die Platte wie eine
unversteifte Platte der Breite 220; wenn t_2 vergrößert
wird, tritt lokales Ausbeulen auf. Übersteigt t_2 den Wert
0.8, so nimmt die kritische Spannung mit zunehmendem t_2 entsprechend langsamer zu, da die Platte als die schwächere
Komponente die Beulspannung bestimmt.

Eine weitere Untersuchung wurde durchgeführt um herauszu-
finden, wie die kritische Spannung einer Platte, bei der
Größe und Abstand der Steifen vorgegeben sind, durch eine

wachsende Anzahl von Steifen beeinflußt wird; dabei vergrößert sich natürlich gleichzeitig die Gesamtbreite der Platte. Man erwartet von zwei Platten, die auf dieser Weise
mit 5 oder 6 Steifen ausgesteift sind, ein ähnliches Beulverhalten mit nahezu gleicher kritischer Spannung. Ver-
gleichberechnungen bestätigen diese Erwartung und deshalb
war es überflüssig, ausgesteifte Platten mit einer großen
Anzahl von Steifen in die Tabellen aufzunehmen. Fig. 24
zeigt, daß die lokale Beulspannung bei vier Platten mit
unterschiedlicher Geometrie nahezu konstant ist, wenn mehr
als 4 Streifen vorhanden sind. Deshalb können die Ergebnisse für n = 5 verwendet werden, wenn die Anzahl der Streifen
5 oder mehr beträgt.

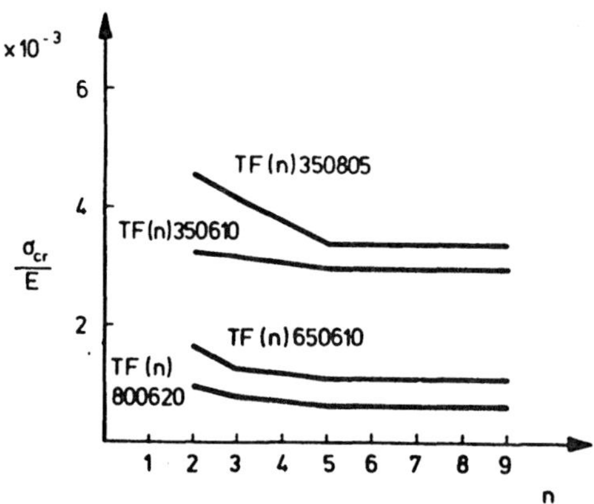

Fig. 24 : Einfluß der Anzahl der Steifen auf σ_{cr}

Schließlich wurde noch der Einfluß der Randbedingungen untersucht um zu sehen, ob einige Fälle von den Tabellen aus-
geschlossen werden können. Unter der Annahme, daß für jede
ausgesteifte Platte die Längsränder frei drehbar und längs-
verschieblich gelagert sind, bestehen vier mögliche Kombinationen von Randbedingungen in x- und z-Richtung, nämlich
jeweils frei oder unverschieblich in beiden Richtungen.
Wenn eine Platte in Längsrichtung (y-Richtung) durch
Druckkräfte beansprucht wird, dann ist die Querdehnung
durch die Randbedingungen in den Fällen 2 und 4

(Fig. 25) behindert. Dies erfordert jedoch Randkräfte, die nur in einem sehr steifen Tragwerk auftreten können. Die Randbedingungen 2 und 4 sind deshalb im allgemeinen von geringem praktischen Interesse und wurden nicht in die Tafeln aufgenommen. Weitere Vergleichsberechnungen zeigten, dass die Breite des Randstreifens einen relativ grossen Einfluss auf σ_{cr} und L besitzt. Deshalb wurden die Randbedingungen der Fälle 1, 3 und 5 in die Tafeln aufgenommen.

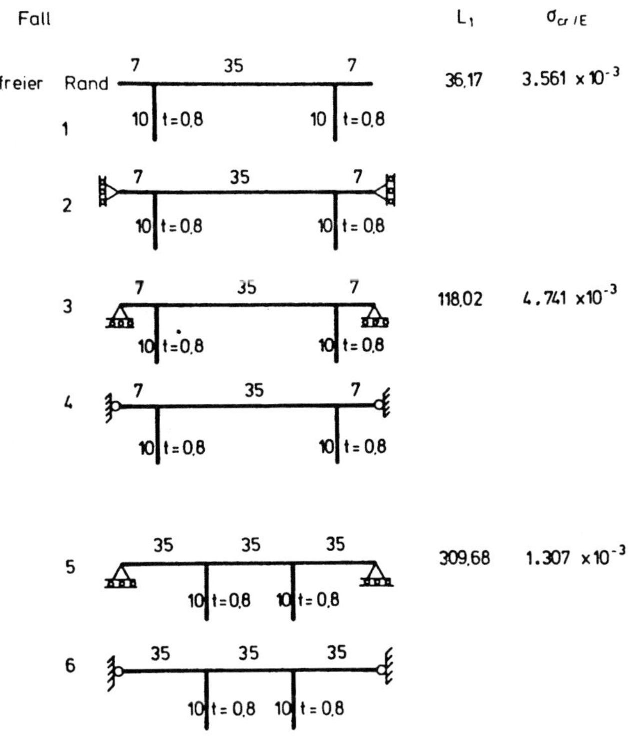

Fig. 25: Einfluß der Randbedingungen auf σ_{cr}

Aus diesen Untersuchungen wird gefolgert:
1. Für eine Platte mit frei verschieblichen Längsrändern sollte die Breite des Randstreifens "a" 0.2 S nicht übersteigen. Für diese Geometrie gelten die Werte von

σ_{cr} und L in den Tafeln. Fur a < 0.2 können die in den Tafeln angegebenen Werte von σ_{cr} benutzt werden, sie liegen jedoch etwas auf der unsicheren Seite.

2. Für eine Platte mit gelenkig gelagerten Längsrändern hat die Breite des Randstreifens "a" wenig Einfluss auf σ_{cr}, wenn die Anzahl der Streifen gross ist. Es ist aber nötig, zwei Werte "a" aufzunehmen, nämlich a = S und a = 0.2 S, um jene Fälle zu erfassen, in denen der Einfluss grösser ist. Für Werte zwischen a = S und a = 0.2 S kann linear interpoliert werden.

3. Die Höhe der Steifen hat grossen Einfluss auf σ_{cr}. Für sehr geringe und sehr grosse Höhen kann der Wert von σ_{cr} sehr klein sein, im ersten Fall verhält sich die Platte wie eine unversteifte und auch im zweiten Fall ist die aussteifende Wirkung gering. Zwischen beiden Extremfällen liegt die optimale Steifenhöhe. Die Tabellen überdecken einen weiten Bereich von Höhen zur Berechnung von σ_{cr}, da bei praktischen Anwendungen die optimale Geometrie nicht immer gefunden werden kann.

4. Die Dicke t_2 der Aussteifungen hat zunächst starken Einfluss, bei zunehmendem t_2 aber steigt σ_{cr} entsprechend langsamer an. Bei kleinem t_2 verhält sich die Platte wie eine unversteifte, wenn aber t_2 gross ist, bestimmt der Plattenstreifen zwischen den Aussteifungen das Beulverhalten; er beult wie eine Platte mit eingespannten Längsrändern aus. Die Tafeln enthalten eine umfassende Zusammenstellung von t_2-Werten.

5. Die in den Tafeln dargestellten Ergebnisse sind für Fälle der Randbedingungen gedacht, die in Fig. 26 mit 1, 3 und 5 bezeichnet sind. In den Tafeln werden diese Falle jeweils mit F, Q und P gekennzeichnet. Die Tafeln sollten für Platten mit den Randbedingungen 2, 4 und 6 (Fig. 25) nicht verwendet werden; die Ergebnisse liegen hierfür auf der unsicheren Seite.

6. Ein Abstand de Quersteifen, der grösser ist als L_{euler} sollte nicht ohne weitgehende, gründliche Untersuchungen

ausgeführt werden. In diesem Bereich tritt nämlich eine starke Interaktion der beiden möglichen Beulformen auf. Diese gegenseitige Beeinflussung ist auch dann von grossem Einfluss wenn die Beulspannung σ_{cr}/E wesentlich kleiner ist als die Eulerspannung

$$\frac{\sigma_E}{E} = \frac{\pi^2\, I}{(L_{euler})^2\, A} \quad .$$

8.4 Gebrauch der Tabellen für Bemessungsvorschriften

Für die Berechnung und Bemessung ausgesteifter Platten
ist es erforderlich, die kritische Belastung im elastischen
Bereich zu berechnen, so daß die Grenzlast abgeschätzt
werden kann. Die Tafeln in diesem Buch können hierfür
anstelle der Näherungen benutzt werden, wie sie in den
einschlägigen Bemessungsvorschriften angegeben sind. In
diesem Abschnitt wird ein Kastenträger untersucht, dessen
oberer Flansch eine ausgesteifte Platte ist. Im ersten
Beispiel werden die Tafeln in Verbindung mit den Regeln der
deutschen DASt-Richtlinie 012[10] angewendet. Zum Vergleich
wird im zweiten Beispiel die kritische Spannung mit Hilfe
der "Merrison Rules"[6] berechnet. Es zeigt sich, daß die
vorliegenden Tafeln leicht zu handhaben sind; sie geben
genaue Werte der kritischen Spannungen und Aufschluß über
den Abstand der Quersteifen.

Beispiel 8.1

Der obere Flansch des Kastenträgers, der in Fig. 26 dargestellt ist, soll mit Benutzung der vorliegenden Tafeln und
der diesbezüglichen Vorschriften der DASt-Richtlinie
012[10] nachgewiesen werden.

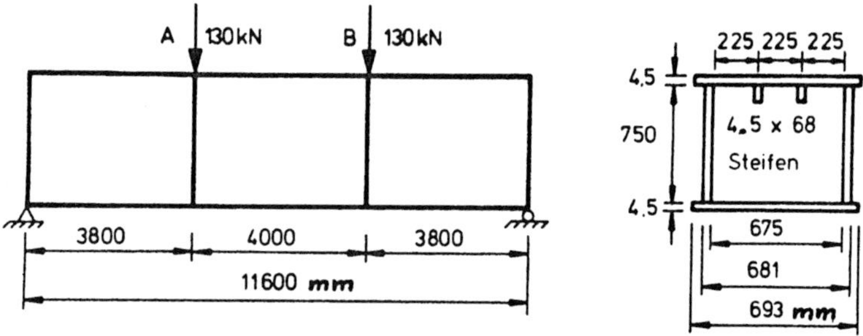

Fig. 26 : Kastenträger. Beispiel 8.1

Die angeführten Vorschriften und Tafeln in diesem
Beispiel beziehen sich auf Richtlinie 012 wenn nicht anders
angegeben.

Querschnittswerte

Querschnittsfläche $= 11349 \text{ mm}^2$

Abstand der neutralen Achse vom oberen

Plattenrand $=\quad 362 \text{ mm}$

I_{NA} $= 1164.958 \times 10^6 \text{ mm}^4$

$Z_{\text{oberer Flansch}}$ $=\quad 3.218 \times 10^6 \text{ mm}^3$

Spannung im oberen Flansch

$$\text{Axialspannung} = \frac{\text{Biegemoment}}{Z}$$

$$\sigma_1 = \frac{130000 \times 3800}{3.218 \times 10^6} = \quad 153.51 \text{ MPa}$$

Untersuchung auf lokales Ausbeulen des oberen Flansches

Stahlgüte St 52 , $\sigma_F = 360$ MPa (= Fließspannung)

Für den nächsten Schritt müssen jene ausgesteifte Platten in den Tabellen dieses Buches gefunden werden, die den skalierten Werten des oberen Flansches entsprechen. Die Plattendicke ist 4.5 mm. Demzufolge hat die Einheitsplatte der vorliegenden Tafeln folgende Abmessungen:

$$\text{Plattendicke} \quad = \frac{4.5}{4.5} = \quad 1 \text{ mm}$$

$$\text{Abstand der Steifen} = \frac{225}{4.5} = \quad 50 \text{ mm}$$

$$\text{Höhe der Steifen} = \frac{68}{4.5} = \quad 15 \text{ mm}$$

$$\text{Dicke der Steifen} = \frac{4.5}{4.5} = \quad 1 \text{ mm}.$$

Daher sind die tabellierten Werte von TP 02501015 maßgebend, d.h.

$$\frac{\sigma_{cr}}{E} = 1.6095 \times 10^{-3}, \ L_1 = 49.1 \text{ mm}, \ L_{culer} = 326 \text{ mm}$$

Für den Flansch des Kastenträgers ergeben sich folgende Größen:

$$\sigma_{cr} = 210000 \times 1.6095 \times 10^{-3} \qquad\qquad = 338 \text{ MPa}$$
$$L_1 = 49.1 \times 4.5 \qquad\qquad = 221 \text{ mm}$$
$$L_{euler} = 326 \times 4.5 \qquad\qquad = 1467 \text{ mm}.$$

Diese letzten Zahlen bedeuten, daß bei einer äußeren Spannung von 338 MPa lokales Ausbeulen mit einer Wellenlänge von 221 mm auftritt. Bei einer gleichen Spannung könnte auch Eulerknicken mit größerer Länge auftreten, sofern dies nicht durch Quersteifen mit Abständen von weniger als 1467 mm verhindert wird. Dies kann durch 2 Quersteifen im Abstand von 4000/3 = 1333 erreicht werden.

Nach Richtlinie 012 Abschnitt 6.1.3 gilt

$$\sigma_{vki} = \sigma_{ki} = 338$$

$$\frac{\sigma_1}{\sigma_{ki}} = \frac{153.51}{338} = 0.454 \quad (= \frac{1}{\nu_B} = \frac{1}{\text{Sicherheitsfaktor}})$$

$$\frac{\sigma_F}{\sigma_{ki}} = \frac{360}{338} = 1.06$$

Nach Abschnitt 6.2 ist die reduzierte Beulspannung

$$\sigma_{vk} = \sigma_F \, (1.474 - 0.677\sqrt{\frac{\sigma_F}{\sigma_{ki}}} \,)$$

$$= 360 \, (1.474 - 0.677\sqrt{1.06}) \qquad\qquad = 279.71 \text{ MPa}$$

Bezogen auf die Fließspannung entspricht dies

$$\bar{\sigma}_{vk} = \frac{279.71}{360} = 0.777$$

Nach Abschnitt 6.3.3.2 ist die relative Grenzspannung $\sigma_G = \sigma_{vk}$. Der damit erreichte rechnerische Beulsicherheitsfaktor (Abschnitt 6.4) beträgt:

$$\nu_B^* = \frac{\sigma_G}{\sigma_1} = \frac{279.71}{153.51} = 1.82$$

Der erforderliche Beulsicherheitsfaktor (Abschn.8 Tab. 7) ist ausreichend:

$$v_B = 1.32 + 0.19(1 + \psi_B) = 1.32 + 0.19(1+1.0) = 1.70 < 1.82$$

Beispiel 8.2

Im folgenden werden die "Merrison Rules" zur Berechnung der kritischen Spannung des in Beispiel 8.1 untersuchten Flansches verwendet. In der folgenden Berechnung beziehen sich die Angaben zu Abschnitt, Tafel und Bezeichnungen auf die "Merrison Rules".

Vorverformung (Tafel 23.1)

$$\text{Für } t < 25 \text{ mm}: \quad \Delta_x = \frac{G}{30t}\left(1 + \frac{b}{5000}\right)$$

$$= \frac{225}{30 \times 4.5}\left(1 + \frac{225}{5000}\right) = 1.74 \text{ mm}$$

$$\text{Eigenspannung (Abschn. 7.3.2)} = \frac{10\,A}{bt} \quad (A = \text{Fläche der Schweißnaht})$$

$$= \frac{10 \times 10.1}{225 \times 4.5} = 0.1 \text{ MPa}$$

Die Eigenspannung wird vernachlässigt mit: $\Sigma\delta = \delta_o$. Nach Abschnitt 18.1.2 ist

$$\delta_o = \frac{1.2\,b\,\Delta_x}{G} \sqrt[3]{\frac{1}{1+N}}$$

$$= \frac{1.2 \times 225 \times 1.74}{225} \sqrt[3]{\frac{1}{1+2}} = 1.45 \text{ mm}$$

$$\frac{\Sigma\delta}{t} = \frac{1.45}{4.5} = 0.32$$

Für eine frei verschiebliche Platte wird Fig. 18.4 benutzt, demnach ist:

$$\sigma_{ecr} = 4 \times \frac{\pi^2 E}{12(1-\mu^2)}\left(\frac{t}{b}\right)^2 = 4\,\frac{\pi^2 \times 206000}{10.92}\cdot\left(\frac{4.5}{225}\right)^2 = 298 \text{ MPa}$$

$$\frac{\sigma_e}{\sigma_{ecr}} = \frac{153.51}{298} \qquad\qquad = 0.52$$

Daraus ergibt sich $K_{bt} = 0.77$ und die mitwirkende Plattenbreite beträgt $0.77 \times 225 = 173$ mm. I_{ox} für diesen Querschnitt ergibt sich zu 408314 mm^4 (Abschn. 20.1.4).

$$\mu_x = \mu_y = 0.3$$

$$D_x = \frac{E\, I_{ox}}{b} + (1-K_{bt})\, \frac{Et^3}{12(1-\mu_x\mu_y)}$$

$$= 206000\left[\frac{408314}{225} + (1-0.77)\, \frac{4.5^3}{12(1-0.3^2)}\right] = 374 \times 10^6 \text{Nmm}.$$

Für Steifen in nur einer Richtung (Abschn. 20.1.4) gilt

$$H = \frac{G\, t^3}{6} + \frac{G\, J_x}{2b}$$

$$= 80\,000\left[\frac{4.5^3}{6} + \frac{68 \times 4.5^3}{3} \times \frac{1}{2\times225}\right] = 1.582 \times 10^6 \text{Nmm}.$$

Um die kleinste Beulspannung zu finden (Abschn. 20.1.5.1a),
nimmt man zuerst $L = 4400$ mm an; dies ergibt

$$\bar{\phi} = \frac{4400}{675} = 6.518$$

$$t_{eff} = t\, (1 + \frac{N\, A_s}{b_s t})$$

$$= 4.5\, (1 + \frac{2 \times 68 \times 4.5}{675 \times 4.5}) \qquad = 5.41 \text{ mm}$$

$$D_y = D = \frac{Et^3}{12(1-\mu^2)} = \frac{206000 \times 4.5^3}{12 \times 0.91} = 1.7206 \times 10^6 \text{Nmm}$$

$$\sigma^*_{cr1} = \frac{\pi^2}{b_s^2\, t_{eff}}\left[\frac{D_x}{\bar{\phi}^2} + 2H + D_y\bar{\phi}^2\right]$$

$$= \frac{\pi^2}{675^2 \times 5.41}\left[\frac{374 \times 10^6}{6.518^2} + 2\times1.582\times10^6 + 1.7206\times10^6 \times 6.518^2\right]$$

$$= 340.6 \text{ MPa}$$

Man wählt nun L = 2200 mm (d.h. zwei Beullängen) :

$$\bar{\phi} \;=\; \frac{2200}{675} \;=\; 3.26$$

$$\sigma^{*}_{cr1} = \frac{\pi^2}{675^2 \times 5.41}\left[\frac{374 \times 10^6}{3.26^2} + 2 \times 1.582 \times 10^6 + 1.7206 \times 10^6 \times 3.26^2\right]$$

$$= 226.7 \text{ MPa} \qquad < 340.6 \text{ MPa}$$

Dann wird die Berechnung mit L = 1467 durchgeführt (d.h.mit drei Beullängen):

$$\bar{\phi} \;=\; \frac{1467}{675} \;=\; 2.173$$

$$\sigma^{*}_{cr1} = \frac{\pi^2}{675^2 \times 5.41}\left[\frac{374 \times 10^6}{2.173^2} + 2 \times 1.582 \times 10^6 + 1.7206 \times 10^6 \times 2.173^2\right]$$

$$= 362.3 \text{ MPa} \qquad > 226.7 \text{ MPa}$$

Die kritische Spannung, wie sie sich aus den "Merrison Rules" ergibt, beträgt 226.7 MPa mit einer Beullänge von 2200 mm (vgl. 338 MPa bei Benutzung der Tafeln dieses Buches). Jedoch kann durch die Anordnung von zwei Quersteifen wie in Beispiel 8.1 gefordert, die kritische Spannung auf 362.3 MPa erhöht werden; dies stimmt mit den Beulergebnissen der Tafeln in diesem Buch näherungsweise überein.

9 Liste der Bezeichnungen

$\underline{A}$ = Allgemeine symmetrische Matrix

a = Breite des Randstreifens (Fig.20)

a_{ij} = Element einer allgemeinen symmetrischen Matrix A

b = Plattenbreite

$\underline{C}$ = Transformationsmatrix (Drehung) für Verformungen
       (Gleichung (27))

$$D \; = \; \frac{Et^3}{12(1-\mu^2)} \; = \; \text{Plattensteifigkeit}$$

d = Höhe der Steife

E = Elastizitätsmodul

G = Schubmodul

h = Halbe Bandbreite der Steifigkeitsmatrix

I_s = Flächenträgheitsmoment der Steife

$\underline{K}$ = Steifigkeitsmatrix

$\underline{K}^G$ = Geometrische Steifigkeitsmatrix

$\underline{K}_m^O$ = Vollständige Steifigkeitsmatrix eines Elements in-
       folge des m-ten Gliedes der Fourierreihe

$\underline{K}_m^G$ = Geometrische Steifigkeitsmatrix eines Elements in-
       folge des m-ten Gliedes der Fourierreihe

$$k \;=\; \text{Beulfaktor} \;=\; \frac{\sigma_{cr}}{E}\,\frac{12(1-\mu^2)}{\pi^2}\left(\frac{b}{t}\right)^2$$

k = S/a = Faktor für relative Breite des Randstreifens

$\underline{k}^{(o)}$, $\underline{k}^{(g)}$ = Elementsteifigkeitsmatrizen nach Theorie 1. oder 2. Ordnung

$\underline{k}^{(1)}$,$\underline{k}^{(2)}$,$\underline{k}^{(3)}$,$\underline{k}^{(4)}$ = Elementsteifigkeitsmatrizen (Gleichungen (18), (20), (22) und (23)).

L = Länge der Platte

L_1 = Länge der lokalen Ausbeulung

L_{euler} = Erste Eulerlänge (Abschnitt 5.2 und Fig. 8)

M = Biege- oder Torsionsmomente pro Längeneinheit (Fig,4)

m = Anzahl der Glieder der Fourierreihe

N = Anzahl der Elemente

n = Anzahl der Steifen

r = Faktor für die relative Breite des Flansches einer Steife für L- und J-Platten (Fig. 18)

r = Trägheitsradius

$\underline{r}_m$ = Vektor der Verschiebungen in der Plattenebene (Gleichung (15))

$\underline{r}$ = Eigenfunktion (Beulform)

S   = Abstand der Steife

t   = Dicke des Elements (Fig. 2)

t_1 = Dicke der Platte (Fig. 20)

t_2 = Dicke der Steife (Fig.18)

U   = Gesamte potentielle Energie eines Elementstreifens
       (Gleichung (2))

U_m = Gesamte potentielle Energie des Elementes aus dem m-
       ten Glied der Fourierreihe

$U_m^{(1)}$, $U_m^{(2)}$ = Scheiben- und Plattenanteile der gesamten
       potentiellen Energie nach Theorie 1. Ordnung
       für das m-te Glied der Fourierreihe

$U_m^{(3)}$, $U_m^{(4)}$ = Scheiben- und Plattenanteile der gesamten po-
       tentiellen Energie nach Theorie 2. Ordnung
       für das m-te Glied der Fourierreihe

U_1 to U_6 = Funktionen nach Gleichung (19)

u,v,w = Verschiebungen in x-, y- oder z-Richtung

u_i, v_i, w_i, θ_i = Verschiebungen und Verdrehungen des
       i-ten Elementrandes (Fig. 2)

u_j, v_j, w_j, θ_j = Verschiebungen (Fig. 2) und Verdrehungen
       des j-ten Elementrandes.

$\bar{u}_i$, $\bar{v}_i$, $\bar{w}_i$ = Komponenten von u_i, v_i oder w_i nach der
       Transformation von lokale in globale
       Koordinaten

u_{im}, v_{im}, w_{im} = Maximalwerte des m-ten Gliedes der Verformung in der Fourierreihe am i-ten Rand (Fig. 2)

u_{jm}, v_{jm}, w_{jm} = Maximalwerte des m-ten Gliedes der Verformung in der Fourierreihe am j-ten Rand

W_1 to W_6 = Funktionen nach Gleichung (21)

$\underline{X}$ = Eigenvektor (Abschnitt 6.1)

x, y, z = Achsen der Elementkoordinaten (Fig. 2)

Z = Widerstandsmoment eines Balkens (Beisp. 8.1)

α = Neigungswinkel der globalen Achsen zu den Elementachsen

$\beta = \dfrac{m\,\pi}{L}$

γ_{xy} = Zusätzliche Schubverzerrungen in der Plattenebene nach dem Ausbeulen

δ = Gesamtverformung in der Mitte einer Stütze oder Platte (Fig. 1)

δ_o = Vorverformung in der Mitte einer Stütze oder Platte (Fig. 1)

$\underline{\delta}_m$ = Vektor der Verschiebungen senkrecht zur Ebene (Gleichung (16)) oder Vektor aller Elementknotenverschiebungen (Gleichung (24))

δ_{ij} = Kronecker-Delta (= 1 wenn i = j; = 0 wenn i $\neq$ j)

ε_x = Zusatzdehnungen in der Plattenebene in x-Richtung
nach dem Ausbeulen

ε_y = Zusatzdehnungen in der Plattenebene   in y-Richtung
nach dem Ausbeulen

$\bar{\varepsilon}_y$ = Lastdehnung in y-Richtung (Gleichung (2))

Θ_i, Θ_j = Maximalwerte des m-ten Gliedes der Verdrehung in
der Fourierreihe am i-ten oder j-ten Rand (Fig.2)

λ = $\dfrac{1}{\sigma_{cr}}$

μ =   Poisson'sche Zahl

Σ =   Summenzeichen

σ =   Normalspannung

σ_e = Eulerspannung

σ_F = Fließspannung

σ_{cr} = Kritischer Wert der Normalspannung

σ_x = Spannungen in der Plattenebene in x-Richtung

σ_y = Zusatzspannungen in der Plattenebene in y-Richtung

$\bar{\sigma}_y$ = Normalspannung in y-Richtung beim Ausbeulen (Fig. 2)

In Beispiel 8.1 wird generell die Bezeichnungsweise der
DAStRichtlinie 012 verwendet und in Beispiel 8.2 die der
"Merrison Rules".

C Tables · Tabellen

Contents · Inhalt   Page/Seite

<u>Deutsche Übersetzung der Tafelköpfe</u>
(Translation of German headings)

edges free out-of-plane ... Ränder frei senkrecht zur
                             Plattenebene
     free in-plane   ...   frei in der Plattenebene
     pinned   ...   gelenkig gelagert

section properties   ... Querschnittswerte
buckling properties   ... Beulwerte

to scale multiply by   ... für die Skalierung ist zu
                           multiplizieren mit ...
rad. gyr.   ... Trägheitsradius (für Knicken
                           senkrecht zur
                           neutralen Achse - N.A.)

<u>Symbols . Bezeichnungen</u>

A Cross-sectional area
               Querschnittsfläche

I_{NA} Moment of inertia
               Flächenträgheitsmoment, bezogen auf N.A.

Z_{plate} Section modulus, plate side
               Widerstandsmoment, Plattenoberkante

Z_{stiff} Section modulus, stiffener side
               Widerstandsmoment, Steifenunterkante

T-PLATE n rectangular stiffeners
edges free out-of-plane
free in-plane

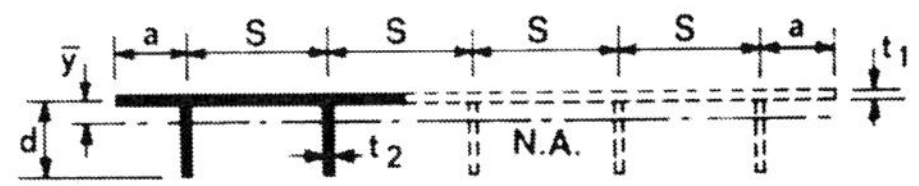

CODE	n	S	t_2	d	a	$\bar{y}$	rad. gyr.	A	I_{NA}	Z_{plate}	$Z_{stiff.}$	σ_{cr}/E	L_1	L_{euler}
To scale multiply by	–	t_1	t_1	t_1	t_1	t_1	t_1	t_1^2	t_1^4	t_1^3	t_1^3	10^{-3}	t_1	t_1
TF02350405	2	35	0.4	5	7	0.19	0.770	53.0	31.45	166.67	6.54	EULER	–	–
TF02350605	2	35	0.6	5	7	0.27	0.914	55.0	45.91	168.33	9.71	EULER	–	–
TF02350805	2	35	0.8	5	7	0.35	1.023	57.0	59.65	170.00	12.83	EULER	–	–
TF02351005	2	35	1.0	5	7	0.42	1.110	59.0	72.74	171.67	15.90	EULER	–	–
TF02350410	2	35	0.4	10	7	0.70	2.046	57.0	238.60	340.00	25.66	1.9036	17.1	145
TF02350610	2	35	0.6	10	7	0.98	2.364	61.0	340.98	346.67	37.82	3.2530	33.9	125
TF02350810	2	35	0.8	10	7	1.23	2.587	65.0	434.87	353.33	49.59	3.5613	36.2	131
TF02351010	2	35	1.0	10	7	1.45	2.750	69.0	521.74	360.00	61.02	3.9813	37.4	131
TF02350415	2	35	0.4	15	7	1.48	3.546	61.0	767.21	520.00	56.73	0.8440	25.6	387
TF02350615	2	35	0.6	15	7	2.01	4.011	67.0	1077.99	535.00	83.02	1.7378	30.1	303
TF02350815	2	35	0.8	15	7	2.47	4.310	73.0	1356.16	550.00	108.20	2.4147	38.5	276
TF02351015	2	35	1.0	15	7	2.85	4.513	79.0	1609.18	565.00	132.42	2.9076	43.2	263
TF02350420	2	35	0.4	20	7	2.46	5.173	65.0	1739.49	706.67	99.18	0.4725	34.6	>500
TF02350620	2	35	0.6	20	7	3.29	5.747	73.0	2410.96	733.33	144.26	0.9812	38.3	>500
TF02350820	2	35	0.8	20	7	3.95	6.098	81.0	3002.47	760.00	187.08	1.4922	44.7	484
TF02351020	2	35	1.0	20	7	4.49	6.303	89.0	3535.58	786.67	228.02	1.9104	50.5	413
TF02350425	2	35	0.4	25	7	3.62	6.875	69.0	3260.87	900.00	152.54	0.3014	43.3	>500
TF02350625	2	35	0.6	25	7	4.75	7.522	79.0	4469.94	941.67	220.70	0.6275	46.9	>500
TF02350825	2	35	0.8	25	7	5.62	7.879	89.0	5524.34	983.33	285.02	0.9704	52.9	>500
TF02351025	2	35	1.0	25	7	6.31	8.085	99.0	6470.96	1025.00	346.28	1.2534	59.7	>500
TF02350430	2	35	0.4	30	7	4.93	8.620	73.0	5424.66	1100.00	216.39	0.2090	51.7	>500
TF02350630	2	35	0.6	30	7	6.35	9.311	85.0	7369.41	1160.00	311.64	0.4358	55.5	>500
TF02350830	2	35	0.8	30	7	7.42	9.662	97.0	9055.67	1220.00	401.10	0.6676	61.6	>500
TF02351030	2	35	1.0	30	7	8.26	9.847	109.0	10568.81	1280.00	486.09	0.9242	68.7	>500
TF02500405	2	50	0.4	5	10	0.14	0.657	74.0	31.98	236.67	6.57	EULER	–	–
TF02500605	2	50	0.6	5	10	0.20	0.787	76.0	47.04	238.33	9.79	EULER	–	–
TF02500805	2	50	0.8	5	10	0.26	0.888	78.0	61.54	240.00	12.97	EULER	–	–
TF02501005	2	50	1.0	5	10	0.31	0.972	80.0	75.52	241.67	16.11	EULER	–	–
TF02500410	2	50	0.4	10	10	0.51	1.776	78.0	246.15	480.00	25.85	1.7465	50.4	120
TF02500610	2	50	0.6	10	10	0.73	2.084	82.0	356.10	486.67	38.42	1.8117	50.4	143
TF02500810	2	50	0.8	10	10	0.93	2.310	86.0	458.91	493.33	50.60	1.9229	50.4	156
TF02501010	2	50	1.0	10	10	1.11	2.485	90.0	555.56	500.00	62.50	2.1098	50.7	160
TF02500415	2	50	0.4	15	10	1.10	3.126	82.0	861.22	730.00	57.63	0.8459	25.7	339
TF02500615	2	50	0.6	15	10	1.53	3.604	88.0	1142.90	745.00	84.87	1.5381	46.6	286
TF02500815	2	50	0.8	15	10	1.91	3.935	94.0	1455.32	760.00	111.22	1.7109	51.4	297
TF02501015	2	50	1.0	15	10	2.25	4.176	100.0	1743.75	775.00	136.76	1.9221	53.4	297
TF02500420	2	50	0.4	20	10	1.86	4.620	86.0	1835.66	986.67	101.20	0.4750	34.1	>500
TF02500620	2	50	0.6	20	10	2.55	5.246	94.0	2587.23	1013.33	148.29	0.9742	41.3	>500
TF02500820	2	50	0.8	20	10	3.14	5.656	102.0	3262.75	1040.00	193.49	1.3051	54.1	495
TF02501020	2	50	1.0	20	10	3.64	5.938	110.0	3878.79	1066.67	237.04	1.5441	59.5	478
TF02500425	2	50	0.4	25	10	2.78	6.211	90.0	3472.22	1250.00	156.25	0.3031	43.3	>500
TF02500625	2	50	0.6	25	10	3.75	6.960	100.0	4843.75	1291.67	227.94	0.6280	48.9	>500
TF02500825	2	50	0.8	25	10	4.55	7.423	110.0	6060.61	1333.33	296.30	0.9230	58.6	>500
TF02501025	2	50	1.0	25	10	5.21	7.725	120.0	7161.46	1375.00	361.84	1.1573	67.1	>500
TF02500430	2	50	0.4	30	10	3.83	7.869	94.0	5821.28	1520.00	222.44	0.2098	51.9	>500
TF02500630	2	50	0.6	30	10	5.09	8.714	106.0	8049.06	1580.00	323.16	0.4361	57.0	>500
TF02500830	2	50	0.8	30	10	6.10	9.209	118.0	10006.78	1640.00	418.72	0.6677	66.0	>500
TF02501030	2	50	1.0	30	10	6.92	9.515	130.0	11769.23	1700.00	510.00	0.8608	75.1	>500
TF02650405	2	65	0.4	5	13	0.11	0.583	95.0	32.28	306.67	6.59	EULER	–	–
TF02650605	2	65	0.6	5	13	0.15	0.701	97.0	47.68	308.33	9.84	EULER	–	–
TF02650805	2	65	0.8	5	13	0.20	0.795	99.0	62.63	310.00	13.05	EULER	–	–
TF02651005	2	65	1.0	5	13	0.25	0.874	101.0	77.15	311.67	16.23	EULER	–	–
TF02650410	2	65	0.4	10	13	0.40	1.591	99.0	250.51	620.00	26.11	1.0580	68.2	131
TF02650610	2	65	0.6	10	13	0.58	1.883	103.0	365.05	626.67	38.76	1.0954	66.0	163
TF02650810	2	65	0.8	10	13	0.75	2.104	107.0	473.52	633.33	51.18	1.1557	65.5	181
TF02651010	2	65	1.0	10	13	0.90	2.279	111.0	576.58	640.00	63.37	1.2538	65.5	189
TF02650415	2	65	0.4	15	13	0.87	2.824	103.0	821.36	940.00	58.14	0.8482	25.7	301
TF02650615	2	65	0.6	15	13	1.24	3.294	109.0	1182.80	955.00	85.95	1.0518	63.9	314
TF02650815	2	65	0.8	15	13	1.57	3.633	115.0	1518.26	970.00	113.01	1.1209	64.9	337
TF02651015	2	65	1.0	15	13	1.86	3.891	121.0	1831.61	985.00	139.39	1.2372	65.6	343
TF02650420	2	65	0.4	20	13	1.50	4.207	107.0	1894.08	1266.67	102.36	0.4757	34.2	>500
TF02650620	2	65	0.6	20	13	2.09	4.845	115.0	2699.13	1293.33	150.68	0.8880	59.3	>500
TF02650820	2	65	0.8	20	13	2.60	5.284	123.0	3434.15	1320.00	197.38	0.9986	67.1	>500
TF02651020	2	65	1.0	20	13	3.05	5.603	131.0	4111.96	1346.67	242.64	1.1252	69.7	>500

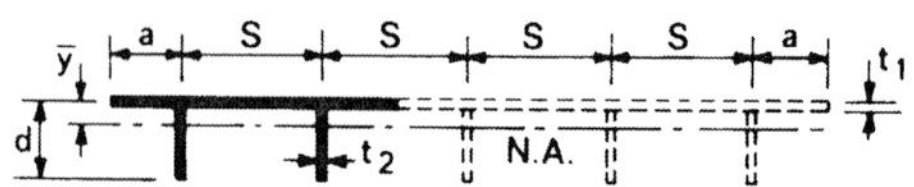

T - PLATE n rectangular stiffeners
edges free out-of-plane
free in-plane

CODE		DIMENSIONS				SECTION PROPERTIES						BUCKLING PROPERTIES		
	n	S	t_2	d	a	$\bar{y}$	rad. gyr.	A	I_{NA}	Z_{plate}	$Z_{stiff.}$	σ_{cr}/E	L_1	L_{euler}
To scale multiply by	−	t_1	t_1	t_1	t_1	t_1	t_1	t_1^2	t_1^4	t_1^3	t_1^3	10^{-3}	t_1	t_1
TF02650425	2	65	0.4	25	13	2.25	5.698	111.0	3603.60	1600.00	158.42	0.3041	43.2	>500
TF02650625	2	65	0.6	25	13	3.10	6.484	121.0	5087.61	1641.67	232.31	0.6215	52.7	>500
TF02650825	2	65	0.8	25	13	3.82	7.003	131.0	6424.94	1683.33	303.30	0.8123	69.3	>500
TF02651025	2	65	1.0	25	13	4.43	7.364	141.0	7646.28	1725.00	371.77	0.9520	75.9	>500
TF02650430	2	65	0.4	30	13	3.13	7.267	115.0	6073.04	1940.00	226.02	0.2108	51.9	>500
TF02650630	2	65	0.6	30	13	4.25	8.153	127.0	8503.54	2000.00	330.28	0.4357	59.1	>500
TF02650830	2	65	0.8	30	13	5.19	8.762	139.0	10670.50	2060.00	429.91	0.6249	73.5	>500
TF02651030	2	65	1.0	30	13	5.96	9.148	151.0	12635.76	2120.00	525.62	0.7681	83.1	>500
TF02800405	2	80	0.4	5	16	0.09	0.529	116.0	32.47	376.67	6.61	EULER	−	−
TF02800605	2	80	0.6	5	16	0.13	0.638	118.0	48.09	378.33	9.87	EULER	−	−
TF02800805	2	80	0.8	5	16	0.17	0.726	120.0	63.33	380.00	13.10	EULER	−	−
TF02801005	2	80	1.0	5	16	0.20	0.801	122.0	78.21	381.67	16.31	EULER	−	−
TF02800410	2	80	0.4	10	16	0.33	1.453	120.0	253.33	760.00	26.21	0.6987	86.7	139
TF02800610	2	80	0.6	10	16	0.48	1.730	124.0	370.97	766.67	38.98	0.7249	82.2	180
TF02800810	2	80	0.8	10	16	0.63	1.943	128.0	483.33	773.33	51.56	0.7614	81.1	202
TF02801010	2	80	1.0	10	16	0.76	2.116	132.0	590.91	780.00	63.93	0.8185	80.9	214
TF02800415	2	80	0.4	15	16	0.73	2.594	124.0	834.68	1150.00	58.47	0.7016	78.3	297
TF02800615	2	80	0.6	15	16	1.04	3.051	130.0	1209.81	1165.00	86.65	0.7224	78.8	349
TF02800815	2	80	0.8	15	16	1.32	3.389	136.0	1561.76	1180.00	114.19	0.7638	79.1	379
TF02801015	2	80	1.0	15	16	1.58	3.652	142.0	1893.49	1195.00	141.14	0.8353	79.3	391
TF02800420	2	80	0.4	20	16	1.25	3.886	128.0	1933.33	1546.67	103.11	0.4767	34.2	>500
TF02800620	2	80	0.6	20	16	1.76	4.518	136.0	2776.47	1573.33	152.26	0.6775	77.9	>500
TF02800820	2	80	0.8	20	16	2.22	4.969	144.0	3555.56	1600.00	200.00	0.7262	80.3	>500
TF02801020	2	80	1.0	20	16	2.63	5.307	152.0	4280.70	1626.67	246.46	0.8048	81.4	>500
TF02800425	2	80	0.4	25	16	1.99	5.289	132.0	3693.18	1950.00	159.84	0.3045	43.2	>500
TF02800625	2	80	0.6	25	16	2.64	6.086	142.0	5259.68	1991.67	235.24	0.5763	72.0	>500
TF02800825	2	80	0.8	25	16	3.29	6.634	152.0	6688.60	2033.33	308.08	0.6530	82.4	>500
TF02801025	2	80	1.0	25	16	3.86	7.030	162.0	8005.40	2075.00	379.65	0.7372	96.3	>500
TF02800430	2	80	0.4	30	16	2.65	6.777	136.0	6247.06	2360.00	228.39	0.2112	51.3	>500
TF02800630	2	80	0.6	30	16	3.65	7.724	146.0	8829.73	2420.00	335.08	0.4304	64.3	>500
TF02800830	2	80	0.8	30	16	4.50	8.352	160.0	11160.00	2480.00	437.65	0.5528	85.0	>500
TF02801030	2	80	1.0	30	16	5.23	8.790	172.0	13290.70	2540.00	536.62	0.6442	92.5	>500
TF03350405	3	35	0.4	5	7	0.17	0.726	90.0	47.50	285.00	9.83	EULER	−	−
TF03350605	3	35	0.6	5	7	0.24	0.865	93.0	69.56	287.50	14.62	EULER	−	−
TF03350805	3	35	0.8	5	7	0.31	0.972	96.0	90.63	290.00	19.33	EULER	−	−
TF03351005	3	35	1.0	5	7	0.38	1.058	99.0	110.80	292.50	23.98	EULER	−	−
TF03350410	3	35	0.4	10	7	0.63	1.943	96.0	362.50	580.00	36.67	1.9063	17.1	135
TF03350610	3	35	0.6	10	7	0.88	2.259	102.0	520.59	590.00	57.10	3.0931	33.6	119
TF03350810	3	35	0.8	10	7	1.11	2.485	108.0	666.67	600.00	75.00	3.3411	34.8	127
TF03351010	3	35	1.0	10	7	1.32	2.653	114.0	802.63	610.00	92.42	3.6912	35.8	130
TF03350415	3	35	0.4	15	7	1.32	3.389	102.0	1171.32	885.00	85.65	0.8442	25.8	369
TF03350615	3	35	0.6	15	7	1.82	3.862	111.0	1655.57	907.50	125.65	1.7479	29.9	289
TF03350815	3	35	0.8	15	7	2.25	4.176	120.0	2092.50	930.00	164.12	2.4076	37.7	265
TF03351015	3	35	1.0	15	7	2.62	4.395	129.0	2452.01	952.50	201.23	2.8693	41.2	255
TF03350420	3	35	0.4	20	7	2.22	4.969	108.0	2666.67	1200.00	150.00	0.4733	34.4	>500
TF03350620	3	35	0.6	20	7	3.00	5.568	120.0	3720.00	1240.00	218.82	0.9892	37.7	>500
TF03350820	3	35	0.8	20	7	3.64	5.938	132.0	4654.55	1280.00	284.44	1.5056	43.5	485
TF03351020	3	35	1.0	20	7	4.17	6.180	144.0	5500.00	1320.00	347.37	1.9442	45.2	443
TF03350425	3	35	0.4	25	7	3.29	6.634	114.0	5016.45	1525.00	231.06	0.3022	43.1	>500
TF03350625	3	35	0.6	25	7	4.36	7.325	129.0	6922.24	1587.50	335.39	0.6337	46.2	>500
TF03350825	3	35	0.8	25	7	5.21	7.725	144.0	8593.75	1650.00	434.21	0.9919	51.4	>500
TF03351025	3	35	1.0	25	7	5.90	7.969	159.0	10097.29	1712.50	528.55	1.3281	57.3	>500
TF03350430	3	35	0.4	30	7	4.50	8.352	120.0	8370.00	1860.00	328.24	0.2097	51.5	>500
TF03350630	3	35	0.6	30	7	5.87	9.107	138.0	11445.65	1950.00	474.32	0.4408	55.0	>500
TF03350830	3	35	0.8	30	7	6.92	9.515	156.0	14123.08	2040.00	612.00	0.6988	59.8	>500
TF03351030	3	35	1.0	30	7	7.76	9.746	174.0	16525.86	2130.00	743.02	0.9523	66.1	>500
TF03500405	3	50	0.4	5	10	0.12	0.619	126.0	48.21	405.00	9.88	EULER	−	−
TF03500605	3	50	0.6	5	10	0.17	0.742	129.0	71.08	407.50	14.73	EULER	−	−
TF03500805	3	50	0.8	5	10	0.23	0.840	132.0	93.18	410.00	19.52	EULER	−	−
TF03501005	3	50	1.0	5	10	0.28	0.921	135.0	114.58	412.50	24.26	EULER	−	−
TF03500410	3	50	0.4	10	10	0.45	1.680	132.0	372.73	820.00	39.05	1.6230	48.6	119
TF03500610	3	50	0.6	10	10	0.65	1.981	138.0	541.30	830.00	57.91	1.6699	48.7	141
TF03500810	3	50	0.8	10	10	0.83	2.205	144.0	700.00	840.00	76.36	1.7568	48.7	155
TF03501010	3	50	1.0	10	10	1.00	2.380	150.0	850.00	850.00	94.44	1.9047	48.6	161

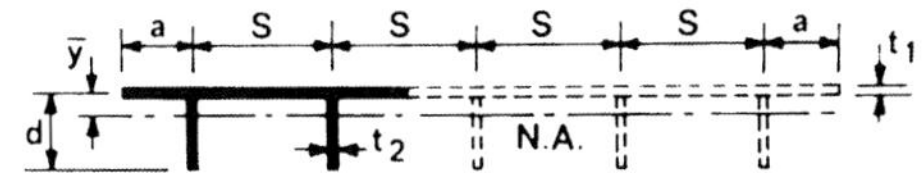

T - PLATE — n rectangular stiffeners, edges free out-of-plane, free in-plane

CODE	n	S	t_2	d	a	$\bar{y}$	rad. gyr.	A	I_{NA}	Z_{plate}	$Z_{stiff.}$	σ_{cr}/E	L_1	L_{euler}
To scale multiply by	–	t_1	t_1	t_1	t_1	t_1	t_1	t_1^2	t_1^4	t_1^3	t_1^3	10^{-3}	t_1	t_1
TF03500415	3	50	0.4	15	10	0.98	2.971	138.0	1217.93	1245.00	86.86	0.8471	25.7	317
TF03500615	3	50	0.6	15	10	1.38	3.446	147.0	1746.05	1267.50	128.17	1.4732	47.1	274
TF03500815	3	50	0.8	15	10	1.73	3.783	156.0	2232.69	1290.00	168.26	1.6108	49.9	289
TF03501015	3	50	1.0	15	10	2.05	4.034	165.0	2624.66	1312.50	207.24	1.7871	50.8	293
TF03500420	3	50	0.4	20	10	1.67	4.410	144.0	2800.00	1680.00	152.73	0.4753	34.3	>500
TF03500620	3	50	0.6	20	10	2.31	5.044	156.0	3969.23	1720.00	224.35	0.9785	41.3	>500
TF03500820	3	50	0.8	20	10	2.86	5.471	168.0	5028.57	1760.00	293.33	1.2884	52.7	478
TF03501020	3	50	1.0	20	10	3.33	5.774	180.0	6000.00	1800.00	360.00	1.5054	56.7	466
TF03500425	3	50	0.4	25	10	2.50	5.951	150.0	5312.50	2125.00	236.11	0.3034	43.1	>500
TF03500625	3	50	0.6	25	10	3.41	6.723	165.0	7457.39	2187.50	345.39	0.6325	48.3	>500
TF03500825	3	50	0.8	25	10	4.17	7.217	180.0	9375.00	2250.00	450.00	0.9324	57.2	>500
TF03501025	3	50	1.0	25	10	4.81	7.551	195.0	11117.79	2312.50	550.60	1.1647	64.0	>500
TF03500430	3	50	0.4	30	10	3.46	7.566	156.0	8930.77	2580.00	336.52	0.2103	51.7	>500
TF03500630	3	50	0.6	30	10	4.66	8.452	174.0	12429.31	2670.00	490.41	0.4398	56.2	>500
TF03500830	3	50	0.8	30	10	5.63	8.992	192.0	15525.00	2760.00	636.92	0.6750	64.1	>500
TF03501030	3	50	1.0	30	10	6.43	9.340	210.0	18321.43	2850.00	777.27	0.8803	72.1	>500
TF03650405	3	65	0.4	5	13	0.09	0.548	162.0	48.61	525.00	9.91	EULER	–	–
TF03650605	3	65	0.6	5	13	0.14	0.660	165.0	71.93	527.50	14.79	EULER	–	–
TF03650805	3	65	0.8	5	13	0.18	0.751	168.0	94.64	530.00	19.63	EULER	–	–
TF03651005	3	65	1.0	5	13	0.22	0.826	171.0	116.78	532.50	24.43	EULER	–	–
TF03650410	3	65	0.4	10	13	0.36	1.501	168.0	376.57	1060.00	39.26	0.9797	63.4	137
TF03650610	3	65	0.6	10	13	0.52	1.783	174.0	553.45	1070.00	58.36	1.0045	62.9	164
TF03650810	3	65	0.8	10	13	0.67	2.000	180.0	720.00	1080.00	77.14	1.0591	62.6	181
TF03651010	3	65	1.0	10	13	0.81	2.174	186.0	879.03	1090.00	95.61	1.1260	62.3	191
TF03650415	3	65	0.4	15	13	0.78	2.675	174.0	1245.26	1605.00	87.55	0.8495	25.7	276
TF03650615	3	65	0.6	15	13	1.11	3.137	183.0	1800.92	1627.50	129.62	0.9719	62.4	304
TF03650815	3	65	0.8	15	13	1.41	3.476	192.0	2320.31	1650.00	170.69	1.0269	62.8	330
TF03651015	3	65	1.0	15	13	1.68	3.738	201.0	2808.30	1672.50	210.82	1.1201	62.9	341
TF03650420	3	65	0.4	20	13	1.33	4.000	180.0	2830.00	2160.00	154.29	0.4764	34.2	>500
TF03650620	3	65	0.6	20	13	1.88	4.635	192.0	4125.00	2200.00	227.59	0.8552	60.8	492
TF03650820	3	65	0.8	20	13	2.35	5.083	204.0	5270.59	2240.00	298.67	0.9429	64.9	>500
TF03651020	3	65	1.0	20	13	2.78	5.415	216.0	6333.33	2280.00	367.74	1.0499	66.6	>500
TF03650425	3	65	0.4	25	13	2.02	5.435	186.0	5493.95	2725.00	239.04	0.3044	43.3	>500
TF03650625	3	65	0.6	25	13	2.80	6.230	201.0	7800.84	2787.50	351.37	0.6233	53.2	>500
TF03650825	3	65	0.8	25	13	3.47	6.769	216.0	9895.83	2850.00	459.68	0.7967	67.8	>500
TF03651025	3	65	1.0	25	13	4.06	7.153	231.0	11820.21	2912.50	564.44	0.9215	72.4	>500
TF03650430	3	65	0.4	30	13	2.91	6.953	192.0	9281.25	3300.00	341.38	0.2109	51.6	>500
TF03650630	3	65	0.6	30	13	3.86	7.891	210.0	13075.71	3390.00	500.16	0.4385	58.4	>500
TF03650830	3	65	0.8	30	13	4.74	8.503	228.0	16484.21	3480.00	652.50	0.6276	71.6	>500
TF03651030	3	65	1.0	30	13	5.49	9.024	246.0	18591.46	3570.00	799.25	0.7656	79.6	>500
TF03800405	3	80	0.4	5	16	0.08	0.497	198.0	48.86	645.00	9.92	EULER	–	–
TF03800605	3	80	0.6	5	16	0.11	0.601	201.0	72.48	647.50	14.83	EULER	–	–
TF03800805	3	80	0.8	5	16	0.15	0.685	204.0	95.59	650.00	19.70	EULER	–	–
TF03801005	3	80	1.0	5	16	0.18	0.756	207.0	118.21	652.50	24.53	EULER	–	–
TF03800410	3	80	0.4	10	16	0.29	1.369	204.0	382.35	1300.00	39.39	0.6439	78.9	154
TF03800610	3	80	0.6	10	16	0.43	1.635	210.0	561.43	1310.00	58.66	0.6650	77.6	184
TF03800810	3	80	0.8	10	16	0.56	1.843	216.0	733.33	1320.00	77.65	0.6918	77.1	205
TF03801010	3	80	1.0	10	16	0.68	2.012	222.0	898.65	1330.00	96.38	0.7354	76.7	218
TF03800415	3	80	0.4	15	16	0.64	2.453	210.0	1263.21	1965.00	87.99	0.6447	76.3	286
TF03800615	3	80	0.6	15	16	0.92	2.897	219.0	1837.76	1987.50	130.57	0.6604	76.5	339
TF03800815	3	80	0.8	15	16	1.18	3.231	228.0	2360.26	2010.00	172.29	0.6928	76.5	372
TF03801015	3	80	1.0	15	16	1.42	3.495	237.0	2894.38	2032.50	213.20	0.7488	76.3	388
TF03800420	3	80	0.4	20	16	1.11	3.685	216.0	2933.33	2640.00	155.29	0.4774	34.2	>500
TF03800620	3	80	0.6	20	16	1.58	4.308	228.0	4231.53	2680.00	229.71	0.6298	76.6	>500
TF03800820	3	80	0.8	20	16	2.00	4.761	240.0	5440.00	2720.00	302.22	0.6686	77.7	>500
TF03801020	3	80	1.0	20	16	2.38	5.107	252.0	6571.43	2760.00	372.97	0.7322	78.0	>500
TF03800425	3	80	0.4	25	16	1.69	5.030	222.0	5616.55	3325.00	240.94	0.3049	43.2	>500
TF03800625	3	80	0.6	25	16	2.37	5.924	237.0	8039.95	3387.50	355.33	0.5570	74.1	>500
TF03800825	3	80	0.8	25	16	2.98	6.383	252.0	10267.66	3450.00	466.22	0.6179	80.2	>500
TF03801025	3	80	1.0	25	16	3.51	6.796	267.0	12333.22	3512.50	573.94	0.6885	82.0	>500
TF03800430	3	80	0.4	30	16	2.37	6.462	228.0	9521.05	4020.00	344.57	0.2114	51.4	>500
TF03800630	3	80	0.6	30	16	3.29	7.417	246.0	13532.93	4110.00	506.71	0.4310	65.5	>500
TF03800830	3	80	0.8	30	16	4.09	8.067	264.0	17181.82	4200.00	663.16	0.5398	82.9	>500
TF03801030	3	80	1.0	30	16	4.79	8.534	282.0	20537.23	4290.00	814.56	0.6206	87.8	>500

T - PLATE

n rectangular stiffeners
edges free out-of-plane
free in-plane

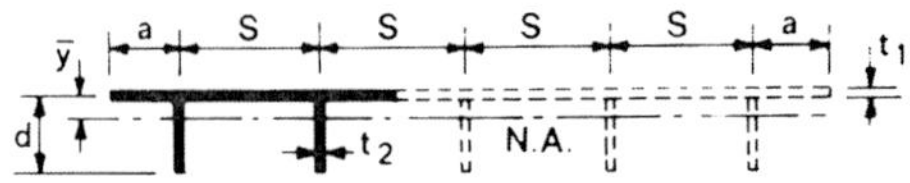

CODE	n	S	t_2	d	a	$\bar{y}$	rad. gyr.	A	I_{NA}	Z_{plate}	$Z_{stiff.}$	σ_{cr}/E	L_1	L_{euler}
To scale multiply by	–	t_1	t_1	t_1	t_1	t_1	t_1	t_1^2	t_1^4	t_1^3	t_1^3	10^{-3}	t_1	t_1
TF05350405	5	35	0.4	5	7	0.15	0.696	164.0	79.52	521.67	16.40	EULER	–	–
TF05350605	5	35	0.6	5	7	0.22	0.831	169.0	116.68	525.83	24.42	EULER	–	–
TF05350805	5	35	0.8	5	7	0.29	0.936	174.0	152.30	530.00	32.32	3.3669	34.6	47
TF05351005	5	35	1.0	5	7	0.35	1.021	179.0	186.51	534.17	40.10	3.5878	34.3	50
TF05350410	5	35	0.4	10	7	0.57	1.871	174.0	609.20	1060.00	64.63	1.9077	17.1	129
TF05350610	5	35	0.6	10	7	0.82	2.184	184.0	877.72	1076.67	95.56	2.9994	33.4	119
TF05350810	5	35	0.8	10	7	1.03	2.410	194.0	1127.15	1093.33	125.67	3.2135	34.3	127
TF05351010	5	35	1.0	10	7	1.23	2.582	204.0	1360.29	1110.00	155.03	3.5219	34.6	131
TF05350415	5	35	0.4	15	7	1.22	3.276	184.0	1974.86	1615.00	143.34	0.8443	25.8	354
TF05350615	5	35	0.6	15	7	1.70	3.753	199.0	2802.61	1652.50	210.66	1.7542	29.7	277
TF05350815	5	35	0.8	15	7	2.10	4.075	214.0	3553.74	1690.00	275.54	2.4016	37.1	256
TF05351015	5	35	1.0	15	7	2.46	4.305	229.0	4243.31	1727.50	338.28	2.8382	40.0	250
TF05350420	5	35	0.4	20	7	2.06	4.821	194.0	4508.59	2186.67	251.34	0.4735	34.4	>500
TF05350620	5	35	0.6	20	7	2.80	5.433	214.0	6317.76	2253.33	367.39	0.9932	37.3	>500
TF05350820	5	35	0.8	20	7	3.42	5.822	234.0	7931.62	2320.00	478.35	1.5202	43.1	470
TF05351020	5	35	1.0	20	7	3.94	6.082	254.0	9396.33	2386.67	584.97	1.9650	47.7	431
TF05350425	5	35	0.4	25	7	3.06	6.456	204.0	8501.84	2775.00	387.57	0.3026	43.0	>500
TF05350625	5	35	0.6	25	7	4.09	7.174	229.0	11786.98	2879.17	563.81	0.6367	45.8	>500
TF05350825	5	35	0.8	25	7	4.92	7.603	254.0	14681.76	2983.33	731.21	1.0028	50.6	>500
TF05351025	5	35	1.0	25	7	5.60	7.872	279.0	17291.11	3087.50	891.31	1.3487	56.1	>500
TF05350430	5	35	0.4	30	7	4.21	8.150	214.0	14214.95	3380.00	551.09	0.2100	51.4	>500
TF05350630	5	35	0.6	30	7	5.53	8.947	244.0	19530.74	3530.00	798.24	0.4432	54.5	>500
TF05350830	5	35	0.8	30	7	6.57	9.393	274.0	24175.18	3690.00	1031.78	0.7069	59.0	>500
TF05351030	5	35	1.0	30	7	7.40	9.656	304.0	28347.04	3830.00	1254.37	0.9682	64.8	>500
TF05500405	5	50	0.4	5	10	0.11	0.592	230.0	80.62	741.67	16.48	EULER	–	–
TF05500605	5	50	0.6	5	10	0.16	0.712	235.0	119.02	745.83	24.59	EULER	–	–
TF05500805	5	50	0.8	5	10	0.21	0.807	240.0	156.25	750.00	32.61	EULER	–	–
TF05501005	5	50	1.0	5	10	0.26	0.886	245.0	192.39	754.17	40.55	EULER	–	–
TF05500410	5	50	0.4	10	10	0.42	1.614	240.0	625.00	1500.00	65.22	1.5566	47.6	121
TF05500610	5	50	0.6	10	10	0.60	1.908	250.0	910.00	1516.67	96.81	1.5943	49.1	143
TF05500810	5	50	0.8	10	10	0.77	2.130	260.0	1179.49	1533.33	127.78	1.6694	47.7	157
TF05501010	5	50	1.0	10	10	0.93	2.306	270.0	1435.19	1550.00	158.16	1.7962	47.6	164
TF05500415	5	50	0.4	15	10	0.90	2.862	250.0	2047.59	2275.00	145.21	0.8476	25.7	302
TF05500615	5	50	0.6	15	10	1.27	3.334	265.0	2945.17	2312.50	214.56	1.4333	47.3	268
TF05500815	5	50	0.8	15	10	1.61	3.673	280.0	3776.79	2350.00	282.00	1.5514	49.2	285
TF05501015	5	50	1.0	15	10	1.91	3.928	295.0	4552.44	2387.50	347.69	1.7071	49.8	292
TF05500420	5	50	0.4	20	10	1.54	4.260	260.0	4717.95	3066.67	255.56	0.4753	34.3	>500
TF05500620	5	50	0.6	20	10	2.14	4.897	280.0	6714.29	3133.33	376.00	0.9814	41.3	486
TF05500820	5	50	0.8	20	10	2.67	5.333	300.0	8533.33	3200.00	492.31	1.2765	52.0	464
TF05501020	5	50	1.0	20	10	3.13	5.648	320.0	10208.33	3266.67	604.94	1.4771	55.4	457
TF05500425	5	50	0.4	25	10	2.31	5.764	270.0	8969.91	3875.00	395.41	0.3035	43.0	>500
TF05500625	5	50	0.6	25	10	3.18	6.547	295.0	12645.66	3979.17	579.49	0.6349	47.5	>500
TF05500825	5	50	0.8	25	10	3.91	7.060	320.0	15950.52	4083.33	756.17	0.9385	56.4	>500
TF05501025	5	50	1.0	25	10	4.53	7.414	345.0	18965.13	4187.50	926.44	1.1684	62.4	>500
TF05500430	5	50	0.4	30	10	3.21	7.345	280.0	15107.14	4700.00	564.00	0.2104	51.5	>500
TF05500630	5	50	0.6	30	10	4.35	8.254	310.0	21120.97	4850.00	823.58	0.4417	55.7	>500
TF05500830	5	50	0.8	30	10	5.29	8.824	340.0	26470.59	5000.00	1071.43	0.6818	63.2	>500
TF05501030	5	50	1.0	30	10	6.08	9.200	370.0	31317.57	5150.00	1309.32	0.8912	70.4	>500
TF05650405	5	65	0.4	5	13	0.08	0.524	296.0	81.22	961.67	16.52	EULER	–	–
TF05650605	5	65	0.6	5	13	0.12	0.632	301.0	120.33	965.83	24.68	EULER	–	–
TF05650805	5	65	0.8	5	13	0.16	0.720	306.0	158.50	970.00	32.77	EULER	–	–
TF05651005	5	65	1.0	5	13	0.20	0.793	311.0	195.77	974.17	40.79	EULER	–	–
TF05650410	5	65	0.4	10	13	0.33	1.439	306.0	633.99	1940.00	65.54	0.9377	62.1	141
TF05650610	5	65	0.6	10	13	0.47	1.714	316.0	928.80	1956.67	97.51	0.9567	61.9	167
TF05650810	5	65	0.8	10	13	0.61	1.927	326.0	1210.63	1973.33	128.98	0.9951	61.7	184
TF05651010	5	65	1.0	10	13	0.74	2.099	336.0	1480.65	1990.00	159.97	1.0599	61.5	195
TF05650415	5	65	0.4	15	13	0.71	2.572	316.0	2089.79	2935.00	146.26	0.8502	25.7	266
TF05650615	5	65	0.6	15	13	1.02	3.026	331.0	3030.87	2972.50	216.99	0.9286	62.0	301
TF05650815	5	65	0.8	15	13	1.30	3.364	346.0	3914.54	3010.00	285.76	0.9760	62.2	320
TF05651015	5	65	1.0	15	13	1.56	3.627	361.0	4748.53	3047.50	353.27	1.0571	62.2	341
TF05650420	5	65	0.4	20	13	1.23	3.854	326.0	4842.54	3946.67	257.95	0.4767	34.2	>500
TF05650620	5	65	0.6	20	13	1.73	4.485	346.0	6959.54	4013.33	381.01	0.8342	61.2	477
TF05650820	5	65	0.8	20	13	2.19	4.936	366.0	8918.03	4080.00	500.61	0.9094	64.1	>500
TF05651020	5	65	1.0	20	13	2.59	5.275	386.0	10742.66	4146.67	617.06	1.0030	65.0	>500

T - PLATE n rectangular stiffeners
edges free out-of-plane
free in-plane

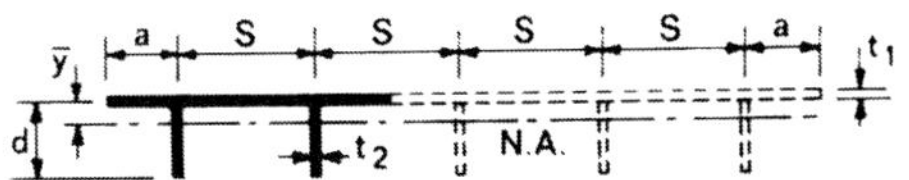

| CODE | n | DIMENSIONS | | | | | SECTION PROPERTIES | | | | | BUCKLING PROPERTIES | | |
| | | S | t_2 | d | a | $\bar{y}$ | rad. gyr. | A | I_{NA} | Z_{plate} | $Z_{stiff.}$ | σ_{cr}/E | L_1 | L_{euler} |
To scale multiply by	–	t_1	t_1	t_1	t_1	t_1	t_1	t_1^2	t_1^4	t_1^3	t_1^3	10^{-3}	t_1	t_1
TF05650425	5	65	0.4	25	13	1.86	5.248	336.0	9254.09	4975.00	399.92	0.3044	42.9	>500
TF05650625	5	65	0.6	25	13	2.60	6.045	361.0	13190.36	5079.17	588.78	0.6245	53.5	>500
TF05650825	5	65	0.8	25	13	3.24	6.594	386.0	16785.41	5183.33	771.33	0.7859	67.1	>500
TF05651025	5	65	1.0	25	13	3.80	6.993	411.0	20101.51	5287.50	948.26	0.8999	70.3	>500
TF05650430	5	65	0.4	30	13	2.60	6.727	346.0	15658.96	6020.00	571.52	0.2109	51.5	>500
TF05650630	5	65	0.6	30	13	3.59	7.676	376.0	22152.93	6170.00	838.82	0.4402	57.9	>500
TF05650830	5	65	0.8	30	13	4.43	8.307	406.0	28019.70	6320.00	1095.95	0.6292	70.6	>500
TF05651030	5	65	1.0	30	13	5.16	8.751	436.0	33388.76	6470.00	1344.18	0.7626	77.2	>500
TF05800405	5	80	0.4	5	16	0.07	0.475	362.0	51.61	1181.67	16.55	EULER	–	–
TF05800605	5	80	0.6	5	16	0.10	0.575	367.0	121.17	1185.83	24.74	EULER	–	–
TF05800805	5	80	0.8	5	16	0.13	0.656	372.0	159.95	1190.00	32.87	EULER	–	–
TF05801005	5	80	1.0	5	16	0.17	0.725	377.0	197.97	1194.17	40.95	EULER	–	–
TF05800410	5	80	0.4	10	16	0.27	1.311	372.0	639.78	2380.00	65.75	0.6215	76.6	159
TF05800610	5	80	0.6	10	16	0.39	1.570	382.0	941.10	2396.67	97.96	0.6333	76.2	188
TF05800810	5	80	0.8	10	16	0.51	1.772	392.0	1231.29	2413.33	129.75	0.6556	75.9	209
TF05801010	5	80	1.0	10	16	0.62	1.939	402.0	1511.19	2430.00	161.14	0.6926	75.7	223
TF05800415	5	80	0.4	15	16	0.59	2.354	382.0	2117.47	3595.00	146.93	0.6147	75.7	286
TF05800615	5	80	0.6	15	16	0.85	2.789	397.0	3088.08	3632.50	218.24	0.6277	75.9	337
TF05800815	5	80	0.8	15	16	1.09	3.119	412.0	4008.50	3670.00	288.22	0.6556	75.9	371
TF05801015	5	80	1.0	15	16	1.32	3.382	427.0	4884.00	3707.50	356.95	0.7038	75.7	390
TF05800420	5	80	0.4	20	16	1.02	3.545	392.0	4925.17	4826.67	259.50	0.4778	34.2	496
TF05800620	5	80	0.6	20	16	1.46	4.159	412.0	7126.21	4893.33	384.29	0.6033	76.2	>500
TF05800820	5	80	0.8	20	16	1.85	4.611	432.0	9185.19	4960.00	506.12	0.6368	76.9	>500
TF05801020	5	80	1.0	20	16	2.21	4.960	452.0	11120.94	5026.67	625.21	0.6923	77.1	>500
TF05800425	5	80	0.4	25	16	1.55	4.847	402.0	9444.96	6075.00	402.85	0.3051	43.2	>500
TF05800625	5	80	0.6	25	16	2.20	5.637	427.0	13566.67	6179.17	594.91	0.5444	74.7	>500
TF05800825	5	80	0.8	25	16	2.77	6.200	452.0	17376.47	6283.33	781.51	0.5965	79.3	>500
TF05801025	5	80	1.0	25	16	3.28	6.623	477.0	20923.41	6387.50	963.13	0.6590	80.3	>500
TF05800430	5	80	0.4	30	16	2.18	6.238	412.0	16033.98	7340.00	576.44	0.2114	51.5	>500
TF05800630	5	80	0.6	30	16	3.05	7.194	442.0	22876.70	7490.00	848.99	0.4313	66.5	>500
TF05800830	5	80	0.8	30	16	3.81	7.857	472.0	29135.59	7640.00	1112.62	0.5310	81.8	>500
TF05801030	5	80	1.0	30	16	4.48	8.340	502.0	34915.34	7790.00	1368.27	0.6043	85.7	>500

T-PLATE n rectangular stiffeners / edges pinned / free in-plane

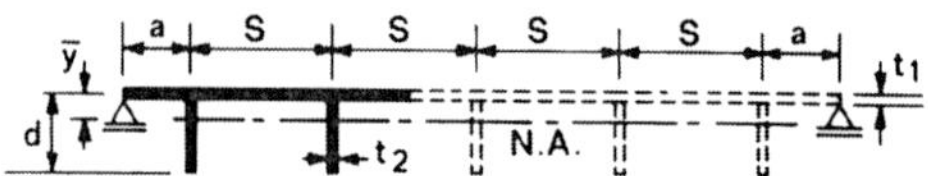

CODE	n	S	t_2	d	a	$\bar{y}$	rad. gyr.	A	I_{NA}	Z_{plate}	$Z_{stiff.}$	σ_{cr}/E	L_1	L_{euler}
To scale multiply by	−	t_1	t_1	t_1	t_1	t_1	t_1	t_1^2	t_1^4	t_1^3	t_1^3	10^{-3}	t_1	t_1
TP01350405	1	35	0.4	5	35	0.07	0.476	72.0	16.32	235.00	3.31	1.2140	108.9	>500
TP01350605	1	35	0.6	5	35	0.10	0.576	73.0	24.23	235.83	4.95	1.3379	118.5	>500
TP01350805	1	35	0.8	5	35	0.14	0.657	74.0	31.98	236.67	6.57	1.4319	126.1	>500
TP01351005	1	35	1.0	5	35	0.17	0.726	75.0	39.58	237.50	8.19	1.5058	132.4	>500
TP01350410	1	35	0.4	10	35	0.27	1.315	74.0	127.93	473.33	13.15	1.9190	17.0	>500
TP01350610	1	35	0.6	10	35	0.39	1.573	76.0	188.16	476.67	19.59	2.7530	192.0	>500
TP01350810	1	35	0.8	10	35	0.51	1.776	78.0	246.15	480.00	25.95	2.9562	205.4	>500
TP01351010	1	35	1.0	10	35	0.63	1.943	80.0	302.08	483.33	32.22	3.0962	216.4	>500
TP01350415	1	35	0.4	15	35	0.59	2.360	76.0	423.36	715.00	29.38	0.8498	25.6	>500
TP01350615	1	35	0.6	15	35	0.85	2.795	79.0	617.33	722.50	43.64	1.8061	27.8	>500
TP01350815	1	35	0.8	15	35	1.10	3.126	82.0	801.22	730.00	57.63	2.5579	34.8	>500
TP01351015	1	35	1.0	15	35	1.32	3.389	85.0	976.10	737.50	71.37	2.9185	36.9	>500
TP01350420	1	35	0.4	20	35	1.03	3.553	78.0	984.62	960.00	51.89	0.4774	34.0	>500
TP01350620	1	35	0.6	20	35	1.46	4.168	82.0	1424.39	973.33	76.84	1.0237	35.7	>500
TP01350820	1	35	0.8	20	35	1.86	4.620	86.0	1835.66	986.67	101.20	1.6311	39.2	>500
TP01351020	1	35	1.0	20	35	2.22	4.969	90.0	2222.22	1000.00	125.00	2.1487	43.0	>500
TP01350425	1	35	0.4	25	35	1.56	4.858	80.0	1888.02	1208.33	80.56	0.3055	42.8	>500
TP01350625	1	35	0.6	25	35	2.21	5.648	85.0	2711.40	1229.17	118.95	0.6580	44.0	>500
TP01350825	1	35	0.8	25	35	2.78	6.211	90.0	3472.22	1250.00	156.25	1.0752	46.8	>500
TP01351025	1	35	1.0	25	35	3.29	6.634	95.0	4180.37	1270.83	192.55	1.4890	50.5	>500
TP01350430	1	35	0.4	30	35	2.20	6.252	82.0	3204.88	1460.00	115.26	0.2123	50.8	>500
TP01350630	1	35	0.6	30	35	3.07	7.208	88.0	4571.59	1490.00	169.75	0.4588	52.3	>500
TP01350830	1	35	0.8	30	35	3.83	7.869	94.0	5821.28	1520.00	222.44	0.7581	55.2	>500
TP01351030	1	35	1.0	30	35	4.50	8.352	100.0	6975.00	1550.00	273.53	1.0710	58.6	>500
TP01500405	1	50	0.4	5	50	0.05	0.401	102.0	16.42	335.00	3.32	0.5494	145.7	>500
TP01500605	1	50	0.6	5	50	0.07	0.487	103.0	24.45	335.83	4.96	0.6036	157.8	>500
TP01500805	1	50	0.8	5	50	0.10	0.558	104.0	32.37	336.67	6.60	0.6466	167.5	>500
TP01501005	1	50	1.0	5	50	0.12	0.619	105.0	40.18	337.50	8.23	0.6819	175.6	>500
TP01500410	1	50	0.4	10	50	0.19	1.116	104.0	129.49	673.33	13.20	1.0796	230.6	>500
TP01500610	1	50	0.6	10	50	0.28	1.344	106.0	191.51	676.67	19.71	1.2259	254.0	>500
TP01500810	1	50	0.8	10	50	0.37	1.527	108.0	251.85	680.00	26.15	1.3314	271.6	>500
TP01501010	1	50	1.0	10	50	0.45	1.680	110.0	310.61	683.33	32.54	1.4105	286.1	>500
TP01500415	1	50	0.4	15	50	0.42	2.016	106.0	430.90	1015.00	29.56	0.8526	25.6	>500
TP01500615	1	50	0.6	15	50	0.62	2.410	109.0	633.20	1022.50	44.03	1.4527	47.3	>500
TP01500815	1	50	0.8	15	50	0.80	2.718	112.0	827.68	1030.00	58.30	1.5163	48.7	>500
TP01501015	1	50	1.0	15	50	0.98	2.971	115.0	1014.95	1037.50	72.38	1.5927	48.8	>500
TP01500420	1	50	0.4	20	50	0.74	3.054	108.0	1007.41	1360.00	52.31	0.4783	34.1	>500
TP01500620	1	50	0.6	20	50	1.07	3.625	112.0	1471.43	1373.33	77.74	1.0126	38.1	>500
TP01500820	1	50	0.8	20	50	1.38	4.061	116.0	1912.64	1386.67	102.72	1.3412	49.5	>500
TP01501020	1	50	1.0	20	50	1.67	4.410	120.0	2333.33	1400.00	127.27	1.4837	51.4	>500
TP01500425	1	50	0.4	25	50	1.14	4.201	110.0	1941.29	1708.33	81.35	0.3057	42.9	>500
TP01500625	1	50	0.6	25	50	1.63	4.951	115.0	2819.29	1729.17	120.64	0.6534	45.2	>500
TP01500825	1	50	0.8	25	50	2.08	5.512	120.0	3645.83	1750.00	159.09	1.0083	51.7	>500
TP01501025	1	50	1.0	25	50	2.50	5.951	125.0	4427.08	1770.83	196.76	1.2536	56.7	>500
TP01500430	1	50	0.4	30	50	1.61	5.437	112.0	3310.71	2060.00	116.60	0.2122	51.0	>500
TP01500630	1	50	0.6	30	50	2.29	6.366	118.0	4782.20	2090.00	172.57	0.4555	53.3	>500
TP01500830	1	50	0.8	30	50	2.90	7.045	124.0	6154.84	2120.00	227.14	0.7313	58.0	>500
TP01501030	1	50	1.0	30	50	3.46	7.566	130.0	7442.31	2150.00	280.43	0.9781	63.3	>500
TP01650405	1	65	0.4	5	65	0.04	0.353	132.0	16.48	435.00	3.32	0.3074	180.7	>500
TP01650605	1	65	0.6	5	65	0.06	0.430	133.0	24.58	435.83	4.97	0.3363	195.0	>500
TP01650805	1	65	0.8	5	65	0.07	0.493	134.0	32.59	436.67	6.62	0.3598	206.6	>500
TP01651005	1	65	1.0	5	65	0.09	0.548	135.0	40.51	437.50	8.25	0.3795	216.4	>500
TP01650410	1	65	0.4	10	65	0.15	0.986	134.0	130.35	873.33	13.23	0.5880	282.8	>500
TP01650610	1	65	0.6	10	65	0.22	1.192	136.0	193.38	876.67	19.77	0.6707	311.2	>500
TP01650810	1	65	0.8	10	65	0.29	1.360	138.0	255.07	880.00	26.27	0.7327	332.9	>500
TP01651010	1	65	1.0	10	65	0.36	1.501	140.0	315.48	883.33	32.72	0.7811	350.9	>500
TP01650415	1	65	0.4	15	65	0.33	1.789	136.0	435.11	1315.00	29.66	0.8551	25.5	>500
TP01650615	1	65	0.6	15	65	0.49	2.149	139.0	642.22	1322.50	44.25	0.9007	62.0	>500
TP01650815	1	65	0.8	15	65	0.63	2.436	142.0	842.96	1330.00	58.68	0.9226	62.1	>500
TP01651015	1	65	1.0	15	65	0.78	2.675	145.0	1037.72	1337.50	72.95	0.9585	61.9	>500
TP01650420	1	65	0.4	20	65	0.58	2.719	138.0	1020.29	1760.00	52.54	0.4795	34.1	>500
TP01650620	1	65	0.6	20	65	0.85	3.249	142.0	1498.59	1773.33	78.24	0.8509	61.3	>500
TP01650820	1	65	0.8	20	65	1.10	3.662	146.0	1957.59	1786.67	103.57	0.8927	63.0	>500
TP01651020	1	65	1.0	20	65	1.33	4.000	150.0	2400.00	1800.00	128.57	0.9394	63.2	>500

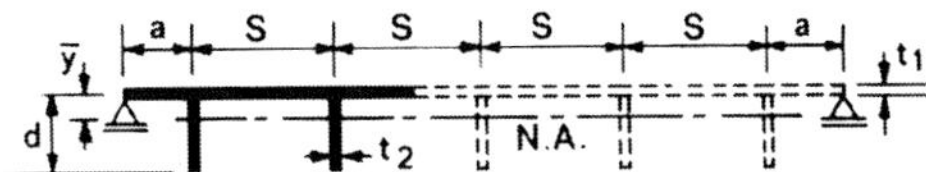

CODE	n	S	t₂	d	a	ȳ	rad. gyr.	A	I_{NA}	Z_{plate}	$Z_{stiff.}$	σ_{cr}/E	L_1	L_{euler}
To scale multiply by	−	t_1	t_1	t_1	t_1	t_1	t_1	t_1^2	t_1^4	t_1^3	t_1^3	10^{-3}	t_1	t_1
TP01650425	1	65	0.4	25	65	0.89	3.753	140.0	1971.73	2208.33	81.79	0.3063	43.0	>500
TP01650625	1	65	0.6	25	65	1.29	4.459	145.0	2882.54	2229.17	121.59	0.6462	48.9	>500
TP01650825	1	65	0.8	25	65	1.67	5.000	150.0	3750.00	2250.00	160.71	0.8174	64.2	>500
TP01651025	1	65	1.0	25	65	2.02	5.435	155.0	4578.29	2270.83	199.20	0.8922	65.9	>500
TP01650430	1	65	0.4	30	65	1.27	4.873	142.0	3371.83	2660.00	117.35	0.2124	51.1	>500
TP01650630	1	65	0.6	30	65	1.82	5.758	148.0	4907.43	2690.00	174.17	0.4529	55.1	>500
TP01650830	1	65	0.8	30	65	2.34	6.426	154.0	6358.44	2720.00	229.86	0.6748	65.4	>500
TP01651030	1	65	1.0	30	65	2.81	6.953	160.0	7734.38	2750.00	284.48	0.8021	70.7	>500
TP01800405	1	80	0.4	5	80	0.03	0.319	162.0	16.51	535.00	3.32	0.1946	214.7	>500
TP01800605	1	80	0.6	5	80	0.05	0.389	163.0	24.65	535.83	4.98	0.2119	230.8	>500
TP01800805	1	80	0.8	5	80	0.06	0.447	164.0	32.72	536.67	6.63	0.2263	244.1	>500
TP01801005	1	80	1.0	5	80	0.08	0.497	165.0	40.72	537.50	8.27	0.2385	255.2	>500
TP01800410	1	80	0.4	10	80	0.12	0.893	164.0	130.89	1073.33	13.25	0.3630	332.3	>500
TP01800610	1	80	0.6	10	80	0.18	1.083	166.0	194.58	1076.67	19.82	0.4148	365.5	>500
TP01800810	1	80	0.8	10	80	0.24	1.237	168.0	257.14	1080.00	26.34	0.4547	391.0	>500
TP01801010	1	80	1.0	10	80	0.29	1.369	170.0	318.63	1083.33	32.83	0.4865	412.0	>500
TP01800415	1	80	0.4	15	80	0.27	1.624	166.0	437.80	1615.00	29.72	0.5839	444.8	>500
TP01800615	1	80	0.6	15	80	0.40	1.958	169.0	648.04	1622.50	44.38	0.6010	76.1	>500
TP01800815	1	80	0.8	15	80	0.52	2.227	172.0	852.91	1630.00	58.92	0.6134	76.1	>500
TP01801015	1	80	1.0	15	80	0.64	2.453	175.0	1052.68	1637.50	73.32	0.6340	75.9	>500
TP01800420	1	80	0.4	20	80	0.48	2.474	168.0	1028.57	2160.00	52.68	0.4806	34.0	>500
TP01800620	1	80	0.6	20	80	0.70	2.969	172.0	1516.28	2173.33	78.55	0.5900	76.3	>500
TP01800820	1	80	0.8	20	80	0.91	3.361	176.0	1987.88	2186.67	104.13	0.6058	76.6	>500
TP01801020	1	80	1.0	20	80	1.11	3.685	180.0	2444.44	2200.00	129.41	0.6310	76.4	>500
TP01800425	1	80	0.4	25	80	0.74	3.423	170.0	1991.42	2708.33	82.07	0.3069	43.0	>500
TP01800625	1	80	0.6	25	80	1.07	4.088	175.0	2924.11	2729.17	122.20	0.5576	74.9	>500
TP01800825	1	80	0.8	25	80	1.39	4.606	180.0	3819.44	2750.00	161.76	0.5874	77.6	>500
TP01801025	1	80	1.0	25	80	1.69	5.030	185.0	4680.46	2770.83	200.79	0.6189	77.9	>500
TP01800430	1	80	0.4	30	80	1.05	4.454	172.0	3411.63	3260.00	117.83	0.2127	51.1	>500
TP01800630	1	80	0.6	30	80	1.52	5.295	178.0	4990.45	3290.00	175.21	0.4476	59.4	>500
TP01800830	1	80	0.8	30	80	1.96	5.942	184.0	6495.65	3320.00	231.63	0.5484	79.0	>500
TP01801030	1	80	1.0	30	80	2.37	6.462	190.0	7934.21	3350.00	287.14	0.5943	80.6	>500
TP02350405	2	35	0.4	5	35	0.09	0.545	109.0	32.42	353.33	6.60	0.5393	164.7	>500
TP02350605	2	35	0.6	5	35	0.14	0.657	111.0	47.97	355.00	9.86	0.5946	179.1	>500
TP02350805	2	35	0.8	5	35	0.18	0.747	113.0	63.13	356.67	13.09	0.6371	190.4	>500
TP02351005	2	35	1.0	5	35	0.22	0.823	115.0	77.90	358.33	16.29	0.6715	199.7	>500
TP02350410	2	35	0.4	10	35	0.35	1.495	113.0	252.51	713.33	26.18	1.0869	264.3	>500
TP02350610	2	35	0.6	10	35	0.51	1.776	117.0	369.23	720.00	38.92	1.2186	290.3	>500
TP02350810	2	35	0.8	10	35	0.66	1.993	121.0	480.44	726.67	51.45	1.3070	309.7	>500
TP02351010	2	35	1.0	10	35	0.80	2.166	125.0	586.67	733.33	63.77	1.3690	325.7	>500
TP02350415	2	35	0.4	15	35	0.77	2.665	117.0	830.77	1080.00	58.38	0.8497	25.5	>500
TP02350615	2	35	0.6	15	35	1.10	3.126	123.0	1201.83	1095.00	86.45	1.7904	28.6	>500
TP02350815	2	35	0.8	15	35	1.40	3.465	129.0	1548.84	1110.00	113.85	2.0329	414.2	>500
TP02351015	2	35	1.0	15	35	1.67	3.727	135.0	1875.00	1125.00	140.63	2.0950	434.6	>500
TP02350420	2	35	0.4	20	35	1.32	3.985	121.0	1921.76	1453.33	102.89	0.4769	34.1	>500
TP02350620	2	35	0.6	20	35	1.86	4.620	129.0	2753.49	1480.00	151.79	1.0164	36.2	>500
TP02350820	2	35	0.8	20	35	2.34	5.068	137.0	3519.22	1506.67	199.23	1.5946	40.2	>500
TP02351020	2	35	1.0	20	35	2.76	5.401	145.0	4229.89	1533.33	245.33	2.0803	44.2	>500
TP02350425	2	35	0.4	25	35	2.00	5.416	125.0	3666.67	1833.33	159.42	0.3050	42.9	>500
TP02350625	2	35	0.6	25	35	2.78	6.211	135.0	5208.33	1875.00	234.37	0.6532	44.5	>500
TP02350825	2	35	0.8	25	35	3.45	6.751	145.0	6609.20	1916.67	306.67	1.0554	47.8	>500
TP02351025	2	35	1.0	25	35	4.03	7.138	155.0	7896.51	1958.33	376.60	1.4460	51.8	>500
TP02350430	2	35	0.4	30	35	2.79	6.930	129.0	6195.35	2220.00	227.69	0.2118	50.9	>500
TP02350630	2	35	0.6	30	35	3.83	7.869	141.0	8731.91	2280.00	333.66	0.4553	52.7	>500
TP02350830	2	35	0.8	30	35	4.71	8.484	153.0	11011.76	2340.00	435.35	0.7453	55.9	>500
TP02351030	2	35	1.0	30	35	5.45	8.907	165.0	13090.91	2400.00	533.33	1.0429	59.9	>500
TP02500405	2	50	0.4	5	50	0.06	0.461	154.0	32.68	503.33	6.62	0.2437	219.5	>500
TP02500605	2	50	0.6	5	50	0.10	0.558	156.0	48.56	505.00	9.90	0.2679	237.9	>500
TP02500805	2	50	0.8	5	50	0.13	0.637	158.0	64.14	506.67	13.16	0.2874	252.6	>500
TP02501005	2	50	1.0	5	50	0.16	0.705	160.0	79.43	508.33	16.40	0.3036	265.0	>500
TP02500410	2	50	0.4	10	50	0.25	1.274	158.0	256.54	1013.33	26.32	0.4785	348.3	>500
TP02500610	2	50	0.6	10	50	0.37	1.527	162.0	377.78	1020.00	39.23	0.5426	382.8	>500
TP02500810	2	50	0.8	10	50	0.48	1.726	166.0	494.78	1026.67	51.98	0.5890	409.0	>500
TP02501010	2	50	1.0	10	50	0.59	1.891	170.0	607.84	1033.33	64.52	0.6242	430.3	>500

n rectangular stiffeners
edges pinned
free in-plane

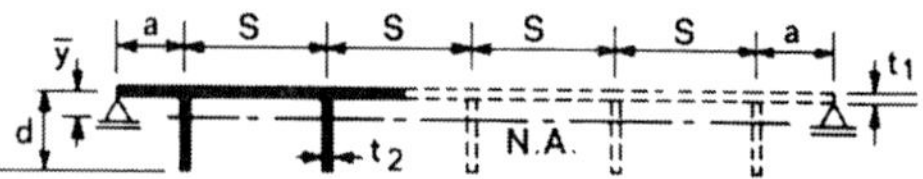

CODE	n	S	t_2	d	a	$\bar{y}$	rad. gyr.	A	I_{NA}	Z_{plate}	$Z_{stiff.}$	$\sigma_{cr/E}$	L_1	L_{euler}
To scale multiply by	–	t_1	t_1	t_1	t_1	t_1	t_1	t_1^2	t_1^4	t_1^3	t_1^3	10^{-3}	t_1	t_1
TP02500415	2	50	0.4	15	50	0.56	2.291	162.0	850.00	1530.00	58.85	0.7716	466.8	>500
TP02500615	2	50	0.6	15	50	0.80	2.718	168.0	1241.52	1545.00	87.45	1.4312	47.3	289
TP02500815	2	50	0.8	15	50	1.03	3.045	174.0	1613.79	1560.00	115.56	1.5107	49.0	313
TP02501015	2	50	1.0	15	50	1.25	3.307	180.0	1968.75	1575.00	143.16	1.6095	49.1	326
TP02500420	2	50	0.4	20	50	0.96	3.453	166.0	1979.12	2053.33	103.97	0.4781	34.0	>500
TP02500620	2	50	0.6	20	50	1.38	4.061	174.0	2868.97	2080.00	154.07	1.0021	39.4	>500
TP02500820	2	50	0.8	20	50	1.76	4.511	182.0	3704.03	2106.67	203.05	1.3078	50.3	>500
TP02501020	2	50	1.0	20	50	2.11	4.862	190.0	4491.23	2133.33	250.98	1.4697	52.4	>500
TP02500425	2	50	0.4	25	50	1.47	4.727	170.0	3799.02	2583.33	161.46	0.3055	43.0	>500
TP02500625	2	50	0.6	25	50	2.08	5.512	180.0	5468.75	2625.00	238.64	0.6487	46.0	>500
TP02500825	2	50	0.8	25	50	2.63	6.077	190.0	7017.54	2666.67	313.73	0.9816	53.2	>500
TP02501025	2	50	1.0	25	50	3.13	6.505	200.0	8463.54	2708.33	386.90	1.2174	58.2	>500
TP02500430	2	50	0.4	30	50	2.07	6.091	174.0	6455.17	3120.00	231.11	0.2119	51.2	>500
TP02500630	2	50	0.6	30	50	2.90	7.045	186.0	9232.26	3180.00	340.71	0.4522	54.3	>500
TP02500830	2	50	0.8	30	50	3.64	7.714	198.0	11781.82	3240.00	446.90	0.7158	59.4	>500
TP02501030	2	50	1.0	30	50	4.29	8.207	210.0	14142.86	3300.00	550.00	0.9471	65.1	>500
TP02650405	2	65	0.4	5	65	0.05	0.406	199.0	32.83	653.33	6.63	0.1363	272.1	>500
TP02650605	2	65	0.6	5	65	0.07	0.493	201.0	48.88	655.00	9.92	0.1492	293.6	>500
TP02650805	2	65	0.8	5	65	0.10	0.565	203.0	64.70	656.67	13.20	0.1598	311.2	>500
TP02651005	2	65	1.0	5	65	0.12	0.626	205.0	80.28	658.33	16.46	0.1688	326.0	>500
TP02650410	2	65	0.4	10	65	0.20	1.129	203.0	258.78	1313.33	26.40	0.2604	426.7	>500
TP02650610	2	65	0.6	10	65	0.29	1.360	207.0	382.61	1320.00	39.40	0.2968	468.8	>500
TP02650810	2	65	0.8	10	65	0.38	1.544	211.0	503.00	1326.67	52.28	0.9380	61.8	195
TP02651010	2	65	1.0	10	65	0.47	1.698	215.0	620.16	1333.33	65.04	0.9744	61.6	210
TP02650415	2	65	0.4	15	65	0.43	2.039	207.0	860.87	1980.00	59.10	0.8538	25.6	278
TP02650615	2	65	0.6	15	65	0.63	2.436	213.0	1264.44	1995.00	88.01	0.8982	62.1	324
TP02650815	2	65	0.8	15	65	0.82	2.747	219.0	1652.05	2010.00	116.52	0.9271	62.2	360
TP02651015	2	65	1.0	15	65	1.00	3.000	225.0	2025.00	2025.00	144.64	0.9746	62.0	381
TP02650420	2	65	0.4	20	65	0.76	3.088	211.0	2012.01	2653.33	104.56	0.4789	34.0	>500
TP02650620	2	65	0.6	20	65	1.10	3.662	219.0	2936.99	2680.00	155.36	0.8366	61.4	>500
TP02650820	2	65	0.8	20	65	1.41	4.100	227.0	3815.57	2706.67	205.24	0.8882	63.3	>500
TP02651020	2	65	1.0	20	65	1.70	4.449	235.0	4652.48	2733.33	254.26	0.9483	63.7	>500
TP02650425	2	65	0.4	25	65	1.16	4.246	215.0	3875.97	3333.33	162.60	0.3061	43.0	>500
TP02650625	2	65	0.6	25	65	1.67	5.000	225.0	5625.00	3375.00	241.07	0.6393	50.8	>500
TP02650825	2	65	0.8	25	65	2.13	5.562	235.0	7269.50	3416.67	317.83	0.7995	65.0	>500
TP02651025	2	65	1.0	25	65	2.55	6.001	245.0	8822.28	3458.33	392.99	0.8872	67.4	>500
TP02650430	2	65	0.4	30	65	1.64	5.493	219.0	6608.22	4020.00	233.04	0.2123	51.2	>500
TP02650630	2	65	0.6	30	65	2.34	6.426	231.0	9537.66	4080.00	344.79	0.4494	56.0	>500
TP02650830	2	65	0.8	30	65	2.96	7.105	243.0	12266.67	4140.00	453.70	0.6556	67.3	>500
TP02651030	2	65	1.0	30	65	3.53	7.624	255.0	14823.53	4200.00	560.00	0.7834	72.7	>500
TP02800405	2	80	0.4	5	80	0.04	0.367	244.0	32.92	803.33	6.64	0.0863	322.9	>500
TP02800605	2	80	0.6	5	80	0.06	0.447	246.0	49.09	805.00	9.94	0.0940	347.3	>500
TP02800805	2	80	0.8	5	80	0.08	0.512	248.0	65.05	806.67	13.22	0.1004	367.4	>500
TP02801005	2	80	1.0	5	80	0.10	0.569	250.0	80.83	808.33	16.50	0.1060	384.3	>500
TP02800410	2	80	0.4	10	80	0.16	1.024	248.0	260.22	1613.33	26.45	0.5999	76.0	164
TP02800610	2	80	0.6	10	80	0.24	1.237	252.0	385.71	1620.00	39.51	0.6059	76.0	197
TP02800810	2	80	0.8	10	80	0.31	1.409	256.0	508.33	1626.67	52.47	0.6184	75.9	221
TP02801010	2	80	1.0	10	80	0.38	1.554	260.0	628.21	1633.33	65.33	0.6390	75.8	239
TP02800415	2	80	0.4	15	80	0.36	1.856	252.0	867.86	2430.00	59.27	0.5938	76.0	303
TP02800615	2	80	0.6	15	80	0.52	2.227	258.0	1279.36	2445.00	88.37	0.6014	76.1	362
TP02800815	2	80	0.8	15	80	0.68	2.521	264.0	1677.27	2460.00	117.14	0.6179	76.1	405
TP02801015	2	80	1.0	15	80	0.83	2.764	270.0	2062.50	2475.00	145.59	0.6453	75.9	432
TP02800420	2	80	0.4	20	80	0.63	2.818	256.0	2033.33	3253.33	104.95	0.4799	34.0	>500
TP02800620	2	80	0.6	20	80	0.91	3.361	264.0	2981.82	3280.00	156.19	0.5870	76.4	>500
TP02800820	2	80	0.8	20	80	1.18	3.782	272.0	3890.20	3306.67	206.67	0.6078	76.8	>500
TP02801020	2	80	1.0	20	80	1.43	4.124	280.0	4761.90	3333.33	256.41	0.6408	76.6	>500
TP02800425	2	80	0.4	25	80	0.96	3.886	260.0	3926.28	4083.33	163.33	0.3065	43.0	>500
TP02800625	2	80	0.6	25	80	1.39	4.606	270.0	5729.17	4125.00	242.65	0.5475	74.9	>500
TP02800825	2	80	0.8	25	80	1.79	5.155	280.0	7440.48	4166.67	320.51	0.5839	78.1	>500
TP02801025	2	80	1.0	25	80	2.16	5.592	290.0	9069.68	4208.33	397.01	0.6243	78.7	>500
TP02800430	2	80	0.4	30	80	1.36	5.041	264.0	6709.09	4920.00	234.29	0.2126	51.0	>500
TP02800630	2	80	0.6	30	80	1.96	5.942	276.0	9743.48	4980.00	347.44	0.4413	62.8	>500
TP02800830	2	80	0.8	30	80	2.50	6.614	288.0	12600.00	5040.00	458.18	0.5375	79.9	>500
TP02801030	2	80	1.0	30	80	3.00	7.141	300.0	15300.00	5100.00	566.67	0.5923	81.9	>500

T-PLATE n rectangular stiffeners / edges pinned / free in-plane

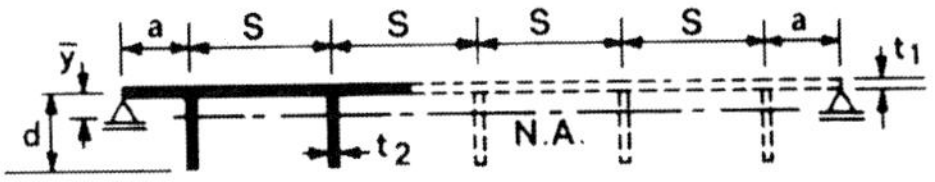

CODE	n	S	t_2	d	a	$\bar{y}$	rad. gyr.	A	I_{NA}	Z_{plate}	$Z_{stiff.}$	σ_{cr}/E	L_1	L_{euler}
To scale multiply by	–	t_1	t_1	t_1	t_1	t_1	t_1	t_1^2	t_1^4	t_1^3	t_1^3	10^{-3}	t_1	t_1
TP03350405	3	35	0.4	5	35	0.10	0.576	146.0	48.46	471.67	9.90	0.3035	219.9	>500
TP03350605	3	35	0.6	5	35	0.15	0.693	149.0	71.60	474.17	14.77	0.3347	239.1	>500
TP03350805	3	35	0.8	5	35	0.20	0.787	152.0	94.08	476.67	19.59	0.3589	254.3	>500
TP03351005	3	35	1.0	5	35	0.24	0.865	155.0	115.93	479.17	24.36	0.3787	266.9	>500
TP03350410	3	35	0.4	10	35	0.39	1.573	152.0	376.32	953.33	39.18	0.6117	353.1	>500
TP03350610	3	35	0.6	10	35	0.57	1.864	158.0	548.73	963.33	58.19	0.6855	387.5	>500
TP03350810	3	35	0.8	10	35	0.73	2.084	164.0	712.20	973.33	76.84	0.7352	413.4	>500
TP03351010	3	35	1.0	10	35	0.88	2.259	170.0	867.65	983.33	95.16	0.7706	434.1	>500
TP03350415	3	35	0.4	15	35	0.85	2.795	158.0	1234.65	1445.00	87.28	0.8496	25.4	>500
TP03350615	3	35	0.6	15	35	1.21	3.264	167.0	1779.45	1467.50	129.06	1.7829	29.0	293
TP03350815	3	35	0.8	15	35	1.53	3.604	176.0	2285.80	1490.00	169.75	2.4551	35.8	264
TP03351015	3	35	1.0	15	35	1.82	3.862	185.0	2759.29	1512.50	209.42	2.8496	38.2	259
TP03350420	3	35	0.4	20	35	1.46	4.168	164.0	2848.78	1946.67	153.68	0.4766	34.1	>500
TP03350620	3	35	0.6	20	35	2.05	4.805	176.0	4063.64	1986.67	226.33	1.0132	36.9	>500
TP03350820	3	35	0.8	20	35	2.55	5.246	188.0	5174.47	2026.67	296.59	1.5786	40.7	>500
TP03351020	3	35	1.0	20	35	3.00	5.568	200.0	6200.00	2066.67	364.71	2.0519	44.7	476
TP03350425	3	35	0.4	25	35	2.21	5.648	170.0	5422.79	2458.33	237.90	0.3047	43.0	>500
TP03350625	3	35	0.6	25	35	3.04	6.437	185.0	7664.70	2520.83	349.04	0.6509	44.7	>500
TP03350825	3	35	0.8	25	35	3.75	6.960	200.0	9687.50	2583.33	455.88	1.0467	48.5	>500
TP03351025	3	35	1.0	25	35	4.36	7.325	215.0	11537.06	2645.83	550.98	1.4276	52.4	>500
TP03350430	3	35	0.4	30	35	3.07	7.208	176.0	9143.18	2980.00	339.49	0.2116	51.0	>500
TP03350630	3	35	0.6	30	35	4.18	8.128	194.0	12818.04	3070.00	496.35	0.4537	52.9	>500
TP03350830	3	35	0.8	30	35	5.09	8.714	212.0	16098.11	3160.00	646.36	0.7355	56.3	>500
TP03351030	3	35	1.0	30	35	5.87	9.107	230.0	19076.09	3250.00	790.54	1.0306	60.8	>500
TP03500405	3	50	0.4	5	50	0.07	0.487	206.0	48.91	671.67	9.93	0.1371	293.1	>500
TP03500605	3	50	0.6	5	50	0.11	0.589	209.0	72.58	674.17	14.83	0.1507	317.6	>500
TP03500805	3	50	0.8	5	50	0.14	0.672	212.0	95.75	676.67	19.71	0.1618	337.2	>500
TP03501005	3	50	1.0	5	50	0.17	0.742	215.0	118.46	679.17	24.55	0.1711	353.4	>500
TP03500410	3	50	0.4	10	50	0.28	1.344	212.0	383.02	1353.33	39.42	0.2691	465.1	>500
TP03500610	3	50	0.6	10	50	0.41	1.607	218.0	562.84	1363.33	58.71	1.5348	47.7	147
TP03500810	3	50	0.8	10	50	0.54	1.812	224.0	735.71	1373.33	77.74	1.5840	47.7	163
TP03501010	3	50	1.0	10	50	0.65	1.981	230.0	902.17	1383.33	96.51	1.6660	47.5	173
TP03500415	3	50	0.4	15	50	0.62	2.410	218.0	1266.40	2045.00	88.06	0.8510	25.5	313
TP03500615	3	50	0.6	15	50	0.89	2.850	227.0	1844.36	2067.50	130.73	1.4212	47.4	275
TP03500815	3	50	0.8	15	50	1.14	3.183	236.0	2391.10	2090.00	172.57	1.5080	49.1	296
TP03501015	3	50	1.0	15	50	1.38	3.446	245.0	2910.08	2112.50	213.62	1.6175	49.3	308
TP03500420	3	50	0.4	20	50	1.07	3.625	224.0	2942.86	2746.67	155.47	0.4780	33.9	>500
TP03500620	3	50	0.6	20	50	1.53	4.244	236.0	4250.85	2786.67	230.09	0.9969	40.1	>500
TP03500820	3	50	0.8	20	50	1.94	4.697	248.0	5470.97	2826.67	302.86	1.2935	50.6	487
TP03501020	3	50	1.0	20	50	2.31	5.044	260.0	6615.38	2866.67	373.91	1.4633	52.9	484
TP03500425	3	50	0.4	25	50	1.63	4.951	230.0	5638.59	3458.33	241.28	0.3054	42.8	>500
TP03500625	3	50	0.6	25	50	2.30	5.744	245.0	8083.55	3520.83	356.04	0.6464	46.4	>500
TP03500825	3	50	0.8	25	50	2.88	6.305	260.0	10336.54	3583.33	467.39	0.9701	53.8	>500
TP03501025	3	50	1.0	25	50	3.41	6.723	275.0	12428.98	3645.83	575.66	1.2023	58.9	>500
TP03500430	3	50	0.4	30	50	2.29	6.366	236.0	9564.41	4180.00	345.14	0.2118	51.2	>500
TP03500630	3	50	0.6	30	50	3.19	7.322	254.0	13616.93	4270.00	507.89	0.4507	54.6	>500
TP03500830	3	50	0.8	30	50	3.97	7.978	272.0	17311.76	4360.00	665.08	0.7089	60.0	>500
TP03501030	3	50	1.0	30	50	4.66	8.452	290.0	20715.52	4450.00	817.35	0.9342	65.9	>500
TP03650405	3	65	0.4	5	65	0.06	0.430	266.0	49.15	871.67	9.94	0.0766	363.1	>500
TP03650605	3	65	0.6	5	65	0.08	0.521	269.0	73.12	874.17	14.87	0.0839	392.1	>500
TP03650805	3	65	0.8	5	65	0.11	0.596	272.0	96.69	876.67	19.77	0.0899	415.2	>500
TP03651005	3	65	1.0	5	65	0.14	0.660	275.0	119.89	879.17	24.65	0.0951	434.8	>500
TP03650410	3	65	0.4	10	65	0.22	1.192	272.0	386.76	1753.33	39.55	0.9055	61.8	144
TP03650610	3	65	0.6	10	65	0.32	1.433	278.0	570.86	1763.33	59.00	0.9169	61.8	172
TP03650810	3	65	0.8	10	65	0.42	1.624	284.0	749.30	1773.33	78.24	0.9417	61.8	191
TP03651010	3	65	1.0	10	65	0.52	1.783	290.0	922.41	1783.33	97.27	0.9825	61.6	205
TP03650415	3	65	0.4	15	65	0.49	2.149	278.0	1284.44	2645.00	88.49	0.8532	25.6	269
TP03650615	3	65	0.6	15	65	0.71	2.561	287.0	1882.12	2667.50	131.67	0.8971	62.1	312
TP03650815	3	65	0.8	15	65	0.91	2.879	296.0	2453.72	2690.00	174.17	0.9294	62.3	343
TP03651015	3	65	1.0	15	65	1.11	3.137	305.0	3001.54	2712.50	216.04	0.9825	62.1	362
TP03650420	3	65	0.4	20	65	0.85	3.249	284.0	2997.18	3546.67	156.47	0.4786	34.0	>500
TP03650620	3	65	0.6	20	65	1.22	3.839	296.0	4362.16	3586.67	232.23	0.8300	61.4	495
TP03650820	3	65	0.8	20	65	1.56	4.284	308.0	5651.95	3626.67	306.48	0.8861	63.5	>500
TP03651020	3	65	1.0	20	65	1.88	4.635	320.0	6875.00	3666.67	379.31	0.9525	63.9	>500

T-PLATE

n rectangular stiffeners
edges pinned
free in-plane

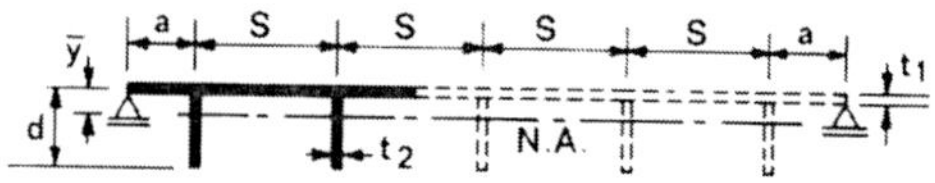

| CODE | n | | DIMENSIONS | | | | | SECTION PROPERTIES | | | | | BUCKLING PROPERTIES | | |
		S	t₂	d	a	ȳ	rad. gyr.	A	I_NA	Z_plate	Z_stiff.	σ_cr/E	L₁	L_euler
To scale multiply by	–	t_1	t_1	t_1	t_1	t_1	t_1	t_1^2	t_1^4	t_1^3	t_1^3	10^{-3}	t_1	t_1
TP03650425	3	65	0.4	25	65	1.29	4.459	290.0	5765.09	4458.33	243.18	0.3060	42.4	>500
TP03650625	3	65	0.6	25	65	1.84	5.228	305.0	8337.60	4520.93	360.07	0.6344	51.9	>500
TP03650825	3	65	0.8	25	65	2.34	5.794	320.0	10742.19	4583.33	474.14	0.7916	65.4	>500
TP03651025	3	65	1.0	25	65	2.80	6.230	335.0	13001.40	4645.83	585.61	0.8848	67.7	>500
TP03650430	3	65	0.4	30	65	1.82	5.758	296.0	9814.86	5380.00	348.35	0.2123	51.3	>500
TP03650630	3	65	0.6	30	65	2.58	6.704	314.0	14110.51	5470.00	514.60	0.4477	56.6	>500
TP03650830	3	65	0.8	30	65	3.25	7.381	332.0	18086.75	5560.00	676.22	0.6474	67.8	>500
TP03651030	3	65	1.0	30	65	3.86	7.891	350.0	21792.86	5650.00	833.61	0.7754	73.5	>500
TP03800405	3	80	0.4	5	80	0.05	0.389	326.0	49.31	1071.67	9.95	0.0485	430.8	>500
TP03800605	3	80	0.6	5	80	0.07	0.473	329.0	73.46	1074.17	14.90	0.0529	463.7	>500
TP03800805	3	80	0.8	5	80	0.09	0.541	332.0	97.29	1076.67	19.82	EULER	–	–
TP03801005	3	80	1.0	5	80	0.11	0.601	335.0	120.80	1079.17	24.71	EULER	–	–
TP03800410	3	80	0.4	10	80	0.18	1.083	332.0	389.16	2153.33	39.63	0.5999	76.0	163
TP03800610	3	80	0.6	10	80	0.27	1.305	338.0	576.04	2163.33	59.18	0.6066	76.0	194
TP03800810	3	80	0.8	10	80	0.35	1.485	344.0	758.14	2173.33	78.55	0.6208	75.9	217
TP03801010	3	80	1.0	10	80	0.43	1.635	350.0	935.71	2183.33	97.76	0.6439	75.7	234
TP03800415	3	80	0.4	15	80	0.40	1.958	338.0	1296.08	3245.00	88.77	0.5931	76.0	295
TP03800615	3	80	0.6	15	80	0.58	2.344	347.0	1906.83	3267.50	132.27	0.6016	76.2	350
TP03800815	3	80	0.8	15	80	0.76	2.647	356.0	2495.22	3290.00	175.21	0.6201	76.2	388
TP03801015	3	80	1.0	15	90	0.92	2.897	365.0	3062.93	3312.50	217.61	0.6569	75.9	413
TP03800420	3	80	0.4	20	80	0.70	2.969	344.0	3032.56	4346.67	157.11	0.4795	34.0	>500
TP03800620	3	80	0.6	20	80	1.01	3.530	356.0	4435.96	4386.67	233.61	0.5855	76.4	>500
TP03800820	3	80	0.8	20	80	1.30	3.961	368.0	5773.91	4426.67	308.84	0.6087	76.9	>500
TP03801020	3	80	1.0	20	80	1.58	4.308	380.0	7052.63	4466.67	382.86	0.6456	76.7	>500
TP03800425	3	80	0.4	25	80	1.07	4.088	350.0	5848.21	5458.33	244.40	0.3063	42.9	>500
TP03800625	3	80	0.6	25	80	1.54	4.828	365.0	8508.13	5520.83	362.68	0.5430	75.0	>500
TP03800825	3	80	0.8	25	80	1.97	5.385	380.0	11019.74	5583.33	478.57	0.5822	78.3	>500
TP03801025	3	80	1.0	25	80	2.37	5.824	395.0	13399.92	5645.83	592.22	0.6268	79.1	>500
TP03800430	3	80	0.4	30	80	1.52	5.295	356.0	9980.90	6580.00	350.41	0.2125	51.0	>500
TP03800630	3	80	0.6	30	80	2.17	6.215	374.0	14445.72	6670.00	518.99	0.4382	64.3	>500
TP03800830	3	80	0.8	30	80	2.76	6.893	392.0	18624.49	6760.00	683.60	0.5327	80.3	>500
TP03801030	3	80	1.0	30	80	3.29	7.417	410.0	22554.88	6850.00	844.52	0.5913	82.6	>500
TP05350405	5	35	0.4	5	35	0.11	0.605	220.0	80.49	708.33	16.47	0.1350	330.2	>500
TP05350605	5	35	0.6	5	35	0.17	0.726	225.0	118.75	712.50	24.57	0.1489	359.2	>500
TP05350805	5	35	0.8	5	35	0.22	0.823	230.0	155.80	716.67	32.58	0.1599	381.4	>500
TP05351005	5	35	1.0	5	35	0.27	0.903	235.0	191.71	720.83	40.50	0.1689	400.3	>500
TP05350410	5	35	0.4	10	35	0.43	1.646	230.0	623.19	1433.33	65.15	1.9142	17.0	128
TP05350610	5	35	0.6	10	35	0.63	1.943	240.0	906.25	1450.00	96.67	2.9433	33.4	120
TP05350810	5	35	0.8	10	35	0.80	2.166	250.0	1173.33	1466.67	127.54	3.1107	34.0	130
TP05351010	5	35	1.0	10	35	0.96	2.342	260.0	1426.28	1483.33	157.80	3.3447	34.1	135
TP05350415	5	35	0.4	15	35	0.94	2.915	240.0	2039.06	2175.00	145.00	0.8496	25.4	361
TP05350615	5	35	0.6	15	35	1.32	3.389	255.0	2928.31	2212.50	214.11	1.7762	29.3	276
TP05350815	5	35	0.8	15	35	1.67	3.727	270.0	3750.00	2250.00	281.25	2.4281	36.1	256
TP05351015	5	35	1.0	15	35	1.97	3.980	285.0	4514.80	2287.50	346.59	2.8299	38.5	251
TP05350420	5	35	0.4	20	35	1.60	4.333	250.0	4693.33	2933.33	255.07	0.4764	34.2	>500
TP05350620	5	35	0.6	20	35	2.22	4.969	270.0	6666.67	3000.00	375.00	1.0101	36.6	>500
TP05350820	5	35	0.8	20	35	2.76	5.401	290.0	8459.77	3066.67	490.67	1.5645	41.1	477
TP05351020	5	35	1.0	20	35	3.23	5.710	310.0	10107.53	3133.33	602.56	2.0271	45.2	435
TP05350425	5	35	0.4	25	35	2.40	5.855	260.0	8914.26	3708.33	394.50	0.3045	43.0	>500
TP05350625	5	35	0.6	25	35	3.29	6.634	285.0	12541.12	3812.50	577.65	0.6490	44.9	>500
TP05350825	5	35	0.8	25	35	4.03	7.138	310.0	15793.01	3916.67	753.21	1.0389	48.9	>500
TP05351025	5	35	1.0	25	35	4.66	7.482	335.0	18753.99	4020.83	922.21	1.4115	52.9	>500
TP05350430	5	35	0.4	30	35	3.33	7.454	270.0	15000.00	4500.00	562.50	0.2114	51.0	>500
TP05350630	5	35	0.6	30	35	4.50	8.352	300.0	20925.00	4650.00	820.59	0.4522	53.1	>500
TP05350830	5	35	0.8	30	35	5.45	8.907	330.0	26181.82	4800.00	1066.67	0.7344	56.7	>500
TP05351030	5	35	1.0	30	35	6.25	9.270	360.0	30937.50	4950.00	1302.63	1.0198	61.4	>500
TP05500405	5	50	0.4	5	50	0.08	0.512	310.0	81.32	1008.33	16.53	0.0609	440.0	>500
TP05500605	5	50	0.6	5	50	0.12	0.619	315.0	120.54	1012.50	24.70	0.0670	477.0	>500
TP05500805	5	50	0.8	5	50	0.16	0.705	320.0	158.85	1016.67	32.80	1.5881	47.6	62
TP05501005	5	50	1.0	5	50	0.19	0.777	325.0	196.31	1020.83	40.83	1.6418	47.5	67
TP05500410	5	50	0.4	10	50	0.31	1.409	320.0	635.42	2033.33	65.59	1.5085	47.5	124
TP05500610	5	50	0.6	10	50	0.45	1.680	330.0	931.82	2050.00	97.62	1.5348	47.7	146
TP05500810	5	50	0.8	10	50	0.59	1.891	340.0	1215.69	2066.67	129.17	1.5892	47.7	161
TP05501010	5	50	1.0	10	50	0.71	2.062	350.0	1488.10	2083.33	160.26	1.6797	47.5	170

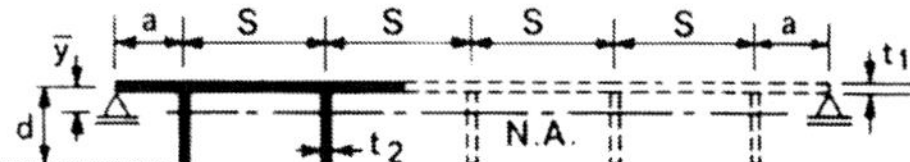

T-PLATE n rectangular stiffeners / edges pinned / free in-plane

CODE	n	S	t_2	d	a	$\bar{y}$	rad. gyr.	A	I_{NA}	Z_{plate}	$Z_{stiff.}$	σ_{cr}/E	L_1	L_{euler}
To scale multiply by	$-$	t_1	t_1	t_1	t_1	t_1	t_1	t_1^2	t_1^4	t_1^3	t_1^3	10^{-3}	t_1	t_1
TP05500415	5	50	0.4	15	50	0.68	2.521	330.0	2996.59	3075.00	146.43	0.8596	25.5	300
TP05500615	5	50	0.6	15	50	0.98	2.971	345.0	3044.84	3112.50	217.15	1.4118	47.4	269
TP05500815	5	50	0.8	15	50	1.25	3.307	360.0	3937.50	3150.00	286.36	1.5054	45.2	289
TP05501015	5	50	1.0	15	50	1.50	3.571	375.0	4781.25	3187.50	354.17	1.6251	49.2	300
TP05500420	5	50	0.4	20	50	1.18	3.782	340.0	4862.75	4133.33	258.33	0.4779	33.9	>500
TP05500620	5	50	0.6	20	50	1.67	4.410	360.0	7000.00	4200.00	381.82	0.9922	40.8	486
TP05500820	5	50	0.8	20	50	2.11	4.862	380.0	8982.46	4266.67	501.96	1.2905	50.9	466
TP05501020	5	50	1.0	20	50	2.50	5.204	400.0	10833.33	4333.33	619.05	1.4571	53.4	465
TP05500425	5	50	0.4	25	50	1.79	5.155	350.0	9300.60	5208.33	400.64	0.3054	42.9	>500
TP05500625	5	50	0.6	25	50	2.50	5.951	375.0	13281.25	5312.50	590.28	0.6445	46.8	>500
TP05500825	5	50	0.8	25	50	3.13	6.505	400.0	16927.08	5416.67	773.81	0.9600	54.6	>500
TP05501025	5	50	1.0	25	50	3.68	6.911	425.0	20297.18	5520.83	951.87	1.1889	59.5	>500
TP05500430	5	50	0.4	30	50	2.50	6.614	360.0	15750.00	6300.00	572.73	0.2117	51.3	>500
TP05500630	5	50	0.6	30	50	3.46	7.566	390.0	22326.92	6450.00	841.30	0.4494	54.9	>500
TP05500830	5	50	0.8	30	50	4.29	8.207	420.0	28285.71	6600.00	1100.00	0.7029	60.9	>500
TP05501030	5	50	1.0	30	50	5.00	8.660	450.0	33750.00	6750.00	1350.00	0.9229	66.8	>500
TP05650405	5	65	0.4	5	65	0.06	0.452	400.0	81.77	1308.33	16.56	EULER	-	-
TP05650605	5	65	0.6	5	65	0.09	0.548	405.0	121.53	1312.50	24.76	EULER	-	-
TP05650805	5	65	0.8	5	65	0.12	0.626	410.0	160.57	1316.67	32.92	EULER	-	-
TP05651005	5	65	1.0	5	65	0.15	0.693	415.0	190.92	1320.83	41.02	EULER	-	-
TP05650410	5	65	0.4	10	65	0.24	1.252	410.0	642.28	2633.33	65.83	0.9052	61.8	144
TP05650610	5	65	0.6	10	65	0.36	1.501	420.0	946.43	2650.00	98.15	0.9179	61.8	171
TP05650810	5	65	0.8	10	65	0.47	1.698	430.0	1240.31	2666.67	130.08	0.9453	61.8	189
TP05651010	5	65	1.0	10	65	0.57	1.861	440.0	1524.62	2683.33	161.65	0.9905	61.5	202
TP05650415	5	65	0.4	15	65	0.54	2.252	420.0	2129.46	3975.00	147.22	0.8527	25.6	265
TP05650615	5	65	0.6	15	65	0.78	2.675	435.0	3113.15	4012.50	218.86	0.8959	62.1	306
TP05650815	5	65	0.8	15	65	1.00	3.000	450.0	4050.00	4050.00	289.29	0.9316	62.3	336
TP05651015	5	65	1.0	15	65	1.21	3.261	465.0	4944.56	4087.50	358.55	0.9901	62.1	353
TP05650420	5	65	0.4	20	65	0.93	3.397	430.0	4961.24	5333.33	260.16	0.4784	34.0	>500
TP05650620	5	65	0.6	20	65	1.33	4.000	450.0	7200.00	5400.00	385.71	0.8239	61.5	480
TP05650820	5	65	0.8	20	65	1.70	4.449	470.0	9304.96	5466.67	508.53	0.8840	63.7	>500
TP05651020	5	65	1.0	20	65	2.04	4.801	490.0	11292.52	5533.33	628.79	0.9566	64.2	>500
TP05650425	5	65	0.4	25	65	1.42	4.654	440.0	9528.88	6708.33	404.12	0.3059	42.4	>500
TP05650625	5	65	0.6	25	65	2.02	5.435	465.0	13734.88	6812.50	597.59	0.6308	52.9	>500
TP05650825	5	65	0.8	25	65	2.55	6.001	490.0	17644.56	6916.67	785.98	0.7844	65.7	>500
TP05651025	5	65	1.0	25	65	3.03	6.431	515.0	21301.07	7020.83	969.73	0.8824	68.3	>500
TP05650430	5	65	0.4	30	65	2.00	6.000	450.0	16200.00	8100.00	578.57	0.2123	51.3	>500
TP05650630	5	65	0.6	30	65	2.81	6.953	480.0	23203.13	8250.00	853.45	0.4462	57.1	>500
TP05650830	5	65	0.8	30	65	3.53	7.624	510.0	29647.06	8400.00	1120.00	0.6402	68.5	>500
TP05651030	5	65	1.0	30	65	4.17	8.122	540.0	35625.00	8550.00	1379.03	0.7681	74.0	>500
TP05800405	5	80	0.4	5	80	0.05	0.409	490.0	82.06	1608.33	16.58	EULER	-	-
TP05800605	5	80	0.6	5	80	0.08	0.497	495.0	122.16	1612.50	24.81	EULER	-	-
TP05800805	5	80	0.8	5	80	0.10	0.569	500.0	161.67	1616.67	32.99	EULER	-	-
TP05801005	5	80	1.0	5	80	0.12	0.630	505.0	200.60	1620.83	41.14	EULER	-	-
TP05800410	5	80	0.4	10	80	0.20	1.137	500.0	646.67	3233.33	65.99	0.5999	76.0	162
TP05800610	5	80	0.6	10	80	0.29	1.369	510.0	955.88	3250.00	98.48	0.6074	76.0	193
TP05800810	5	80	0.8	10	80	0.38	1.554	520.0	1256.41	3266.67	130.67	0.6231	75.9	215
TP05801010	5	80	1.0	10	80	0.47	1.709	530.0	1548.74	3283.33	162.54	0.6488	75.7	231
TP05800415	5	80	0.4	15	80	0.44	2.054	510.0	2150.74	4875.00	147.73	0.5924	76.0	291
TP05800615	5	80	0.6	15	80	0.64	2.453	525.0	3158.04	4912.50	219.96	0.6018	76.2	344
TP05800815	5	80	0.8	15	80	0.83	2.764	540.0	4125.00	4950.00	291.18	0.6223	76.2	380
TP05801015	5	80	1.0	15	80	1.01	3.018	555.0	5054.90	4987.50	361.41	0.6564	75.8	403
TP05800420	5	80	0.4	20	80	0.77	3.109	520.0	5025.64	6533.33	261.33	0.4792	34.0	492
TP05800620	5	80	0.6	20	80	1.11	3.685	540.0	7333.33	6600.00	388.24	0.5841	76.5	>500
TP05800820	5	80	0.8	20	80	1.43	4.124	560.0	9523.81	6666.67	512.82	0.6096	77.0	>500
TP05801020	5	80	1.0	20	80	1.72	4.474	580.0	11609.20	6733.33	635.22	0.6503	76.8	>500
TP05800425	5	80	0.4	25	80	1.18	4.274	530.0	9679.64	8208.33	406.35	0.3062	42.9	>500
TP05800625	5	80	0.6	25	80	1.69	5.030	555.0	14041.39	8312.50	602.36	0.5387	75.1	>500
TP05800825	5	80	0.8	25	80	2.16	5.592	580.0	18139.37	8416.67	794.03	0.5806	78.7	>500
TP05801025	5	80	1.0	25	80	2.58	6.031	605.0	22006.28	8520.83	981.66	0.6293	79.4	>500
TP05800430	5	80	0.4	30	80	1.67	5.528	540.0	16500.00	9900.00	582.35	0.2124	50.9	>500
TP05800630	5	80	0.6	30	80	2.37	6.462	570.0	23802.63	10050.00	861.43	0.4353	65.7	>500
TP05800830	5	80	0.8	30	80	3.00	7.141	600.0	30600.00	10200.00	1133.33	0.5282	80.6	>500
TP05801030	5	80	1.0	30	80	3.57	7.660	630.0	36964.29	10350.00	1398.65	0.5903	83.2	>500

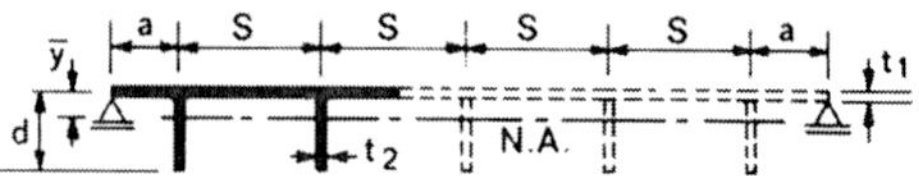

T-PLATE   n rectangular stiffeners / edges pinned / free in-plane

CODE	n	S	t_2	d	a	$\bar{y}$	rad. gyr.	A	I_{NA}	Z_{plate}	$Z_{stiff.}$	σ_{cr}/E	L_1	L_{euler}
To scale multiply by	—	t_1	t_1	t_1	t_1	t_1	t_1	t_1^2	t_1^4	t_1^3	t_1^3	10^{-3}	t_1	t_1
TQ02350405	2	35	0.4	5	7	0.19	0.770	53.0	31.45	166.67	6.54	2.1742	66.6	>500
TQ02350605	2	35	0.6	5	7	0.27	0.914	55.0	45.91	168.33	9.71	2.3892	71.6	>500
TQ02350805	2	35	0.8	5	7	0.35	1.023	57.0	59.65	170.00	12.83	2.5879	75.8	>500
TQ02351005	2	35	1.0	5	7	0.42	1.110	59.0	72.74	171.67	15.90	2.7899	79.5	>500
TQ02350410	2	35	0.4	10	7	0.70	2.046	57.0	238.60	340.00	25.66	1.9146	16.9	>500
TQ02350610	2	35	0.6	10	7	0.98	2.364	61.0	340.98	346.67	37.82	4.0517	19.1	>500
TQ02350810	2	35	0.8	10	7	1.23	2.587	65.0	434.87	353.33	49.59	4.7413	118.0	>500
TQ02351010	2	35	1.0	10	7	1.45	2.750	69.0	521.74	360.00	61.02	5.0692	124.2	>500
TQ02350415	2	35	0.4	15	7	1.48	3.546	61.0	767.21	520.00	56.73	0.8526	25.3	>500
TQ02350615	2	35	0.6	15	7	2.01	4.011	67.0	1077.99	535.00	83.02	1.8498	26.3	>500
TQ02350815	2	35	0.8	15	7	2.47	4.310	73.0	1356.16	550.00	108.20	3.0228	27.5	>500
TQ02351015	2	35	1.0	15	7	2.85	4.513	79.0	1609.18	565.00	132.42	4.1282	28.9	>500
TQ02350420	2	35	0.4	20	7	2.46	5.173	65.0	1739.49	706.67	99.18	0.4801	33.7	>500
TQ02350620	2	35	0.6	20	7	3.29	5.747	73.0	2410.96	733.33	144.26	1.0495	34.3	>500
TQ02350820	2	35	0.8	20	7	3.95	6.088	81.0	3002.47	760.00	187.08	1.7653	35.9	>500
TQ02351020	2	35	1.0	20	7	4.49	6.303	89.0	3535.58	786.67	228.02	2.5390	36.9	>500
TQ02350425	2	35	0.4	25	7	3.62	6.875	69.0	3260.87	900.00	152.54	0.3076	42.0	>500
TQ02350625	2	35	0.6	25	7	4.75	7.522	79.0	4469.94	941.67	220.70	0.6756	43.0	>500
TQ02350825	2	35	0.8	25	7	5.62	7.879	89.0	5524.34	983.33	285.02	1.1486	43.7	>500
TQ02351025	2	35	1.0	25	7	6.31	8.085	99.0	6470.96	1025.00	346.28	1.6821	45.2	>500
TQ02350430	2	35	0.4	30	7	4.93	8.620	73.0	5424.66	1100.00	216.39	0.2139	50.4	>500
TQ02350630	2	35	0.6	30	7	6.35	9.311	85.0	7369.41	1160.00	311.64	0.4712	50.9	>500
TQ02350830	2	35	0.8	30	7	7.42	9.662	97.0	9055.67	1220.00	401.10	0.8063	52.0	>500
TQ02351030	2	35	1.0	30	7	8.26	9.847	109.0	10568.81	1280.00	486.08	1.1921	53.5	>500
TQ02500405	2	50	0.4	5	10	0.14	0.657	74.0	31.98	236.67	6.57	1.0026	90.3	>500
TQ02500605	2	50	0.6	5	10	0.20	0.787	76.0	47.04	238.33	9.79	1.0936	96.7	>500
TQ02500805	2	50	0.8	5	10	0.26	0.888	78.0	61.54	240.00	12.97	1.1786	102.0	>500
TQ02501005	2	50	1.0	5	10	0.31	0.972	80.0	75.52	241.67	16.11	1.2645	106.5	>500
TQ02500410	2	50	0.4	10	10	0.51	1.776	78.0	246.15	480.00	25.95	1.7815	135.6	>500
TQ02500610	2	50	0.6	10	10	0.73	2.084	82.0	356.10	486.67	38.42	2.0255	148.7	>500
TQ02500810	2	50	0.8	10	10	0.93	2.310	86.0	458.91	493.33	50.60	2.2235	158.8	>500
TQ02501010	2	50	1.0	10	10	1.11	2.485	90.0	555.56	500.00	62.50	2.4016	167.1	>500
TQ02500415	2	50	0.4	15	10	1.10	3.126	82.0	801.22	730.00	57.63	0.8512	25.4	>500
TQ02500615	2	50	0.6	15	10	1.53	3.604	88.0	1142.90	745.00	84.87	1.8176	27.7	>500
TQ02500815	2	50	0.8	15	10	1.91	3.935	94.0	1455.32	760.00	111.22	2.5743	31.4	>500
TQ02501015	2	50	1.0	15	10	2.25	4.176	100.0	1743.75	775.00	136.76	2.9906	32.4	>500
TQ02500420	2	50	0.4	20	10	1.86	4.620	86.0	1835.66	986.67	101.20	0.4795	33.7	>500
TQ02500620	2	50	0.6	20	10	2.55	5.246	94.0	2587.23	1013.33	148.29	1.0393	35.0	>500
TQ02500820	2	50	0.8	20	10	3.14	5.656	102.0	3262.75	1040.00	193.49	1.6822	37.0	>500
TQ02501020	2	50	1.0	20	10	3.64	5.938	110.0	3878.75	1066.67	237.04	2.2555	38.8	>500
TQ02500425	2	50	0.4	25	10	2.78	6.211	90.0	3472.22	1250.00	156.25	0.3072	42.1	>500
TQ02500625	2	50	0.6	25	10	3.75	6.960	100.0	4843.75	1291.67	227.94	0.6698	43.1	>500
TQ02500825	2	50	0.8	25	10	4.55	7.423	110.0	6060.61	1333.33	296.30	1.1171	44.7	>500
TQ02501025	2	50	1.0	25	10	5.21	7.725	120.0	7161.46	1375.00	361.84	1.5829	46.7	>500
TQ02500430	2	50	0.4	30	10	3.83	7.869	94.0	5821.28	1520.00	222.44	0.2135	50.5	>500
TQ02500630	2	50	0.6	30	10	5.09	8.714	106.0	8049.06	1580.00	323.18	0.4675	51.3	>500
TQ02500830	2	50	0.8	30	10	6.10	9.209	118.0	10006.78	1640.00	418.72	0.7893	52.8	>500
TQ02501030	2	50	1.0	30	10	6.92	9.515	130.0	11769.23	1700.00	510.00	1.1428	55.1	>500
TQ02650405	2	65	0.4	5	13	0.11	0.583	95.0	32.28	306.67	6.59	0.5631	113.3	>500
TQ02650605	2	65	0.6	5	13	0.15	0.701	97.0	47.68	308.33	9.84	0.6152	120.7	>500
TQ02650805	2	65	0.8	5	13	0.20	0.795	99.0	62.63	310.00	13.05	0.6596	126.9	>500
TQ02651005	2	65	1.0	5	13	0.25	0.874	101.0	77.15	311.67	16.23	0.7043	132.2	>500
TQ02650410	2	65	0.4	10	13	0.40	1.591	99.0	250.51	620.00	26.11	0.9850	167.8	>500
TQ02650610	2	65	0.6	10	13	0.58	1.883	103.0	365.05	626.67	38.76	1.1232	184.0	>500
TQ02650810	2	65	0.8	10	13	0.75	2.104	107.0	473.52	633.33	51.18	1.2379	196.3	>500
TQ02651010	2	65	1.0	10	13	0.90	2.279	111.0	576.58	640.00	63.37	1.3420	206.4	>500
TQ02650415	2	65	0.4	15	13	0.87	2.824	103.0	821.36	940.00	58.14	0.8513	25.5	>500
TQ02650615	2	65	0.6	15	13	1.24	3.294	109.0	1182.80	955.00	85.95	1.6327	36.7	>500
TQ02650815	2	65	0.8	15	13	1.57	3.633	115.0	1518.26	970.00	113.01	1.8032	38.6	>500
TQ02651015	2	65	1.0	15	13	1.86	3.891	121.0	1831.61	985.00	139.39	1.9481	38.8	>500
TQ02650420	2	65	0.4	20	13	1.50	4.207	107.0	1894.08	1266.67	102.36	0.4789	33.8	>500
TQ02650620	2	65	0.6	20	13	2.09	4.845	115.0	2699.13	1293.33	150.68	1.0259	36.9	>500
TQ02650820	2	65	0.8	20	13	2.60	5.284	123.0	3434.15	1320.00	197.38	1.4900	40.8	>500
TQ02651020	2	65	1.0	20	13	3.05	5.603	131.0	4111.96	1346.67	242.64	1.7569	42.0	>500

CODE	n	S	t_2	d	a	$\bar{y}$	rad. gyr.	A	I_{NA}	Z_{plate}	$Z_{stiff.}$	σ_{cr}/E	L_1	L_{euler}
To scale multiply by	−	t_1	t_1	t_1	t_1	t_1	t_1	t_1^2	t_1^4	t_1^3	t_1^3	10^{-3}	t_1	t_1
TQ02650425	2	65	0.4	25	13	2.25	5.698	111.0	3603.60	1600.00	158.42	0.3068	42.1	>500
TQ02650625	2	65	0.6	25	13	3.10	6.484	121.0	5087.81	1641.67	232.31	0.6645	43.9	>500
TQ02650825	2	65	0.8	25	13	3.82	7.003	131.0	6424.94	1683.33	303.30	1.0677	46.6	>500
TQ02651025	2	65	1.0	25	13	4.43	7.364	141.0	7646.28	1725.00	371.77	1.4128	49.0	>500
TQ02650430	2	65	0.4	30	13	3.13	7.267	115.0	6073.04	1940.00	226.02	0.2133	50.6	>500
TQ02650630	2	65	0.6	30	65	2.34	6.426	231.0	9537.66	4080.00	344.79	0.4494	56.0	>500
TQ02650830	2	65	0.8	30	65	2.96	7.105	243.0	12266.67	4140.00	453.70	0.6556	67.3	>500
TQ02651030	2	65	1.0	30	65	3.53	7.624	255.0	14823.53	4200.00	560.00	0.7834	72.7	>500
TQ02800405	2	80	0.4	5	16	0.09	0.529	116.0	32.47	376.67	6.61	0.3631	135.8	>500
TQ02800605	2	80	0.6	5	16	0.13	0.638	118.0	48.09	378.33	9.87	0.3908	143.9	>500
TQ02800805	2	80	0.8	5	16	0.17	0.726	120.0	63.33	380.00	13.10	0.4170	150.9	>500
TQ02801005	2	80	1.0	5	16	0.20	0.801	122.0	78.21	381.57	16.31	0.4434	157.0	>500
TQ02800410	2	80	0.4	10	16	0.33	1.453	120.0	253.33	760.00	26.21	0.6136	198.5	>500
TQ02800610	2	80	0.6	10	16	0.48	1.730	124.0	370.97	766.67	38.98	0.6996	217.2	>500
TQ02800810	2	80	0.8	10	16	0.63	1.943	128.0	483.33	773.33	51.56	0.7720	231.7	>500
TQ02801010	2	80	1.0	10	16	0.76	2.116	132.0	590.91	780.00	63.93	0.8380	243.7	>500
TQ02800415	2	80	0.4	15	16	0.73	2.594	124.0	834.68	1150.00	58.47	0.8522	25.5	>500
TQ02800615	2	80	0.6	15	16	1.04	3.051	130.0	1209.81	1165.00	86.65	1.0907	287.6	>500
TQ02800815	2	80	0.8	15	16	1.32	3.389	136.0	1561.76	1180.00	114.19	1.2033	307.1	>500
TQ02801015	2	80	1.0	15	16	1.58	3.652	142.0	1893.49	1195.00	141.14	1.3001	323.0	>500
TQ02800420	2	80	0.4	20	16	1.25	3.886	128.0	1933.33	1546.67	103.11	0.4788	33.9	>500
TQ02800620	2	80	0.6	20	16	1.76	4.518	136.0	2776.47	1573.33	152.26	0.9838	42.6	>500
TQ02800820	2	80	0.8	20	16	2.22	4.969	144.0	3555.56	1600.00	200.00	1.1637	47.5	>500
TQ02801020	2	80	1.0	20	16	2.63	5.307	152.0	4280.70	1626.67	246.46	1.2728	48.2	>500
TQ02800425	2	80	0.4	25	16	1.89	5.289	132.0	3693.18	1950.00	159.84	0.3065	42.2	>500
TQ02800625	2	80	0.6	25	16	2.64	6.086	142.0	5259.68	1991.67	235.24	0.6576	45.4	>500
TQ02800825	2	80	0.8	25	16	3.29	6.634	152.0	6688.60	2033.33	308.08	0.9679	50.5	>500
TQ02801025	2	80	1.0	25	16	3.86	7.030	162.0	8005.40	2075.00	378.65	1.1521	51.9	>500
TQ02800430	2	80	0.4	30	16	2.65	6.777	136.0	6247.06	2360.00	228.39	0.2130	50.6	>500
TQ02800630	2	80	0.6	30	16	3.65	7.724	148.0	8829.73	2420.00	335.08	0.4611	52.8	>500
TQ02800830	2	80	0.8	30	16	4.50	8.352	160.0	11160.00	2480.00	437.65	0.7367	56.1	>500
TQ02801030	2	80	1.0	30	16	5.23	8.790	172.0	13290.70	2540.00	536.62	0.9653	58.6	>500
TQ03350405	3	35	0.4	5	7	0.17	0.726	90.0	47.50	285.00	9.83	0.8311	129.8	>500
TQ03350605	3	35	0.6	5	7	0.24	0.865	93.0	69.56	287.50	14.62	0.9191	141.0	>500
TQ03350805	3	35	0.8	5	7	0.31	0.972	96.0	90.63	290.00	19.33	0.9919	149.8	>500
TQ03351005	3	35	1.0	5	7	0.38	1.058	99.0	110.80	292.50	23.98	1.0574	157.1	>500
TQ03350410	3	35	0.4	10	7	0.63	1.943	96.0	362.50	580.00	38.67	1.6504	206.7	>500
TQ03350610	3	35	0.6	10	7	0.88	2.259	102.0	520.59	590.00	57.10	1.8541	227.1	>500
TQ03350810	3	35	0.8	10	7	1.11	2.485	108.0	666.67	600.00	75.00	1.9997	242.4	>500
TQ03351010	3	35	1.0	10	7	1.32	2.653	114.0	802.63	610.00	92.42	2.1145	254.8	>500
TQ03350415	3	35	0.4	15	7	1.32	3.389	102.0	1171.32	885.00	85.65	0.8498	25.6	>500
TQ03350615	3	35	0.6	15	7	1.82	3.862	111.0	1655.57	907.50	125.65	1.8076	27.5	>500
TQ03350815	3	35	0.8	15	7	2.25	4.176	120.0	2092.50	930.00	164.12	2.7303	31.2	>500
TQ03351015	3	35	1.0	15	7	2.62	4.395	129.0	2492.01	952.50	201.23	3.1425	339.7	>500
TQ03350420	3	35	0.4	20	7	2.22	4.969	108.0	2666.67	1200.00	150.00	0.4774	34.0	>500
TQ03350620	3	35	0.6	20	7	3.00	5.568	120.0	3720.00	1240.00	218.82	1.0252	35.9	>500
TQ03350820	3	35	0.8	20	7	3.64	5.938	132.0	4654.55	1280.00	284.44	1.6543	38.1	>500
TQ03351020	3	35	1.0	20	7	4.17	6.180	144.0	5500.00	1320.00	347.37	2.2534	40.9	>500
TQ03350425	3	35	0.4	25	7	3.29	6.634	114.0	5016.45	1525.00	231.06	0.3056	42.3	>500
TQ03350625	3	35	0.6	25	7	4.36	7.325	129.0	6922.24	1587.50	335.39	0.6595	43.7	>500
TQ03350825	3	35	0.8	25	7	5.21	7.725	144.0	8593.75	1650.00	434.21	1.0855	46.0	>500
TQ03351025	3	35	1.0	25	7	5.90	7.969	159.0	10097.29	1712.50	528.55	1.5261	49.3	>500
TQ03350430	3	35	0.4	30	7	4.50	8.352	120.0	8370.00	1860.00	328.24	0.2124	50.7	>500
TQ03350630	3	35	0.6	30	7	5.87	9.107	138.0	11445.65	1950.00	474.32	0.4601	52.0	>500
TQ03350830	3	35	0.8	30	7	6.92	9.515	156.0	14123.08	2040.00	612.00	0.7650	54.5	>500
TQ03351030	3	35	1.0	30	7	7.76	9.746	174.0	16525.86	2130.00	743.02	1.0917	57.3	>500
TQ03500405	3	50	0.4	5	10	0.12	0.619	126.0	48.21	405.00	9.88	0.3764	173.4	>500
TQ03500605	3	50	0.6	5	10	0.17	0.742	129.0	71.08	407.50	14.73	0.4146	187.6	>500
TQ03500805	3	50	0.8	5	10	0.23	0.840	132.0	93.18	410.00	19.52	0.4471	199.0	>500
TQ03501005	3	50	1.0	5	10	0.28	0.921	135.0	114.58	412.50	24.26	0.4766	208.6	>500
TQ03500410	3	50	0.4	10	10	0.45	1.680	132.0	372.73	820.00	39.05	0.7308	273.5	>500
TQ03500610	3	50	0.6	10	10	0.65	1.981	138.0	541.30	830.00	57.91	0.8305	300.6	>500
TQ03500810	3	50	0.8	10	10	0.83	2.205	144.0	700.00	840.00	76.36	0.9060	320.9	>500
TQ03501010	3	50	1.0	10	10	1.00	2.380	150.0	850.00	850.00	94.44	0.9680	337.7	>500

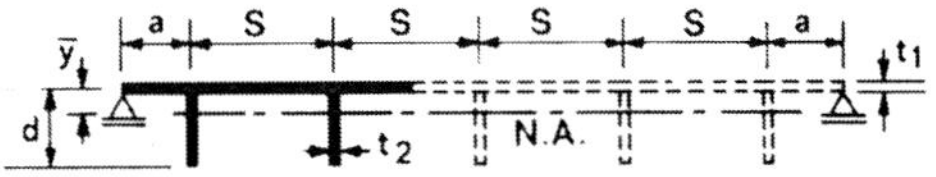

T-PLATE — n rectangular stiffeners / edges pinned / free in-plane

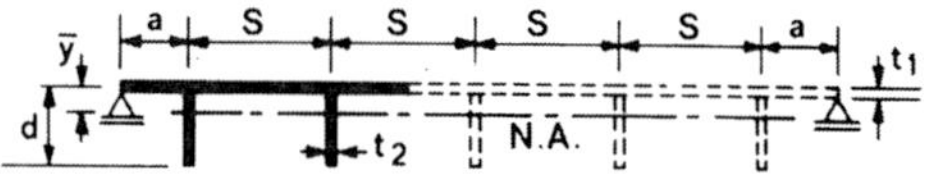

| CODE | n | DIMENSIONS | | | | SECTION PROPERTIES | | | | | | BUCKLING PROPERTIES | | |
| | | S | t_2 | d | a | $\bar{y}$ | rad. gyr. | A | I_{NA} | Z_{plate} | $Z_{stiff.}$ | σ_{cr}/E | L_1 | L_{euler} |
To scale multiply by	–	t_1	t_1	t_1	t_1	t_1	t_1	t_1^2	t_1^4	t_1^3	t_1^3	10^{-3}	t_1	t_1
TQ03500415	3	50	0.4	15	10	0.98	2.971	138.0	1217.93	1245.00	86.86	0.8508	25.4	>500
TQ03500615	3	50	0.6	15	10	1.38	3.446	147.0	1746.05	1267.50	128.17	1.3192	402.0	>500
TQ03500815	3	50	0.8	15	10	1.73	3.783	156.0	2232.69	1290.00	168.26	1.4233	429.1	>500
TQ03501015	3	50	1.0	15	10	2.05	4.034	165.0	2684.66	1312.50	207.24	1.5016	451.2	>500
TQ03500420	3	50	0.4	20	10	1.67	4.410	144.0	2800.00	1680.00	152.73	0.4783	34.0	>500
TQ03500620	3	50	0.6	20	10	2.31	5.044	156.0	3969.23	1720.00	224.35	1.0143	37.2	>500
TQ03500820	3	50	0.8	20	10	2.86	5.471	168.0	5028.57	1760.00	293.33	1.4867	43.0	>500
TQ03501020	3	50	1.0	20	10	3.33	5.774	180.0	6000.00	1800.00	360.00	1.8000	45.8	>500
TQ03500425	3	50	0.4	25	10	2.50	5.951	150.0	5312.50	2125.00	236.11	0.3057	42.9	>500
TQ03500625	3	50	0.6	25	10	3.41	6.723	165.0	7457.39	2187.50	345.39	0.6542	44.9	>500
TQ03500825	3	50	0.8	25	10	4.17	7.217	180.0	9375.00	2250.00	450.00	1.0354	49.2	>500
TQ03501025	3	50	1.0	25	10	4.81	7.551	195.0	11117.79	2312.50	550.60	1.3669	52.7	>500
TQ03500430	3	50	0.4	30	10	3.46	7.566	156.0	8930.77	2580.00	336.52	0.2122	50.9	>500
TQ03500630	3	50	0.6	30	10	4.66	8.452	174.0	12429.31	2670.00	490.41	0.4563	53.0	>500
TQ03500830	3	50	0.8	30	10	5.63	8.992	192.0	15525.00	2760.00	636.92	0.7410	56.4	>500
TQ03501030	3	50	1.0	30	10	6.43	9.340	210.0	18321.43	2850.00	777.27	1.0194	60.7	>500
TQ03650405	3	65	0.4	5	13	0.09	0.548	162.0	48.61	525.00	9.91	0.2108	215.1	>500
TQ03650605	3	65	0.6	5	13	0.14	0.660	165.0	71.93	527.50	14.79	0.2310	231.9	>500
TQ03650805	3	65	0.8	5	13	0.18	0.751	168.0	94.64	530.00	19.63	0.2484	245.5	>500
TQ03651005	3	65	1.0	5	13	0.22	0.826	171.0	116.78	532.50	24.43	0.2644	257.0	>500
TQ03650410	3	65	0.4	10	13	0.36	1.501	168.0	378.57	1060.00	39.26	0.3987	335.4	>500
TQ03650610	3	65	0.6	10	13	0.52	1.783	174.0	553.45	1070.00	58.36	0.4552	368.4	>500
TQ03650810	3	65	0.8	10	13	0.67	2.000	180.0	720.00	1080.00	77.14	0.4993	393.7	>500
TQ03651010	3	65	1.0	10	13	0.81	2.174	186.0	879.03	1090.00	95.61	0.5363	414.4	>500
TQ03650415	3	65	0.4	15	13	0.78	2.675	174.0	1245.26	1605.00	87.55	0.6388	448.6	>500
TQ03650615	3	65	0.6	15	13	1.11	3.137	183.0	1800.92	1627.50	129.62	1.1857	50.4	275
TQ03650815	3	65	0.8	15	13	1.41	3.476	192.0	2320.31	1650.00	170.69	1.2487	50.7	303
TQ03651015	3	65	1.0	15	13	1.68	3.738	201.0	2808.30	1672.50	210.82	1.3427	50.7	316
TQ03650420	3	65	0.4	20	13	1.33	4.000	180.0	2880.00	2160.00	154.29	0.4786	33.9	>500
TQ03650620	3	65	0.6	20	13	1.88	4.635	192.0	4125.00	2200.00	227.59	0.9698	45.4	>500
TQ03650820	3	65	0.8	20	13	2.35	5.083	204.0	5270.59	2240.00	298.67	1.1404	52.0	>500
TQ03651020	3	65	1.0	20	13	2.78	5.415	216.0	6333.33	2280.00	367.74	1.2629	53.1	>500
TQ03650425	3	65	0.4	25	13	2.02	5.435	186.0	5493.95	2725.00	239.04	0.3062	42.9	>500
TQ03650625	3	65	0.6	25	13	2.80	6.230	201.0	7800.84	2787.50	351.37	0.6479	47.0	>500
TQ03650825	3	65	0.8	25	13	3.47	6.769	216.0	9895.83	2850.00	459.68	0.9289	55.1	>500
TQ03651025	3	65	1.0	25	13	4.06	7.153	231.0	11820.21	2912.50	564.44	1.1064	58.0	>500
TQ03650430	3	65	0.4	30	13	2.81	6.953	192.0	9281.25	3300.00	341.38	0.2124	51.1	>500
TQ03650630	3	65	0.6	30	13	3.86	7.891	210.0	13075.71	3390.00	500.16	0.4534	54.6	>500
TQ03650830	3	65	0.8	30	13	4.74	8.503	228.0	16484.21	3480.00	652.50	0.7045	60.1	>500
TQ03651030	3	65	1.0	30	13	5.49	8.924	246.0	19591.46	3570.00	799.25	0.9064	64.7	>500
TQ03800405	3	80	0.4	5	16	0.08	0.497	198.0	48.86	645.00	9.92	0.1335	255.5	>500
TQ03800605	3	80	0.6	5	16	0.11	0.601	201.0	72.48	647.50	14.83	0.1456	274.6	>500
TQ03800805	3	80	0.8	5	16	0.15	0.685	204.0	95.59	650.00	19.70	0.1561	290.3	>500
TQ03801005	3	80	1.0	5	16	0.18	0.756	207.0	118.21	652.50	24.53	0.1659	303.5	>500
TQ03800410	3	80	0.4	10	16	0.29	1.369	204.0	382.35	1300.00	39.39	0.2463	394.1	>500
TQ03800610	3	80	0.6	10	16	0.43	1.635	210.0	561.43	1310.00	58.66	0.2818	432.7	>500
TQ03800810	3	80	0.8	10	16	0.56	1.843	216.0	733.33	1320.00	77.65	0.3099	462.6	>500
TQ03801010	3	80	1.0	10	16	0.68	2.012	222.0	898.65	1330.00	96.38	0.3339	486.9	>500
TQ03800415	3	80	0.4	15	16	0.64	2.453	210.0	1263.21	1965.00	87.99	0.7859	62.1	254
TQ03800615	3	80	0.6	15	16	0.92	2.897	219.0	1837.76	1987.50	130.57	0.8094	62.1	303
TQ03800815	3	80	0.8	15	16	1.18	3.231	228.0	2380.26	2010.00	172.29	0.8435	61.8	337
TQ03801015	3	80	1.0	15	16	1.42	3.495	237.0	2894.38	2032.50	213.20	0.8991	61.6	356
TQ03800420	3	80	0.4	20	16	1.11	3.685	216.0	2933.33	2640.00	155.29	0.4788	33.9	>500
TQ03800620	3	80	0.6	20	16	1.58	4.308	228.0	4231.58	2680.00	229.71	0.7673	61.3	>500
TQ03800820	3	80	0.8	20	16	2.00	4.761	240.0	5440.00	2720.00	302.22	0.8159	62.2	>500
TQ03801020	3	80	1.0	20	16	2.38	5.107	252.0	6571.43	2760.00	372.97	0.8809	62.5	>500
TQ03800425	3	80	0.4	25	16	1.69	5.030	222.0	5616.55	3325.00	240.94	0.3063	42.3	>500
TQ03800625	3	80	0.6	25	16	2.37	5.824	237.0	8039.95	3387.50	355.33	0.6257	55.2	>500
TQ03800825	3	80	0.8	25	16	2.98	6.383	252.0	10267.86	3450.00	466.22	0.7475	64.0	>500
TQ03801025	3	80	1.0	25	16	3.51	6.796	267.0	12333.22	3512.50	573.94	0.8305	65.4	>500
TQ03800430	3	80	0.4	30	16	2.37	6.462	228.0	9521.05	4020.00	344.57	0.2127	51.0	>500
TQ03800630	3	80	0.6	30	16	3.29	7.417	246.0	13532.93	4110.00	506.71	0.4492	56.8	>500
TQ03800830	3	80	0.8	30	16	4.09	8.067	264.0	17181.82	4200.00	663.16	0.6336	67.0	>500
TQ03801030	3	80	1.0	30	16	4.79	8.534	282.0	20537.23	4290.00	814.56	0.7468	70.3	>500

n rectangular stiffeners
edges pinned
free in-plane

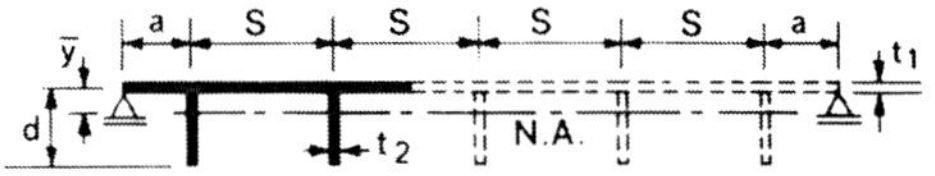

CODE	n	\multicolumn{4}{c}{DIMENSIONS}				$\bar{y}$	rad. gyr.	A	I_{NA}	Z_{plate}	$Z_{stiff.}$	σ_{cr}/E	L_1	L_{euler}
		S	t_2	d	a	$\bar{y}$	rad. gyr.	A	I_{NA}	Z_{plate}	$Z_{stiff.}$	σ_{cr}/E	L_1	L_{euler}
To scale multiply by	−	t_1	t_1	t_1	t_1	t_1	t_1	t_1^2	t_1^4	t_1^3	t_1^3	10^{-3}	t_1	t_1
TG05350405	5	35	0.4	5	7	0.15	0.696	164.0	79.52	521.67	16.40	0.2506	241.5	>500
TG05350605	5	35	0.6	5	7	0.22	0.831	169.0	116.68	525.83	24.42	0.2770	262.5	>500
TG05350805	5	35	0.8	5	7	0.29	0.936	174.0	152.30	530.00	32.32	0.2984	279.2	>500
TG05351005	5	35	1.0	5	7	0.35	1.021	179.0	166.51	534.17	40.10	0.3171	292.6	>500
TG05350410	5	35	0.4	10	7	0.57	1.871	174.0	609.20	1060.00	64.63	0.5038	387.5	>500
TG05350610	5	35	0.6	10	7	0.82	2.184	184.0	877.72	1076.67	95.56	0.5656	425.1	>500
TG05350810	5	35	0.8	10	7	1.03	2.410	194.0	1127.15	1093.33	125.67	0.6092	453.4	>500
TG05351010	5	35	1.0	10	7	1.23	2.582	204.0	1360.29	1110.00	155.03	0.6426	476.3	>500
TG05350415	5	35	0.4	15	7	1.22	3.276	184.0	1974.86	1615.00	143.34	0.8496	25.4	436
TG05350615	5	35	0.6	15	7	1.70	3.753	199.0	2802.61	1652.50	210.66	1.7846	28.7	283
TG05350815	5	35	0.8	15	7	2.10	4.075	214.0	3553.74	1690.00	275.54	2.5220	34.4	255
TG05351015	5	35	1.0	15	7	2.46	4.305	229.0	4243.31	1727.50	338.28	3.0051	36.9	246
TG05350420	5	35	0.4	20	7	2.05	4.821	194.0	4508.59	2186.67	251.34	0.4766	34.1	>500
TG05350620	5	35	0.6	20	7	2.80	5.433	214.0	6317.76	2253.33	367.39	1.0138	36.3	>500
TG05350820	5	35	0.8	20	7	3.42	5.822	234.0	7931.62	2320.00	478.35	1.5897	40.1	>500
TG05351020	5	35	1.0	20	7	3.94	6.082	254.0	9396.33	2386.67	584.97	2.0918	43.8	448
TG05350425	5	35	0.4	25	7	3.06	6.456	204.0	8501.84	2775.00	387.57	0.3048	43.0	>500
TG05350625	5	35	0.6	25	7	4.09	7.174	229.0	11786.98	2879.17	563.81	0.6516	44.5	>500
TG05350825	5	35	0.8	25	7	4.92	7.603	254.0	14681.76	2983.33	731.21	1.0515	47.8	>500
TG05351025	5	35	1.0	25	7	5.60	7.872	279.0	17291.11	3087.50	891.31	1.4432	51.7	>500
TG05350430	5	35	0.4	30	7	4.21	8.150	214.0	14214.95	3380.00	551.09	0.2116	50.9	>500
TG05350630	5	35	0.6	30	7	5.53	8.947	244.0	19520.74	3530.00	798.24	0.4543	52.8	>500
TG05350830	5	35	0.8	30	7	6.57	9.393	274.0	24175.18	3680.00	1031.78	0.7428	55.9	>500
TG05351030	5	35	1.0	30	7	7.40	9.656	304.0	28347.04	3830.00	1254.37	1.0396	59.9	>500
TG05500405	5	50	0.4	5	10	0.11	0.592	230.0	80.62	741.67	16.49	0.1132	321.9	>500
TG05500605	5	50	0.6	5	10	0.16	0.712	235.0	119.02	745.83	24.59	0.1247	348.8	>500
TG05500805	5	50	0.8	5	10	0.21	0.807	240.0	156.25	750.00	32.61	0.1343	370.1	>500
TG05501005	5	50	1.0	5	10	0.26	0.886	245.0	192.39	754.17	40.55	0.1428	387.9	>500
TG05500410	5	50	0.4	10	10	0.42	1.614	240.0	625.00	1500.00	65.22	1.6356	45.4	118
TG05500610	5	50	0.6	10	10	0.60	1.908	250.0	910.00	1516.67	96.81	1.6817	45.2	139
TG05500810	5	50	0.8	10	10	0.77	2.130	260.0	1179.49	1533.33	127.78	1.7577	45.1	153
TG05501010	5	50	1.0	10	10	0.93	2.306	270.0	1435.19	1550.00	158.16	1.8843	45.0	160
TG05500415	5	50	0.4	15	10	0.90	2.862	250.0	2047.50	2275.00	145.21	0.8505	25.5	306
TG05500615	5	50	0.6	15	10	1.27	3.334	265.0	2945.17	2312.50	214.56	1.5100	43.7	262
TG05500815	5	50	0.8	15	10	1.61	3.673	280.0	3776.79	2350.00	282.00	1.6433	45.8	278
TG05501015	5	50	1.0	15	10	1.91	3.928	295.0	4552.44	2387.50	347.69	1.7998	46.4	286
TG05500420	5	50	0.4	20	10	1.54	4.260	260.0	4717.95	3066.67	255.56	0.4780	33.9	>500
TG05500620	5	50	0.6	20	10	2.14	4.897	280.0	6714.29	3133.33	376.00	0.9987	39.5	>500
TG05500820	5	50	0.8	20	10	2.67	5.333	300.0	8533.33	3200.00	492.31	1.3450	48.4	464
TG05501020	5	50	1.0	20	10	3.13	5.648	320.0	10208.33	3266.67	604.94	1.5649	50.9	456
TG05500425	5	50	0.4	25	10	2.31	5.764	270.0	8969.91	3875.00	395.41	0.3054	42.8	>500
TG05500625	5	50	0.6	25	10	3.18	6.547	295.0	12645.66	3979.17	579.49	0.6469	46.2	>500
TG05500825	5	50	0.8	25	10	3.91	7.060	320.0	15950.52	4083.33	756.17	0.9823	52.5	>500
TG05501025	5	50	1.0	25	10	4.53	7.414	345.0	18965.13	4187.50	926.44	1.2418	57.2	>500
TG05500430	5	50	0.4	30	10	3.21	7.345	280.0	15107.14	4700.00	564.00	0.2118	51.2	>500
TG05500630	5	50	0.6	30	10	4.35	8.254	310.0	21120.97	4850.00	823.58	0.4511	54.4	>500
TG05500830	5	50	0.8	30	10	5.29	8.824	340.0	26470.59	5000.00	1071.43	0.7135	59.2	>500
TG05501030	5	50	1.0	30	10	6.08	9.200	370.0	31317.57	5150.00	1309.32	0.9504	64.6	>500
TG05650405	5	65	0.4	5	13	0.08	0.524	296.0	81.22	961.67	16.52	0.0633	398.8	>500
TG05650605	5	65	0.6	5	13	0.12	0.632	301.0	120.33	963.83	24.68	0.0694	430.2	>500
TG05650805	5	65	0.8	5	13	0.16	0.720	306.0	158.50	970.00	32.77	0.0746	455.7	>500
TG05651005	5	65	1.0	5	13	0.20	0.793	311.0	195.77	974.17	40.79	0.0792	477.5	>500
TG05650410	5	65	0.4	10	13	0.33	1.439	306.0	633.99	1940.00	65.54	0.9840	59.8	137
TG05650610	5	65	0.6	10	13	0.47	1.714	316.0	925.80	1956.67	97.51	1.0081	58.9	162
TG05650810	5	65	0.8	10	13	0.61	1.927	326.0	1210.63	1973.33	128.98	1.0479	58.5	179
TG05651010	5	65	1.0	10	13	0.74	2.099	336.0	1480.65	1990.00	159.97	1.1126	58.2	190
TG05650415	5	65	0.4	15	13	0.71	2.572	316.0	2089.79	2935.00	146.26	0.8517	25.5	266
TG05650615	5	65	0.6	15	13	1.02	3.026	331.0	3030.87	2972.50	216.79	0.9856	57.9	292
TG05650815	5	65	0.8	15	13	1.30	3.364	346.0	3914.74	3010.00	285.76	1.0334	58.1	320
TG05651015	5	65	1.0	15	13	1.56	3.627	361.0	4748.53	3047.50	353.27	1.1131	58.2	334
TG05650420	5	65	0.4	20	13	1.23	3.854	326.0	4842.54	3946.67	257.95	0.4794	33.9	>500
TG05650620	5	65	0.6	20	13	1.73	4.485	346.0	6959.54	4013.33	381.01	0.8778	56.1	470
TG05650820	5	65	0.8	20	13	2.19	4.936	366.0	8918.03	4080.00	500.61	0.9647	59.5	497
TG05651020	5	65	1.0	20	13	2.59	5.275	386.0	10742.66	4146.67	617.06	1.0592	60.7	>500

145

T-PLATE n rectangular stiffeners
edges pinned
free in-plane

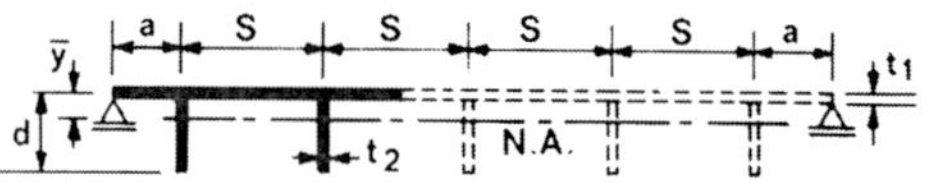

CODE	n	S	t_2	d	a	$\bar{y}$	rad. gyr.	A	I_{NA}	Z_{plate}	$Z_{stiff.}$	σ_{cr}/E	L_1	L_{euler}
To scale multiply by	–	t_1	t_1	t_1	t_1	t_1	t_1	t_1^2	t_1^4	t_1^3	t_1^3	10^{-3}	t_1	t_1
TG05650425	5	65	0.4	25	13	1.86	5.248	336.0	9254.09	4975.00	399.92	0.3059	42.4	>500
TG05650625	5	65	0.6	25	13	2.60	6.045	361.0	13190.36	5079.17	588.78	0.6363	50.5	>500
TG05650825	5	65	0.8	25	13	3.24	6.594	386.0	16785.41	5183.33	771.33	0.8305	61.8	>500
TG05651025	5	65	1.0	25	13	3.80	6.993	411.0	20101.51	5287.50	948.26	0.9536	64.8	>500
TG05650430	5	65	0.4	30	13	2.60	6.727	346.0	15658.96	6020.00	571.52	0.2123	51.3	>500
TG05650630	5	65	0.6	30	13	3.59	7.676	376.0	22152.93	6170.00	838.82	0.4481	56.3	>500
TG05650830	5	65	0.8	30	13	4.43	8.307	406.0	28019.70	6320.00	1095.95	0.6600	65.4	>500
TG05651030	5	65	1.0	30	13	5.16	8.751	436.0	33388.76	6470.00	1344.18	0.8102	70.9	>500
TG05800405	5	80	0.4	5	16	0.07	0.475	362.0	81.61	1181.67	16.55	0.0400	473.3	>500
TG05800605	5	80	0.6	5	16	0.10	0.575	367.0	121.17	1185.83	24.74	EULER	–	–
TG05800805	5	80	0.8	5	16	0.13	0.656	372.0	159.95	1190.00	32.87	EULER	–	–
TG05801005	5	80	1.0	5	16	0.17	0.725	377.0	197.97	1194.17	40.95	EULER	–	–
TG05800410	5	80	0.4	10	16	0.27	1.311	372.0	639.78	2380.00	65.75	0.6507	74.2	154
TG05800610	5	80	0.6	10	16	0.39	1.570	382.0	941.10	2396.67	97.96	0.6664	73.0	183
TG05800810	5	80	0.8	10	16	0.51	1.772	392.0	1231.29	2413.33	129.75	0.6901	72.1	204
TG05801010	5	80	1.0	10	16	0.62	1.939	402.0	1511.19	2430.00	161.14	0.7273	71.8	218
TG05800415	5	80	0.4	15	16	0.59	2.354	382.0	2117.47	3595.00	146.93	0.6518	71.1	277
TG05800615	5	80	0.6	15	16	0.85	2.789	397.0	3088.08	3632.50	218.24	0.6663	71.1	327
TG05800815	5	80	0.8	15	16	1.09	3.119	412.0	4008.50	3670.00	288.22	0.6940	71.0	361
TG05801015	5	80	1.0	15	16	1.32	3.382	427.0	4884.00	3707.50	356.95	0.7414	70.8	380
TG05800420	5	80	0.4	20	16	1.02	3.545	392.0	4925.17	4826.67	259.50	0.4789	34.0	>500
TG05800620	5	80	0.6	20	16	1.46	4.159	412.0	7126.21	4893.33	384.29	0.6413	70.9	>500
TG05800820	5	80	0.8	20	16	1.85	4.611	432.0	9185.19	4960.00	506.12	0.6755	71.6	>500
TG05801020	5	80	1.0	20	16	2.21	4.960	452.0	11120.94	5026.67	625.21	0.7301	71.9	>500
TG05800425	5	80	0.4	25	16	1.55	4.847	402.0	9444.96	6075.00	402.85	0.3062	42.9	>500
TG05800625	5	80	0.6	25	16	2.20	5.637	427.0	13566.67	6179.17	594.91	0.5721	68.4	>500
TG05800825	5	80	0.8	25	16	2.77	6.200	452.0	17376.47	6283.33	781.51	0.6332	73.5	>500
TG05801025	5	80	1.0	25	16	3.28	6.623	477.0	20923.41	6387.50	963.13	0.6965	74.5	>500
TG05800430	5	80	0.4	30	16	2.18	6.238	412.0	16033.98	7340.00	576.44	0.2125	50.9	>500
TG05800630	5	80	0.6	30	16	3.05	7.194	442.0	22876.70	7490.00	848.99	0.4401	61.9	>500
TG05800830	5	80	0.8	30	16	3.81	7.857	472.0	29135.59	7640.00	1112.62	0.5618	75.6	>500
TG05801030	5	80	1.0	30	16	4.48	8.340	502.0	34915.34	7790.00	1368.27	0.6404	79.1	>500

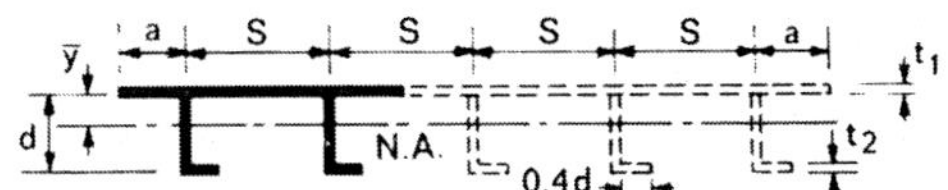

L-PLATE   n angle stiffeners
edges free out-of-plane
free in-plane

CODE	n	S	t_2	d	a	$\bar{y}$	rad. gyr.	A	I_{NA}	Z_{plate}	$Z_{stiff.}$	σ_{cr}/E	L_1	L_{euler}
To scale multiply by	−	t_1	t_1	t_1	t_1	t_1	t_1	t_1^2	t_1^4	t_1^3	t_1^3	10^{-3}	t_1	t_1
LF02350405	2	35	0.4	5	7	0.33	1.111	54.6	67.40	204.44	14.43	EULER	−	−
LF02350605	2	35	0.6	5	7	0.47	1.302	57.4	97.30	206.85	21.48	EULER	−	−
LF02350805	2	35	0.8	5	7	0.60	1.442	60.2	125.14	209.26	28.43	EULER	−	−
LF02351005	2	35	1.0	5	7	0.71	1.549	63.0	151.19	211.67	35.28	EULER	−	−
LF02350410	2	35	0.4	10	7	1.20	2.884	60.2	500.55	418.52	56.86	3.9278	33.1	123
LF02350610	2	35	0.6	10	7	1.64	3.268	65.8	702.74	428.15	84.07	4.4215	30.0	140
LF02350810	2	35	0.8	10	7	2.02	3.516	71.4	882.91	437.78	110.60	5.0310	69.1	141
LF02351010	2	35	1.0	10	7	2.34	3.686	77.0	1045.89	447.41	136.50	5.4028	72.8	143
LF02350415	2	35	0.4	15	7	2.46	4.902	65.8	1581.16	642.22	126.11	3.1312	14.7	230
LF02350615	2	35	0.6	15	7	3.27	5.413	74.2	2174.19	663.89	185.43	4.2226	29.9	214
LF02350815	2	35	0.8	15	7	3.92	5.706	82.6	2689.10	685.56	242.75	4.9287	125.7	198
LF02351015	2	35	1.0	15	7	4.45	5.881	91.0	3147.53	707.22	298.36	5.0726	127.4	196
LF02350420	2	35	0.4	20	7	4.03	7.033	71.4	3531.65	875.56	221.19	1.7623	19.6	433
LF02350620	2	35	0.6	20	7	5.23	7.608	82.6	4760.63	914.07	323.67	3.7102	27.7	176
LF02350820	2	35	0.8	20	7	6.14	7.897	93.8	5849.61	952.59	422.07	4.4753	28.5	147
LF02351020	2	35	1.0	20	7	6.86	8.045	105.0	6796.19	991.11	517.10	5.3736	26.0	123
LF02350425	2	35	0.4	25	7	5.84	9.214	77.0	6536.80	1118.52	341.24	1.1282	24.4	633
LF02350625	2	35	0.6	25	7	7.42	9.802	91.0	8743.13	1178.70	497.27	2.4740	25.8	335
LF02350825	2	35	0.8	25	7	8.57	10.057	105.0	10619.05	1238.89	646.38	3.7702	30.1	195
LF02351025	2	35	1.0	25	7	9.45	10.159	119.0	12281.16	1299.07	789.98	4.7593	28.9	161
LF02350430	2	35	0.4	30	7	7.85	11.412	82.6	10756.42	1371.11	485.51	0.7831	29.4	824
LF02350630	2	35	0.6	30	7	9.78	11.975	99.4	14255.13	1457.78	704.96	1.7248	30.3	464
LF02350830	2	35	0.8	30	7	11.15	12.175	116.2	17225.47	1544.44	913.97	2.8549	32.7	278
LF02351030	2	35	1.0	30	7	12.18	12.222	133.0	19867.67	1631.11	1114.94	3.9326	33.1	212
LF02500405	2	50	0.4	5	10	0.24	0.956	75.6	69.05	290.00	14.50	EULER	−	−
LF02500605	2	50	0.6	5	10	0.34	1.133	78.4	100.70	292.41	21.63	EULER	−	−
LF02500805	2	50	0.8	5	10	0.44	1.269	81.2	130.71	294.81	28.68	EULER	−	−
LF02501005	2	50	1.0	5	10	0.54	1.377	84.0	159.23	297.22	35.67	EULER	−	−
LF02500410	2	50	0.4	10	10	0.89	2.537	81.2	622.82	589.63	57.37	1.9560	48.4	163
LF02500610	2	50	0.6	10	10	1.24	2.931	86.8	745.62	599.26	85.16	2.2063	48.4	181
LF02500810	2	50	0.8	10	10	1.56	3.205	92.4	948.92	608.89	112.41	2.4733	69.9	187
LF02501010	2	50	1.0	10	10	1.84	3.405	98.0	1136.05	618.52	139.17	2.7517	76.9	188
LF02500415	2	50	0.4	15	10	1.87	4.396	86.8	1677.65	898.89	127.74	1.9537	45.8	293
LF02500615	2	50	0.6	15	10	2.55	4.968	95.2	2349.74	920.56	188.77	2.1958	41.2	317
LF02500815	2	50	0.8	15	10	3.13	5.333	103.6	2946.72	942.22	248.20	2.5786	36.2	311
LF02501015	2	50	1.0	15	10	3.62	5.579	112.0	3485.49	963.89	306.18	3.0946	32.0	293
LF02500420	2	50	0.4	20	10	3.12	6.409	92.4	3795.67	1217.78	224.82	1.7635	19.6	440
LF02500620	2	50	0.6	20	10	4.17	7.111	103.6	5238.61	1256.30	330.93	2.1095	42.1	439
LF02500820	2	50	0.8	20	10	5.02	7.523	114.8	6496.63	1294.81	433.61	2.4508	37.6	416
LF02501020	2	50	1.0	20	10	5.71	7.776	126.0	7619.05	1333.33	533.33	2.9281	33.5	376
LF02500425	2	50	0.4	25	10	4.59	8.512	98.0	7100.34	1546.30	347.92	1.1287	24.4	713
LF02500625	2	50	0.6	25	10	6.03	9.298	112.0	9681.92	1606.48	510.29	1.9913	42.8	525
LF02500825	2	50	0.8	25	10	7.14	9.720	126.0	11904.76	1666.67	666.67	2.3163	39.2	480
LF02501025	2	50	1.0	25	10	8.04	9.956	140.0	13876.49	1726.85	817.98	2.7670	35.3	416
LF02500430	2	50	0.4	30	10	6.25	10.666	103.6	11786.87	1884.44	496.39	0.7838	29.3	1004
LF02500630	2	50	0.6	30	10	8.07	11.496	120.4	15912.96	1971.11	725.73	1.6948	34.5	616
LF02500830	2	50	0.8	30	10	9.45	11.903	137.2	19437.90	2057.78	945.70	2.1462	41.0	516
LF02501030	2	50	1.0	30	10	10.52	12.103	154.0	22558.44	2144.44	1158.00	2.5908	37.6	387
LF02650405	2	65	0.4	5	13	0.19	0.851	96.6	69.98	375.56	14.54	EULER	−	−
LF02650605	2	65	0.6	5	13	0.27	1.016	99.4	102.67	377.96	21.71	EULER	−	−
LF02650805	2	65	0.8	5	13	0.35	1.145	102.2	133.99	380.37	28.83	EULER	−	−
LF02651005	2	65	1.0	5	13	0.43	1.250	105.0	164.05	382.78	35.89	EULER	−	−
LF02650410	2	65	0.4	10	13	0.70	2.290	102.2	535.94	760.74	57.66	1.1534	67.0	192
LF02650610	2	65	0.6	10	13	1.00	2.676	107.8	771.80	770.37	85.77	1.2611	73.6	220
LF02650810	2	65	0.8	10	13	1.27	2.955	113.4	990.48	780.00	113.45	1.3847	80.1	233
LF02651010	2	65	1.0	10	13	1.51	3.168	119.0	1194.40	789.63	140.73	1.5426	85.2	236
LF02650415	2	65	0.4	15	13	1.50	4.014	107.8	1736.55	1155.56	128.66	1.1774	58.9	352
LF02650615	2	65	0.6	15	13	2.09	4.603	116.2	2461.83	1177.22	190.71	1.3422	53.0	381
LF02650815	2	65	0.8	15	13	2.60	5.002	124.6	3117.50	1198.89	251.42	1.5936	45.9	379
LF02651015	2	65	1.0	15	13	3.05	5.286	133.0	3716.73	1220.56	310.90	1.9245	40.4	361
LF02650420	2	65	0.4	20	13	2.54	5.911	113.4	3961.90	1560.00	226.91	1.1600	59.1	524
LF02650620	2	65	0.6	20	13	3.47	6.669	124.6	5542.22	1598.52	335.22	1.3021	53.4	559
LF02650820	2	65	0.8	20	13	4.24	7.151	135.8	6943.54	1637.04	440.62	1.5290	46.6	548
LF02651020	2	65	1.0	20	13	4.90	7.472	147.0	8296.80	1675.56	543.42	1.8360	41.2	514

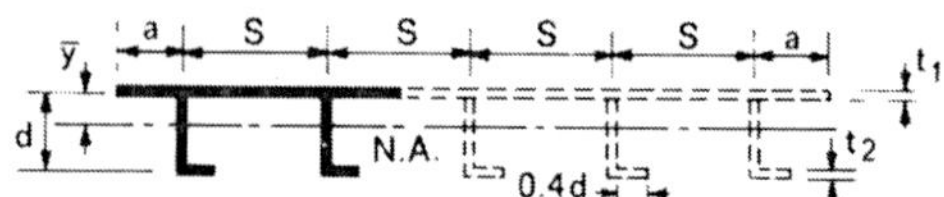

L-PLATE n angle stiffeners, edges free out-of-plane, free in-plane

CODE	n	S	t_2	d	a	$\bar{y}$	rad. gyr.	A	I_{NA}	Z_{plate}	$Z_{stiff.}$	σ_{cr}/E	L_1	L_{euler}
To scale multiply by	—	t_1	t_1	t_1	t_1	t_1	t_1	t_1^2	t_1^4	t_1^3	t_1^3	10^{-3}	t_1	t_1
LF02650425	2	65	0.4	25	13	3.78	7.920	119.0	7464.99	1974.07	351.82	1.1735	60.4	697
LF02650625	2	65	0.6	25	13	5.08	8.811	133.0	10324.25	2034.26	518.16	1.2601	54.3	725
LF02650825	2	65	0.8	25	13	6.12	9.340	147.0	12823.13	2094.44	679.28	1.4663	48.3	690
LF02651025	2	65	1.0	25	13	6.99	9.670	161.0	15055.64	2154.63	835.85	1.7535	43.0	631
LF02650430	2	65	0.4	30	13	5.20	10.004	124.6	12469.98	2397.78	502.83	0.7842	29.4	1040
LF02650630	2	65	0.6	30	13	6.87	10.990	141.4	17078.36	2484.44	738.50	1.2119	55.3	858
LF02650830	2	65	0.8	30	13	8.19	11.539	158.2	21062.96	2571.11	965.84	1.4053	50.0	796
LF02651030	2	65	1.0	30	13	9.26	11.857	175.0	24603.43	2657.78	1186.12	1.6781	44.9	713
LF02800405	2	80	0.4	5	16	0.15	0.775	117.6	70.58	461.11	14.56	EULER	–	–
LF02800605	2	80	0.6	5	16	0.22	0.929	120.4	103.95	463.52	21.77	EULER	–	–
LF02800805	2	80	0.8	5	16	0.29	1.051	123.2	136.15	465.93	28.92	EULER	–	–
LF02801005	2	80	1.0	5	16	0.36	1.152	126.0	167.26	468.33	36.03	EULER	–	–
LF02800410	2	80	0.4	10	16	0.58	2.102	123.2	544.59	931.85	57.84	0.7516	84.7	218
LF02800610	2	80	0.6	10	16	0.84	2.476	128.8	789.44	941.48	86.17	0.8049	89.3	255
LF02800810	2	80	0.8	10	16	1.07	2.754	134.4	1019.05	951.11	114.13	0.8717	92.6	275
LF02801010	2	80	1.0	10	16	1.29	2.970	140.0	1235.24	960.74	141.75	0.9626	96.1	282
LF02800415	2	80	0.4	15	16	1.26	3.714	128.8	1776.24	1412.22	129.25	0.7864	74.1	401
LF02800615	2	80	0.6	15	16	1.77	4.302	137.2	2539.61	1433.89	191.98	0.9041	65.1	436
LF02800815	2	80	0.8	15	16	2.23	4.717	145.6	3239.01	1455.56	253.55	1.0913	57.2	435
LF02801015	2	80	1.0	15	16	2.63	5.023	154.0	3884.90	1477.22	314.06	1.2940	136.7	420
LF02800420	2	80	0.4	20	16	2.14	5.507	134.4	4076.19	1902.22	228.27	0.7779	71.9	605
LF02800620	2	80	0.6	20	16	2.97	6.289	145.6	5758.24	1940.74	338.06	0.8840	63.9	650
LF02800820	2	80	0.8	20	16	3.67	6.810	156.8	7270.75	1979.26	445.33	1.0485	55.7	643
LF02801020	2	80	1.0	20	16	4.29	7.175	168.0	8647.62	2017.78	550.30	1.2657	48.8	613
LF02800425	2	80	0.4	25	16	3.21	7.426	140.0	7720.24	2401.85	354.37	0.7666	72.5	819
LF02800625	2	80	0.6	25	16	4.38	8.371	154.0	10791.40	2462.04	523.43	0.8597	65.2	869
LF02800825	2	80	0.8	25	16	5.36	8.968	168.0	13511.90	2522.22	687.88	1.0093	57.3	849
LF02801025	2	80	1.0	25	16	6.18	9.365	182.0	15962.68	2582.41	848.24	1.2121	50.5	795
LF02800430	2	80	0.4	30	16	4.45	9.433	145.6	12956.04	2911.11	507.10	0.7520	72.6	1035
LF02800630	2	80	0.6	30	16	5.99	10.511	162.4	17942.36	2997.78	747.14	0.8366	66.8	1076
LF02800830	2	80	0.8	30	16	7.23	11.157	179.2	22307.14	3084.44	979.76	0.9745	59.1	1028
LF02801030	2	80	1.0	30	16	8.27	11.564	196.0	26210.20	3171.11	1205.92	1.1661	52.5	943
LF03350405	3	35	0.4	5	7	0.29	1.051	92.4	102.11	349.44	21.69	EULER	–	–
LF03350605	3	35	0.6	5	7	0.42	1.238	96.6	148.02	353.06	32.31	EULER	–	–
LF03350805	3	35	0.8	5	7	0.54	1.377	100.8	191.07	356.67	42.80	3.6909	37.0	62
LF03351005	3	35	1.0	5	7	0.64	1.485	105.0	231.61	360.28	53.16	4.0204	37.1	65
LF03350410	3	35	0.4	10	7	1.07	2.754	100.8	764.29	713.33	85.60	3.5716	32.3	123
LF03350610	3	35	0.6	10	7	1.48	3.144	109.2	1079.67	727.78	126.77	3.9399	30.0	142
LF03350810	3	35	0.8	10	7	1.84	3.405	117.6	1363.27	742.22	167.00	4.5223	27.5	145
LF03351010	3	35	1.0	10	7	2.14	3.587	126.0	1621.43	756.67	206.36	5.3725	63.2	140
LF03350415	3	35	0.4	15	7	2.23	4.717	109.2	2429.26	1091.67	190.16	3.1321	14.7	233
LF03350615	3	35	0.6	15	7	2.99	5.256	121.8	3364.19	1124.17	280.18	3.7914	30.1	246
LF03350815	3	35	0.8	15	7	3.62	5.579	134.4	4182.59	1156.67	367.41	4.3143	27.7	245
LF03351015	3	35	1.0	15	7	4.13	5.782	147.0	4914.41	1189.17	452.22	5.0581	25.0	229
LF03350420	3	35	0.4	20	7	3.67	6.810	117.6	5453.06	1484.44	334.00	1.7627	19.6	470
LF03350620	3	35	0.6	20	7	4.82	7.438	134.4	7435.71	1542.22	489.88	3.4862	30.1	326
LF03350820	3	35	0.8	20	7	5.71	7.776	151.2	9142.86	1500.00	640.00	4.0321	29.0	307
LF03351020	3	35	1.0	20	7	6.43	7.965	168.0	10657.14	1657.78	785.25	4.5159	189.1	276
LF03350425	3	35	0.4	25	7	5.36	8.968	126.0	10133.93	1891.67	515.91	1.1282	24.4	749
LF03350625	3	35	0.6	25	7	6.89	9.637	147.0	13651.15	1981.94	753.70	2.4742	26.0	469
LF03350825	3	35	0.8	25	7	8.04	9.956	168.0	16651.79	2072.22	981.58	3.5769	30.6	225
LF03351025	3	35	1.0	25	7	8.93	10.107	189.0	19308.04	2162.50	1201.39	4.3392	28.9	183
LF03350430	3	35	0.4	30	7	7.23	11.157	134.4	16730.36	2313.33	734.82	0.7833	29.3	1049
LF03350630	3	35	0.6	30	7	9.14	11.826	159.6	22320.68	2443.33	1069.78	1.7271	30.3	644
LF03350830	3	35	0.8	30	7	10.52	12.103	184.8	27070.13	2573.33	1389.60	2.8296	32.8	326
LF03351030	3	35	1.0	30	7	11.57	12.205	210.0	31281.43	2703.33	1697.44	3.7698	32.5	231
LF03500405	3	50	0.4	5	10	0.21	0.901	128.4	104.32	496.11	21.78	EULER	–	–
LF03500605	3	50	0.6	5	10	0.31	1.073	132.6	152.63	499.72	32.51	EULER	–	–
LF03500805	3	50	0.8	5	10	0.39	1.205	136.8	198.68	503.33	43.14	EULER	–	–
LF03501005	3	50	1.0	5	10	0.48	1.312	141.0	242.69	506.94	53.68	EULER	–	–
LF03500410	3	50	0.4	10	10	0.79	2.410	136.8	794.74	1006.67	86.29	1.7743	46.4	161
LF03500610	3	50	0.6	10	10	1.12	2.801	145.2	1139.26	1021.11	128.23	1.9633	45.2	182
LF03500810	3	50	0.8	10	10	1.41	3.079	153.6	1456.25	1035.56	169.45	2.2363	48.6	189
LF03501010	3	50	1.0	10	10	1.67	3.287	162.0	1750.00	1050.00	210.00	2.5281	63.7	190

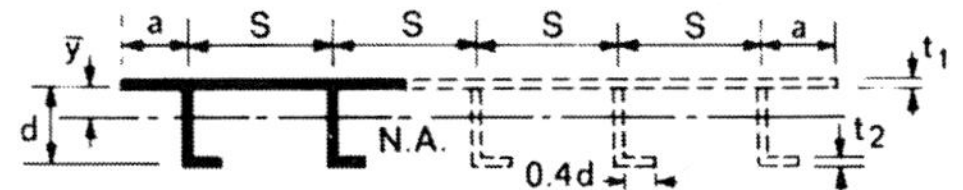

CODE	n	S	t_2	d	a	$\bar{y}$	rad. gyr.	A	I_{NA}	Z_{plate}	$Z_{stiff.}$	$\sigma_{cr/E}$	L_1	L_{euler}
To scale multiply by	–	t_1	t_1	t_1	t_1	t_1	t_1	t_1^2	t_1^4	t_1^3	t_1^3	10^{-3}	t_1	t_1
LF03500415	3	50	0.4	15	10	1.67	4.202	145.2	2563.33	1531.67	192.35	1.7627	45.4	252
LF03500615	3	50	0.6	15	10	2.31	4.785	157.8	3613.05	1564.17	284.71	1.9463	41.8	324
LF03500815	3	50	0.8	15	10	2.85	5.170	170.4	4553.87	1596.67	374.87	2.2441	37.6	327
LF03501015	3	50	1.0	15	10	3.32	5.436	183.0	5408.30	1629.17	463.03	2.6510	33.7	316
LF03500420	3	50	0.4	20	10	2.81	6.158	153.6	5825.00	2071.11	338.91	1.7229	45.5	437
LF03500620	3	50	0.6	20	10	3.80	6.893	170.4	8095.77	2128.89	499.83	1.8938	42.7	476
LF03500820	3	50	0.8	20	10	4.62	7.342	187.2	10092.31	2186.67	656.00	2.1496	38.7	473
LF03501020	3	50	1.0	20	10	5.29	7.632	204.0	11882.35	2244.44	808.00	2.5254	35.0	448
LF03500425	3	50	0.4	25	10	4.17	8.217	162.0	10937.50	2625.00	525.00	1.1290	24.4	732
LF03500625	3	50	0.6	25	10	5.53	9.061	183.0	15023.05	2715.28	771.71	1.8068	43.5	620
LF03500825	3	50	0.8	25	10	6.62	9.540	204.0	18566.18	2805.56	1010.00	2.0559	40.2	600
LF03501025	3	50	1.0	25	10	7.50	9.825	225.0	21718.75	2895.83	1241.07	2.4101	36.6	555
LF03500430	3	50	0.4	30	10	5.70	10.339	170.4	18215.49	3193.33	749.74	0.7840	29.3	1041
LF03500630	3	50	0.6	30	10	7.45	11.254	195.6	24772.09	3323.33	1098.73	1.6448	41.2	763
LF03500830	3	50	0.8	30	10	8.80	11.735	220.8	30404.35	3453.33	1434.46	1.9457	41.7	704
LF03501030	3	50	1.0	30	10	9.88	11.995	246.0	35396.34	3583.33	1759.09	2.2920	38.3	635
LF03650405	3	65	0.4	5	13	0.16	0.801	164.4	105.57	642.78	21.83	EULER	–	–
LF03650605	3	65	0.6	5	13	0.24	0.960	168.6	155.27	646.39	32.62	EULER	–	–
LF03650805	3	65	0.8	5	13	0.31	1.084	172.8	203.13	650.00	43.33	1.0223	68.9	96
LF03651005	3	65	1.0	5	13	0.38	1.187	177.0	249.26	653.61	53.97	1.0864	67.7	102
LF03650410	3	65	0.4	10	13	0.63	2.168	172.8	812.50	1300.00	86.67	1.0501	64.0	190
LF03650610	3	65	0.6	10	13	0.89	2.547	181.2	1175.17	1314.44	129.05	1.1379	65.4	219
LF03650810	3	65	0.8	10	13	1.14	2.826	189.6	1513.92	1328.89	170.86	1.2472	70.5	234
LF03651010	3	65	1.0	10	13	1.36	3.042	198.0	1831.82	1343.33	212.11	1.3842	75.2	240
LF03650415	3	65	0.4	15	13	1.34	3.820	181.2	2644.12	1971.67	193.58	1.0583	58.5	347
LF03650615	3	65	0.6	15	13	1.88	4.410	193.8	3769.45	2004.17	287.32	1.1831	53.5	384
LF03650815	3	65	0.8	15	13	2.35	4.820	206.4	4795.64	2036.67	379.24	1.3763	48.0	390
LF03651015	3	65	1.0	15	13	2.77	5.119	219.0	5739.81	2069.17	469.47	1.6390	43.0	379
LF03650420	3	65	0.4	20	13	2.20	5.651	189.6	6055.70	2657.78	341.71	1.0453	60.4	526
LF03650620	3	65	0.6	20	13	3.14	6.427	206.4	8525.58	2715.56	505.66	1.1528	54.0	577
LF03650820	3	65	0.8	20	13	3.87	6.935	223.2	10735.48	2773.33	665.60	1.3322	50.5	579
LF03651020	3	65	1.0	20	13	4.50	7.286	240.0	12740.00	2831.11	821.94	1.5723	43.5	558
LF03650425	3	65	0.4	25	13	3.41	7.604	198.0	11448.86	3358.33	530.26	1.0262	59.1	718
LF03650625	3	65	0.6	25	13	4.62	8.532	219.0	15943.92	3448.61	782.46	1.1225	55.1	775
LF03650825	3	65	0.8	25	13	5.63	9.107	240.0	19906.25	3538.89	1027.42	1.2832	49.9	770
LF03651025	3	65	1.0	25	13	6.47	9.482	261.0	23464.44	3629.17	1265.99	1.5095	45.0	732
LF03650430	3	65	0.4	30	13	4.71	9.640	206.4	19182.56	4073.33	758.48	0.7844	29.4	1045
LF03650630	3	65	0.6	30	13	6.30	10.689	231.6	26461.40	4203.33	1116.30	1.0900	56.0	970
LF03650830	3	65	0.8	30	13	7.57	11.302	256.8	32803.74	4333.33	1462.50	1.2399	51.4	946
LF03651030	3	65	1.0	30	13	8.62	11.678	282.0	38460.64	4463.33	1798.66	1.4544	46.6	884
LF03800405	3	80	0.4	5	16	0.13	0.729	200.4	106.36	789.44	21.86	EULER	–	–
LF03800605	3	80	0.6	5	16	0.20	0.876	204.6	156.98	793.06	32.69	EULER	–	–
LF03800805	3	80	0.8	5	16	0.26	0.993	208.8	206.03	796.67	43.45	EULER	–	–
LF03801005	3	80	1.0	5	16	0.32	1.091	213.0	253.61	800.28	54.15	0.6995	88.2	118
LF03800410	3	80	0.4	10	16	0.52	1.987	208.8	824.14	1593.33	86.91	0.6867	78.8	217
LF03800610	3	80	0.6	10	16	0.75	2.350	217.2	1199.17	1607.78	129.58	0.7290	81.6	253
LF03800810	3	80	0.8	10	16	0.96	2.624	225.6	1553.19	1622.22	171.76	0.7839	84.7	274
LF03801010	3	80	1.0	10	16	1.15	2.841	234.0	1888.46	1636.67	213.48	0.8584	87.8	285
LF03800415	3	80	0.4	15	16	1.12	3.525	217.2	2698.14	2411.67	194.37	0.7053	71.9	393
LF03800615	3	80	0.6	15	16	1.59	4.107	229.8	3876.84	2444.17	289.02	0.7943	66.7	436
LF03800815	3	80	0.8	15	16	2.00	4.526	242.4	4965.59	2476.67	382.11	0.9304	60.1	444
LF03801015	3	80	1.0	15	16	2.38	4.842	255.0	5977.72	2509.17	473.76	1.1107	53.9	434
LF03800420	3	80	0.4	20	16	1.91	5.248	225.6	6212.77	3244.44	343.53	0.6982	71.7	601
LF03800620	3	80	0.6	20	16	2.67	6.035	242.4	8827.72	3302.22	509.49	0.7791	65.3	660
LF03800820	3	80	0.8	20	16	3.33	6.573	259.2	11200.00	3360.00	672.00	0.9076	58.3	667
LF03801020	3	80	1.0	20	16	3.91	6.961	276.0	13373.91	3417.78	831.35	1.0796	51.8	646
LF03800425	3	80	0.4	25	16	2.88	7.102	234.0	11802.88	4091.67	533.70	0.6898	72.3	826
LF03800625	3	80	0.6	25	16	3.97	8.069	255.0	16604.78	4181.94	789.60	0.7609	66.7	900
LF03800825	3	80	0.8	25	16	4.89	8.701	276.0	20896.74	4272.22	1039.19	0.8777	59.6	902
LF03801025	3	80	1.0	25	16	5.68	9.136	297.0	24786.93	4362.50	1283.09	1.0390	53.3	868
LF03800430	3	80	0.4	30	16	4.01	9.052	242.4	19862.38	4953.33	764.23	0.6799	72.6	1062
LF03800630	3	80	0.6	30	16	5.45	10.173	267.6	27696.19	5083.33	1128.08	0.7442	67.7	1144
LF03800830	3	80	0.8	30	16	6.64	10.873	292.8	34613.11	5213.33	1461.68	0.8517	61.2	1137
LF03801030	3	80	1.0	30	16	7.64	11.331	318.0	40831.13	5343.33	1826.20	1.0027	55.0	1082

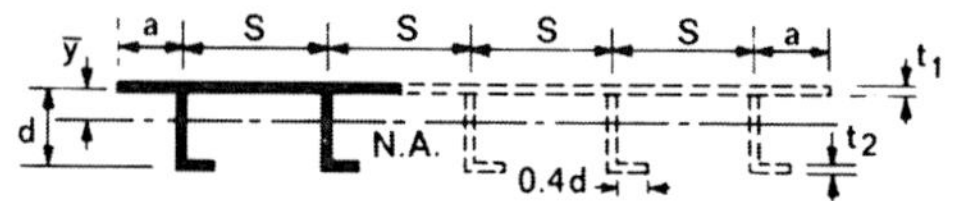

L-PLATE   n angle stiffeners, edges free out-of-plane, free in-plane

CODE	n	DIMENSIONS				SECTION PROPERTIES						BUCKLING PROPERTIES		
		S	t_2	d	a	$\bar{y}$	rad. gyr.	A	I_{NA}	Z_{plate}	$Z_{stiff.}$	σ_{cr}/E	L_1	L_{euler}
To scale multiply by	–	t_1	t_1	t_1	t_1	t_1	t_1	t_1^2	t_1^4	t_1^3	t_1^3	10^{-3}	t_1	t_1
LF05350405	5	35	0.4	5	7	0.27	1.010	168.0	171.28	639.44	36.19	EULER	–	–
LF05350605	5	35	0.6	5	7	0.39	1.193	175.0	248.96	645.46	53.96	EULER	–	–
LF05350805	5	35	0.8	5	7	0.49	1.330	182.0	322.16	651.48	71.50	3.5412	35.1	65
LF05351005	5	35	1.0	5	7	0.60	1.439	189.0	391.37	657.50	88.85	3.8281	35.2	68
LF05350410	5	35	0.4	10	7	0.99	2.661	182.0	1288.64	1302.96	143.01	3.3903	32.1	125
LF05350610	5	35	0.6	10	7	1.38	3.054	196.0	1828.06	1327.04	212.01	3.7026	30.2	145
LF05350810	5	35	0.8	10	7	1.71	3.321	210.0	2316.19	1351.11	279.54	4.2098	28.1	150
LF05351010	5	35	1.0	10	7	2.01	3.512	224.0	2762.65	1375.19	345.72	4.9127	25.8	147
LF05350415	5	35	0.4	15	7	2.07	4.581	196.0	4113.14	1990.56	318.02	3.2938	31.8	216
LF05350615	5	35	0.6	15	7	2.80	5.136	217.0	5724.28	2044.72	469.19	3.5732	30.4	253
LF05350815	5	35	0.8	15	7	3.40	5.478	238.0	7143.28	2098.89	615.98	4.0302	28.2	257
LF05351015	5	35	1.0	15	7	3.91	5.701	259.0	8416.87	2153.06	758.91	4.6897	25.8	248
LF05350420	5	35	0.4	20	7	3.43	6.642	210.0	9264.76	2702.22	559.08	1.7628	19.6	473
LF05350620	5	35	0.6	20	7	4.54	7.305	238.0	12699.16	2798.52	821.30	3.3402	30.8	366
LF05350820	5	35	0.8	20	7	5.41	7.676	266.0	15671.18	2894.81	1074.36	3.7986	29.3	360
LF05351020	5	35	1.0	20	7	6.12	7.892	294.0	18312.93	2991.11	1319.61	4.4177	27.2	337
LF05350425	5	35	0.4	25	7	5.02	8.780	224.0	17266.56	3437.76	864.29	1.1282	24.4	785
LF05350625	5	35	0.6	25	7	6.52	9.501	259.0	23380.19	3588.43	1264.85	2.4741	26.1	541
LF05350825	5	35	0.8	25	7	7.65	9.865	294.0	28613.95	3738.89	1649.51	3.4462	30.8	449
LF05351025	5	35	1.0	25	7	8.55	10.053	329.0	33248.64	3889.35	2021.03	3.7687	257.2	426
LF05350430	5	35	0.4	30	7	6.81	10.957	238.0	28573.11	4197.78	1231.96	0.7834	29.3	1156
LF05350630	5	35	0.6	30	7	8.68	11.697	280.0	38311.07	4414.44	1796.83	1.7285	30.2	779
LF05350830	5	35	0.8	30	7	10.06	12.030	322.0	46598.76	4631.11	2337.20	2.8087	32.9	567
LF05351030	5	35	1.0	30	7	11.13	12.173	364.0	53938.19	4847.78	2857.86	3.1623	342.7	508
LF05500405	5	50	0.4	5	10	0.19	0.864	234.0	174.68	908.33	36.33	EULER	–	–
LF05500605	5	50	0.6	5	10	0.28	1.031	241.0	256.09	914.35	54.26	EULER	–	–
LF05500805	5	50	0.8	5	10	0.36	1.161	248.0	334.01	920.37	72.03	EULER	–	–
LF05501005	5	50	1.0	5	10	0.44	1.266	255.0	408.70	926.39	89.65	EULER	–	–
LF05500410	5	50	0.4	10	10	0.73	2.321	248.0	1336.02	1840.74	144.06	1.6816	46.0	163
LF05500610	5	50	0.6	10	10	1.03	2.708	262.0	1921.76	1864.81	214.26	1.8414	45.0	186
LF05500810	5	50	0.8	10	10	1.30	2.988	276.0	2463.77	1888.89	283.33	2.0784	45.9	194
LF05501010	5	50	1.0	10	10	1.55	3.199	290.0	2968.39	1912.96	351.36	2.3669	55.8	196
LF05500415	5	50	0.4	15	10	1.55	4.062	262.0	4323.95	2797.22	321.38	1.6662	45.5	291
LF05500615	5	50	0.6	15	10	2.15	4.651	283.0	6120.91	2851.39	476.21	1.8239	42.4	326
LF05500815	5	50	0.8	15	10	2.66	5.046	304.0	7741.78	2905.56	627.60	2.0853	38.6	333
LF05501015	5	50	1.0	15	10	3.12	5.326	325.0	9220.67	2959.72	775.85	2.4475	34.9	326
LF05500420	5	50	0.4	20	10	2.61	5.976	276.0	9855.07	3777.78	566.67	1.6334	45.7	436
LF05500620	5	50	0.6	20	10	3.55	6.729	304.0	13763.16	3874.07	836.80	1.7706	43.2	484
LF05500820	5	50	0.8	20	10	4.34	7.202	332.0	17220.88	3970.37	1099.49	2.0035	39.6	490
LF05501020	5	50	1.0	20	10	5.00	7.515	360.0	20333.33	4066.67	1355.56	2.3372	36.0	473
LF05500425	5	50	0.4	25	10	3.88	7.998	290.0	18552.44	4782.41	878.40	1.1292	24.4	722
LF05500625	5	50	0.6	25	10	5.19	8.877	325.0	25612.98	4932.87	1293.08	1.7090	43.9	650
LF05500825	5	50	0.8	25	10	6.25	9.394	360.0	31770.83	5083.33	1694.44	1.9251	40.9	648
LF05501025	5	50	1.0	25	10	7.12	9.713	395.0	37265.95	5233.80	2084.26	2.2382	37.5	616
LF05500430	5	50	0.4	30	10	5.33	10.093	304.0	30967.11	5811.11	1255.20	0.7840	29.3	1102
LF05500630	5	50	0.6	30	10	7.02	11.061	346.0	42333.82	6027.78	1842.45	1.5924	43.3	827
LF05500830	5	50	0.8	30	10	8.35	11.593	388.0	52144.33	6244.44	2408.57	1.8370	42.2	799
LF05501030	5	50	1.0	30	10	9.42	11.896	430.0	60854.65	6461.11	2956.78	2.1405	39.1	749
LF05650405	5	65	0.4	5	13	0.15	0.767	300.0	176.58	1177.22	36.41	EULER	–	–
LF05650605	5	65	0.6	5	13	0.22	0.921	307.0	260.16	1183.24	54.43	0.9515	64.6	88
LF05650805	5	65	0.8	5	13	0.29	1.042	314.0	340.87	1189.26	72.32	0.9872	63.9	99
LF05651005	5	65	1.0	5	13	0.35	1.142	321.0	418.91	1195.28	90.10	1.0399	63.5	106
LF05650410	5	65	0.4	10	13	0.57	2.084	314.0	1363.48	2378.52	144.64	0.9953	51.2	193
LF05650610	5	65	0.6	10	13	0.82	2.456	328.0	1977.74	2402.59	215.51	1.0708	63.1	223
LF05650810	5	65	0.8	10	13	1.05	2.733	342.0	2554.39	2426.67	285.49	1.1684	67.2	239
LF05651010	5	65	1.0	10	13	1.26	2.950	356.0	3097.85	2450.74	354.61	1.2915	70.8	246
LF05650415	5	65	0.4	15	13	1.23	3.683	328.0	4449.92	3603.89	323.27	0.9994	60.4	345
LF05650615	5	65	0.6	15	13	1.74	4.271	349.0	6367.53	3658.06	480.23	1.1059	54.4	386
LF05650815	5	65	0.8	15	13	2.19	4.687	370.0	8126.76	3712.22	634.37	1.2772	49.5	396
LF05651015	5	65	1.0	15	13	2.59	4.994	391.0	9753.12	3766.39	785.88	1.5092	44.6	389
LF05650420	5	65	0.4	20	13	2.11	5.466	342.0	10217.54	4853.33	570.98	0.9861	59.0	521
LF05650620	5	65	0.6	20	13	2.92	6.249	370.0	14447.57	4949.63	845.82	1.0795	54.9	578
LF05650820	5	65	0.8	20	13	3.62	6.773	398.0	18256.62	5045.93	1114.44	1.2352	49.8	589
LF05651020	5	65	1.0	20	13	4.23	7.142	426.0	21727.70	5142.22	1377.38	1.4515	44.9	574

L-PLATE   n angle stiffeners
edges free out-of-plane
free in-plane

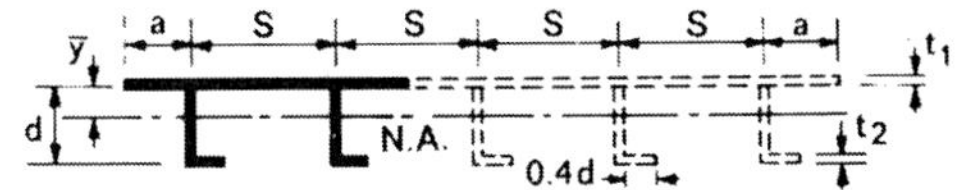

| CODE | n | DIMENSIONS | | | | $\bar{y}$ | rad. gyr. | A | I_{NA} | Z_{plate} | $Z_{stiff.}$ | σ_{cr}/E | L_1 | L_{euler} |
| | | S | t_2 | d | a | | | | | | | | | |
To scale multiply by	–	t_1	t_1	t_1	t_1	t_1	t_1	t_1^2	t_1^4	t_1^3	t_1^3	10^{-3}	t_1	t_1
LF05650425	5	65	0.4	25	13	3.16	7.375	356.0	19361.54	6126.85	886.52	0.9712	59.3	714
LF05650625	5	65	0.6	25	13	4.32	8.324	391.0	27091.99	6277.31	1309.80	1.0558	55.8	785
LF05650825	5	65	0.8	25	13	5.28	8.927	426.0	33949.53	6427.78	1721.73	1.1947	51.1	793
LF05651025	5	65	1.0	25	13	6.10	9.330	461.0	40132.98	6578.24	2123.54	1.3958	46.3	767
LF05650430	5	65	0.4	30	13	4.38	9.373	370.0	32507.03	7424.44	1268.73	0.7845	29.5	1023
LF05650630	5	65	0.6	30	13	5.90	10.459	412.0	45067.72	7641.11	1869.88	1.0274	56.7	1000
LF05650830	5	65	0.8	30	13	7.14	11.114	454.0	56077.53	7857.78	2452.72	1.1582	52.5	1000
LF05651030	5	65	1.0	30	13	8.17	11.529	496.0	65930.44	8074.44	3019.53	1.3481	47.8	958
LF05800405	5	80	0.4	5	16	0.12	0.697	366.0	177.80	1446.11	36.46	EULER	–	–
LF05800605	5	80	0.6	5	16	0.18	0.839	373.0	262.78	1452.13	54.53	EULER	–	–
LF05800805	5	80	0.8	5	16	0.24	0.953	380.0	345.35	1458.15	72.50	EULER	–	–
LF05801005	5	80	1.0	5	16	0.29	1.049	387.0	425.63	1464.17	90.38	0.6729	78.3	122
LF05800410	5	80	0.4	10	16	0.47	1.907	380.0	1381.40	2916.30	145.01	0.6521	76.9	221
LF05800610	5	80	0.6	10	16	0.69	2.261	394.0	2014.97	2940.37	216.32	0.6881	79.5	257
LF05800810	5	80	0.8	10	16	0.88	2.532	408.0	2615.69	2964.44	286.88	0.7360	82.0	279
LF05801010	5	80	1.0	10	16	1.07	2.748	422.0	3186.81	2988.52	356.72	0.8011	84.0	292
LF05800415	5	80	0.4	15	16	1.03	3.392	394.0	4533.69	4410.56	324.48	0.6648	72.1	392
LF05800615	5	80	0.6	15	16	1.46	3.968	415.0	6535.71	4464.72	482.83	0.7412	67.6	438
LF05800815	5	80	0.8	15	16	1.86	4.388	436.0	8395.18	4518.89	638.80	0.8598	62.0	452
LF05801015	5	80	1.0	15	16	2.22	4.709	457.0	10131.77	4573.06	792.51	1.0218	60.4	446
LF05800420	5	80	0.4	20	16	1.76	5.064	408.0	10462.75	5928.89	573.76	0.6581	72.1	594
LF05800620	5	80	0.6	20	16	2.48	5.851	436.0	14924.77	6025.19	851.73	0.7281	66.8	660
LF05800820	5	80	0.8	20	16	3.10	6.399	464.0	18997.70	6121.48	1124.35	0.8414	60.0	674
LF05801020	5	80	1.0	20	16	3.66	6.890	492.0	22747.97	6217.78	1392.04	0.9952	53.7	660
LF05800425	5	80	0.4	25	16	2.67	6.870	422.0	19917.56	7471.30	891.80	0.6509	72.6	817
LF05800625	5	80	0.6	25	16	3.69	7.848	457.0	28143.80	7621.76	1320.84	0.7123	67.6	902
LF05800825	5	80	0.8	25	16	4.57	8.500	492.0	35543.70	7772.22	1740.05	0.8151	61.2	916
LF05801025	5	80	1.0	25	16	5.34	8.957	527.0	42281.88	7922.69	2150.31	0.9581	55.1	892
LF05800430	5	80	0.4	30	16	3.72	8.776	436.0	33580.73	9037.78	1277.59	0.6431	74.2	1056
LF05800630	5	80	0.6	30	16	5.08	9.921	478.0	47046.65	9254.44	1888.19	0.6982	68.5	1157
LF05800830	5	80	0.8	30	16	6.23	10.653	520.0	59012.31	9471.11	2482.72	0.7926	62.6	1168
LF05801030	5	80	1.0	30	16	7.21	11.146	562.0	69814.06	9687.78	3062.88	0.9269	56.7	1130

L-PLATE n angle stiffeners / edges pinned / free in-plane

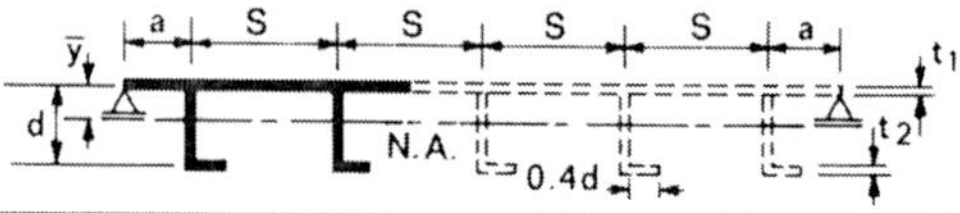

CODE	n	S	t_2	d	a	$\bar{y}$	rad. gyr.	A	I_{NA}	Z_{plate}	$Z_{stiff.}$	σ_{cr}/E	L_1	L_{euler}
To scale multiply by	$-$	t_1	t_1	t_1	t_1	t_1	t_1	t_1^2	t_1^4	t_1^3	t_1^3	10^{-3}	t_1	t_1
LP01350405	1	35	0.4	5	35	0.12	0.699	72.8	35.55	287.59	7.29	1.5201	128.0	1612
LP01350605	1	35	0.6	5	35	0.18	0.842	74.2	52.54	288.80	10.91	1.6937	140.7	1513
LP01350805	1	35	0.8	5	35	0.24	0.956	75.6	69.05	290.00	14.50	1.8180	149.9	1446
LP01351005	1	35	1.0	5	35	0.29	1.051	77.0	85.09	291.20	18.07	1.9105	158.1	1398
LP01350410	1	35	0.4	10	35	0.48	1.911	75.6	276.19	580.00	29.00	3.2220	206.7	1084
LP01350610	1	35	0.6	10	35	0.69	2.267	78.4	402.81	584.81	43.26	3.6001	229.8	1006
LP01350810	1	35	0.8	10	35	0.89	2.537	81.2	522.82	589.63	57.37	3.8206	245.2	959
LP01351010	1	35	1.0	10	35	1.07	2.754	84.0	636.90	594.44	71.33	3.7686	30.1	950
LP01350415	1	35	0.4	15	35	1.03	3.400	78.4	906.31	877.22	64.89	3.8980	139.2	966
LP01350615	1	35	0.6	15	35	1.47	3.977	82.6	1306.28	888.06	96.55	3.2747	32.0	1028
LP01350815	1	35	0.8	15	35	1.87	4.396	86.8	1677.65	898.89	127.74	3.4712	30.8	974
LP01351015	1	35	1.0	15	35	2.23	4.717	91.0	2024.38	909.72	158.47	3.7203	29.5	918
LP01350420	1	35	0.4	20	35	1.77	5.075	81.2	2091.30	1179.26	114.74	3.1790	207.5	1050
LP01350620	1	35	0.6	20	35	2.49	5.862	86.8	2982.49	1198.52	170.32	3.1825	32.2	1016
LP01350820	1	35	0.8	20	35	3.12	6.409	92.4	3795.67	1217.78	224.82	3.3844	31.4	955
LP01351020	1	35	1.0	20	35	3.67	6.810	98.0	4544.22	1237.04	278.33	3.6325	30.1	894
LP01350425	1	35	0.4	25	35	2.68	6.884	84.0	3980.65	1486.11	178.33	1.1298	24.4	1744
LP01350625	1	35	0.6	25	35	3.71	7.861	91.0	5623.28	1516.20	264.11	2.4952	25.5	1122
LP01350825	1	35	0.8	25	35	4.59	8.512	98.0	7100.34	1546.30	347.92	3.2496	32.1	944
LP01351025	1	35	1.0	25	35	5.36	8.968	105.0	8444.94	1576.39	429.92	3.5314	31.0	873
LP01350430	1	35	0.4	30	35	3.73	8.793	86.8	6710.60	1797.78	255.47	0.7846	29.2	>2000
LP01350630	1	35	0.6	30	35	5.11	9.936	95.2	9398.95	1841.11	377.54	1.7406	29.8	1317
LP01350830	1	35	0.8	30	35	6.25	10.666	103.6	11786.87	1884.44	496.39	2.8474	32.8	978
LP01351030	1	35	1.0	30	35	7.23	11.157	112.0	13941.96	1927.78	612.35	3.3751	32.3	859
LP01500405	1	50	0.4	5	50	0.09	0.591	102.8	35.88	409.81	7.30	0.6800	170.8	>2000
LP01500605	1	50	0.6	5	50	0.13	0.715	104.2	53.25	411.02	10.93	0.7593	186.7	>2000
LP01500805	1	50	0.8	5	50	0.17	0.816	105.6	70.27	412.22	14.55	0.8196	199.1	>2000
LP01501005	1	50	1.0	5	50	0.21	0.901	107.0	86.94	413.43	18.15	0.8670	208.9	>2000
LP01500410	1	50	0.4	10	50	0.34	1.631	105.6	281.06	824.44	29.10	1.4388	275.6	>2000
LP01500610	1	50	0.6	10	50	0.50	1.952	108.4	413.10	829.26	43.48	1.6281	304.3	>2000
LP01500810	1	50	0.8	10	50	0.65	2.204	111.2	540.05	834.07	57.74	1.7546	325.9	>2000
LP01501010	1	50	1.0	10	50	0.79	2.410	114.0	662.28	838.89	71.90	1.8423	343.1	1997
LP01500415	1	50	0.4	15	50	0.75	2.928	108.4	929.47	1243.89	65.21	1.5648	46.7	>2000
LP01500615	1	50	0.6	15	50	1.08	3.468	112.6	1353.90	1254.72	97.26	1.6345	45.3	>2000
LP01500815	1	50	0.8	15	50	1.39	3.877	116.8	1755.31	1265.56	128.94	1.7422	43.4	>2000
LP01501015	1	50	1.0	15	50	1.67	4.202	121.0	2136.11	1276.39	160.25	1.8696	41.6	1924
LP01500420	1	50	0.4	20	50	1.29	4.408	111.2	2160.19	1668.15	115.49	1.5505	46.8	>2000
LP01500620	1	50	0.6	20	50	1.85	5.169	116.8	3120.55	1687.41	171.92	1.6122	45.7	>2000
LP01500820	1	50	0.8	20	50	2.35	5.728	122.4	4015.69	1706.67	227.56	1.7118	43.8	>2000
LP01501020	1	50	1.0	20	50	2.81	6.158	128.0	4854.17	1725.93	282.42	1.8365	41.9	1888
LP01500425	1	50	0.4	25	50	1.97	6.026	114.0	4139.25	2097.22	179.76	1.1306	24.4	>2000
LP01500625	1	50	0.6	25	50	2.79	7.003	121.0	5933.63	2127.31	267.15	1.5872	46.0	>2000
LP01500825	1	50	0.8	25	50	3.52	7.698	128.0	7584.64	2157.41	353.03	1.6819	44.4	1974
LP01501025	1	50	1.0	25	50	4.17	8.217	135.0	9114.58	2187.50	437.50	1.8038	42.5	1856
LP01500430	1	50	0.4	30	50	2.77	7.753	116.8	7021.23	2531.11	257.89	0.7850	29.3	>2000
LP01500630	1	50	0.6	30	50	3.88	8.934	125.2	9993.45	2574.44	382.62	1.5417	45.9	>2000
LP01500830	1	50	0.8	30	50	4.85	9.749	133.6	12697.01	2617.78	504.86	1.6491	45.0	1951
LP01501030	1	50	1.0	30	50	5.70	10.339	142.0	15179.58	2661.11	624.78	1.7721	43.3	1826
LP01650405	1	65	0.4	5	65	0.07	0.521	132.8	36.06	532.04	7.31	0.3759	210.2	>2000
LP01650605	1	65	0.6	5	65	0.10	0.632	134.2	53.64	533.24	10.95	0.4154	225.3	>2000
LP01650805	1	65	0.8	5	65	0.13	0.723	135.6	70.94	534.44	14.58	0.4534	244.3	>2000
LP01651005	1	65	1.0	5	65	0.16	0.801	137.0	87.97	535.65	18.19	0.4811	256.7	>2000
LP01650410	1	65	0.4	10	65	0.27	1.447	135.6	283.78	1068.89	29.15	0.7856	339.4	>2000
LP01650610	1	65	0.6	10	65	0.39	1.740	138.4	418.93	1073.70	43.59	0.8961	373.9	>2000
LP01650810	1	65	0.8	10	65	0.51	1.974	141.2	549.95	1078.52	57.95	0.9745	400.2	>2000
LP01651010	1	65	1.0	10	65	0.63	2.168	144.0	677.08	1083.33	72.22	1.0324	422.2	>2000
LP01650415	1	65	0.4	15	65	0.59	2.610	138.4	942.59	1610.56	65.39	0.9313	60.6	>2000
LP01650615	1	65	0.6	15	65	0.85	3.113	142.6	1361.48	1621.39	97.64	0.9787	60.4	>2000
LP01650815	1	65	0.8	15	65	1.10	3.503	146.8	1801.23	1632.22	129.62	1.0450	56.5	>2000
LP01651015	1	65	1.0	15	65	1.34	3.820	151.0	2203.44	1643.06	161.32	1.1205	54.4	>2000
LP01650420	1	65	0.4	20	65	1.02	3.947	141.2	2199.81	2157.04	115.90	0.9256	60.7	>2000
LP01650620	1	65	0.6	20	65	1.47	4.670	146.8	3202.18	2176.30	172.82	0.9672	58.7	>2000
LP01650820	1	65	0.8	20	65	1.89	5.218	152.4	4149.08	2195.56	229.10	1.0320	56.1	>2000
LP01651020	1	65	1.0	20	65	2.28	5.651	158.0	5046.41	2214.81	284.76	1.1091	53.5	>2000

L-PLATE n angle stiffeners
edges pinned
free in-plane

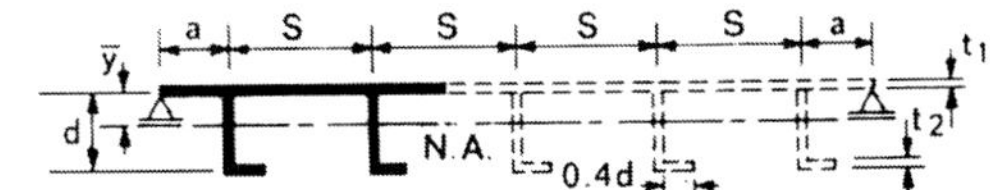

CODE	n	S	t2	d	a	$\bar{y}$	rad. gyr.	A	I_{NA}	Z_{plate}	$Z_{stiff.}$	σ_{cr}/E	L_1	L_{euler}
To scale multiply by	–	t_1	t_1	t_1	t_1	t_1	t_1	t_1^2	t_1^4	t_1^3	t_1^3	10^{-3}	t_1	t_1
LP01650425	1	65	0.4	25	65	1.56	5.421	144.0	4231.77	2708.33	180.56	0.9192	60.9	>2000
LP01650625	1	65	0.6	25	65	2.24	6.367	151.0	6120.65	2738.43	268.86	0.9563	59.2	>2000
LP01650825	1	65	0.8	25	65	2.85	7.064	158.0	7885.02	2768.52	355.95	1.0165	56.7	>2000
LP01651025	1	65	1.0	25	65	3.41	7.604	165.0	9540.72	2798.61	441.89	1.0913	54.1	>2000
LP01650430	1	65	0.4	30	65	2.21	7.006	146.8	7204.90	3264.44	259.24	0.7355	29.4	>2000
LP01650630	1	65	0.6	30	65	3.13	8.169	155.2	10358.12	3307.78	385.51	0.9455	59.6	>2000
LP01650830	1	65	0.8	30	65	3.96	9.007	163.6	13273.35	3351.11	509.75	1.0024	57.4	>2000
LP01651030	1	65	1.0	30	65	4.71	9.640	172.0	15985.47	3394.44	632.07	1.0752	54.8	>2000
LP01800405	1	80	0.4	5	80	0.06	0.471	162.8	36.17	654.26	7.31	0.2354	248.3	>2000
LP01800605	1	80	0.6	5	80	0.08	0.573	164.2	53.89	655.46	10.96	0.2620	270.5	>2000
LP01800805	1	80	0.8	5	80	0.11	0.657	165.6	71.38	656.67	14.59	0.2833	287.6	>2000
LP01801005	1	80	1.0	5	80	0.13	0.729	167.0	88.64	657.87	18.22	0.3010	302.1	>2000
LP01800410	1	80	0.4	10	80	0.22	1.313	165.6	285.51	1313.33	29.19	0.4846	398.6	>2000
LP01800610	1	80	0.6	10	80	0.32	1.584	168.4	422.68	1318.15	43.67	0.5554	439.2	>2000
LP01800810	1	80	0.8	10	80	0.42	1.803	171.2	556.39	1322.96	58.08	0.6075	470.2	>2000
LP01801010	1	80	1.0	10	80	0.52	1.987	174.0	686.78	1327.78	72.42	0.6474	495.4	>2000
LP01800415	1	80	0.4	15	80	0.48	2.376	168.4	951.04	1977.22	65.50	0.6172	74.5	>2000
LP01800615	1	80	0.6	15	80	0.70	2.847	172.6	1399.47	1988.06	97.89	0.6495	74.1	>2000
LP01800815	1	80	0.8	15	80	0.92	3.219	176.8	1831.56	1998.89	130.05	0.6928	70.8	>2000
LP01801015	1	80	1.0	15	80	1.12	3.525	181.0	2248.45	2009.72	161.90	0.7102	67.1	>2000
LP01800420	1	80	0.4	20	80	0.84	3.606	171.2	2225.55	2645.93	116.16	0.6141	74.5	>2000
LP01800620	1	80	0.6	20	80	1.22	4.291	176.8	3256.11	2665.19	173.40	0.6449	71.8	>2000
LP01800820	1	80	0.8	20	80	1.58	4.821	182.4	4238.60	2684.44	230.10	0.6909	68.5	>2000
LP01801020	1	80	1.0	20	80	1.91	5.248	188.0	5177.30	2703.70	286.27	0.7431	65.5	>2000
LP01800425	1	80	0.4	25	80	1.29	4.967	174.0	4292.39	3319.44	181.06	0.6108	74.7	>2000
LP01800625	1	80	0.6	25	80	1.86	5.874	181.0	6245.68	3349.54	269.96	0.6383	72.2	>2000
LP01800825	1	80	0.8	25	80	2.39	6.560	188.0	8089.54	3379.63	357.84	0.6813	68.9	>2000
LP01801025	1	80	1.0	25	80	2.88	7.102	195.0	9835.74	3409.72	444.75	0.7326	65.7	>2000
LP01800430	1	80	0.4	30	80	1.83	6.437	176.8	7326.24	3997.78	260.10	0.6074	74.9	>2000
LP01800630	1	80	0.6	30	80	2.62	7.567	185.2	10604.64	4041.11	387.37	0.6323	72.7	>2000
LP01800830	1	80	0.8	30	80	3.35	8.403	193.6	13671.07	4084.44	512.93	0.6726	69.6	>2000
LP01801030	1	80	1.0	30	80	4.01	9.052	202.0	16551.98	4127.78	636.86	0.7223	66.7	>2000
LP02350405	2	35	0.4	5	35	0.16	0.798	110.6	70.40	432.59	14.55	0.6765	194.2	>2000
LP02350605	2	35	0.6	5	35	0.24	0.956	113.4	103.57	435.00	21.75	0.7531	212.8	>2000
LP02350805	2	35	0.8	5	35	0.31	1.080	116.2	135.51	437.41	28.89	0.8087	226.9	>2000
LP02351005	2	35	1.0	5	35	0.38	1.182	119.0	166.32	439.81	35.98	0.8512	238.0	>2000
LP02350410	2	35	0.4	10	35	0.62	2.160	116.2	542.05	874.81	57.79	1.4396	316.0	>2000
LP02350610	2	35	0.6	10	35	0.89	2.537	121.8	784.24	884.44	86.05	1.5944	347.3	>2000
LP02350810	2	35	0.8	10	35	1.13	2.816	127.4	1010.57	894.07	113.94	1.6856	370.5	>2000
LP02351010	2	35	1.0	10	35	1.35	3.032	133.0	1223.06	903.70	141.45	1.7407	388.7	>2000
LP02350415	2	35	0.4	15	35	1.33	3.806	121.8	1764.53	1326.67	129.08	2.2903	421.0	1898
LP02350615	2	35	0.6	15	35	1.87	4.396	130.2	2516.47	1348.33	191.61	2.4802	462.9	1785
LP02350815	2	35	0.8	15	35	2.34	4.807	138.6	3202.60	1370.00	252.92	2.5610	493.0	1722
LP02351015	2	35	1.0	15	35	2.76	5.107	147.0	3834.18	1391.67	313.13	2.5872	516.4	1683
LP02350420	2	35	0.4	20	35	2.26	5.633	127.4	4042.28	1788.15	227.87	1.7648	19.5	>2000
LP02350620	2	35	0.6	20	35	3.12	6.409	138.6	5693.51	1826.67	337.23	3.1963	31.9	1538
LP02350820	2	35	0.8	20	35	3.85	6.919	149.8	7171.87	1865.19	443.94	3.3510	600.9	1466
LP02351020	2	35	1.0	20	35	4.47	7.272	161.0	8513.46	1903.70	548.27	3.3230	629.0	1441
LP02350425	2	35	0.4	25	35	3.38	7.581	133.0	7644.11	2259.26	353.62	1.1298	24.4	>2000
LP02350625	2	35	0.6	25	35	4.59	8.512	147.0	10650.51	2319.44	521.88	2.4865	25.8	1715
LP02350825	2	35	0.8	25	35	5.59	9.090	161.0	13302.28	2379.63	685.33	3.2805	31.8	1447
LP02351025	2	35	1.0	25	35	6.43	9.467	175.0	15684.52	2439.81	844.55	3.6470	30.4	1335
LP02350430	2	35	0.4	30	35	4.68	9.614	138.6	12810.39	2740.00	505.85	0.7845	29.3	>2000
LP02350630	2	35	0.6	30	35	6.25	10.666	155.4	17680.31	2826.67	744.59	1.7368	30.0	>2000
LP02350830	2	35	0.8	30	35	7.53	11.284	172.2	21926.13	2913.33	975.63	2.8200	32.9	1529
LP02351030	2	35	1.0	30	35	8.57	11.664	189.0	25714.29	3000.00	1200.00	3.4357	32.2	1339
LP02500405	2	50	0.4	5	50	0.12	0.677	155.6	71.25	615.93	14.59	0.3019	257.5	>2000
LP02500605	2	50	0.6	5	50	0.17	0.816	158.4	105.40	618.33	21.82	0.3371	281.5	>2000
LP02500805	2	50	0.8	5	50	0.22	0.927	161.2	138.63	620.74	29.02	0.3640	299.8	>2000
LP02501005	2	50	1.0	5	50	0.27	1.021	164.0	170.99	623.15	36.18	0.3858	315.0	>2000
LP02500410	2	50	0.4	10	50	0.45	1.855	161.2	554.51	1241.48	58.04	0.6388	418.8	>2000
LP02500610	2	50	0.6	10	50	0.65	2.204	166.8	810.07	1251.11	86.62	0.7290	459.6	>2000
LP02500810	2	50	0.8	10	50	0.84	2.471	172.4	1053.05	1250.74	114.90	0.7746	490.7	>2000
LP02501010	2	50	1.0	10	50	1.01	2.686	178.0	1284.64	1270.37	142.92	0.8128	515.7	>2000

n angle stiffeners
edges pinned
free in-plane

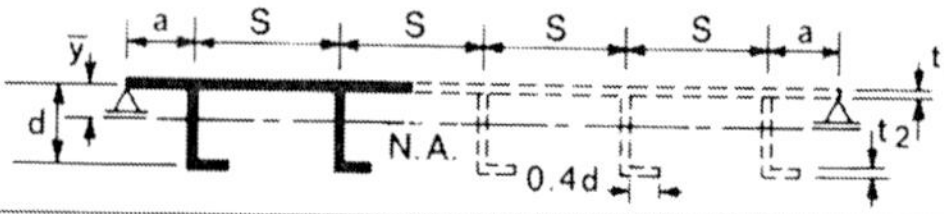

$$N.A. \qquad 0.4d$$

CODE	n	S	t_2	d	a	$\bar{y}$	rad. gyr.	A	I_{NA}	Z_{plate}	$Z_{stiff.}$	σ_{cr}/E	L_1	L_{euler}
To scale multiply by	−	t_1	t_1	t_1	t_1	t_1	t_1	t_1^2	t_1^4	t_1^3	t_1^3	10^{-3}	t_1	t_1
LP02500415	2	50	0.4	15	50	0.97	3.306	166.8	1822.66	1876.67	129.92	1.0375	560.5	>2000
LP02500615	2	50	0.6	15	50	1.39	3.877	175.2	2632.96	1398.33	193.42	1.1516	615.8	>2000
LP02500815	2	50	0.8	15	50	1.76	4.296	183.6	3388.24	1920.00	256.00	1.2174	656.0	>2000
LP02501015	2	50	1.0	15	50	2.11	4.619	192.0	4095.70	1941.67	317.73	1.2556	688.6	>2000
LP02500420	2	50	0.4	20	50	1.67	4.943	172.4	4212.22	2521.48	229.81	1.4591	687.1	>2000
LP02500620	2	50	0.6	20	50	2.35	5.728	183.6	6023.53	2560.00	341.33	1.5911	759.4	>2000
LP02500820	2	50	0.8	20	50	2.96	6.280	194.8	7683.50	2598.52	450.63	1.6523	804.8	>2000
LP02501020	2	50	1.0	20	50	3.50	6.689	206.0	9216.83	2637.04	558.43	1.6768	846.2	>2000
LP02500425	2	50	0.4	25	50	2.53	6.716	178.0	8029.03	3175.93	357.23	1.1303	24.4	>2000
LP02500625	2	50	0.6	25	50	3.52	7.698	192.0	11376.95	3236.11	529.55	1.6033	45.6	>2000
LP02500825	2	50	0.8	25	50	4.37	8.361	206.0	14401.29	3296.30	698.04	1.7272	43.6	>2000
LP02501025	2	50	1.0	25	50	5.11	8.833	220.0	17163.83	3356.48	863.10	1.8832	41.4	>2000
LP02500430	2	50	0.4	30	50	3.53	8.592	183.6	13552.94	3840.00	512.00	0.7849	29.2	>2000
LP02500630	2	50	0.6	30	50	4.85	9.749	200.4	19045.51	3926.67	757.29	1.5428	45.3	>2000
LP02500830	2	50	0.8	30	50	5.97	10.500	217.2	23946.96	4013.33	996.41	1.6934	44.3	>2000
LP02501030	2	50	1.0	30	50	6.92	11.014	234.0	28384.62	4100.00	1230.00	1.8425	42.2	>2000
LP02650405	2	65	0.4	5	65	0.09	0.598	200.6	71.72	799.26	14.61	0.1667	317.0	>2000
LP02650605	2	65	0.6	5	65	0.13	0.723	203.4	106.42	801.67	21.86	0.1860	345.8	>2000
LP02650805	2	65	0.8	5	65	0.17	0.825	206.2	140.38	804.07	29.09	0.2013	369.1	>2000
LP02651005	2	65	1.0	5	65	0.22	0.912	209.0	173.64	806.48	36.29	0.2139	386.9	>2000
LP02650410	2	65	0.4	10	65	0.35	1.650	206.2	561.53	1608.15	58.18	0.3481	512.4	>2000
LP02650610	2	65	0.6	10	65	0.51	1.974	211.8	824.93	1617.78	86.93	0.3962	563.9	>2000
LP02650810	2	65	0.8	10	65	0.66	2.227	217.4	1077.95	1627.41	115.44	0.4304	602.1	>2000
LP02651010	2	65	1.0	10	65	0.81	2.434	223.0	1321.38	1637.04	143.74	0.4558	633.4	>2000
LP02650415	2	65	0.4	15	65	0.76	2.960	211.8	1856.09	2426.67	130.39	0.5656	687.9	>2000
LP02650615	2	65	0.6	15	65	1.10	3.503	220.2	2701.84	2448.33	194.43	0.6420	760.1	>2000
LP02650815	2	65	0.8	15	65	1.42	3.913	228.6	3500.79	2470.00	257.74	0.6885	807.4	>2000
LP02651015	2	65	1.0	15	65	1.71	4.239	237.0	4257.91	2491.67	320.36	0.7194	848.3	>2000
LP02650420	2	65	0.4	20	65	1.32	4.453	217.4	4311.81	3254.81	230.88	0.8112	847.3	>2000
LP02650620	2	65	0.6	20	65	1.89	5.218	228.6	6223.62	3293.33	343.65	0.9013	930.8	>2000
LP02650820	2	65	0.8	20	65	2.40	5.777	239.8	8003.11	3331.85	454.77	0.9525	991.9	>2000
LP02651020	2	65	1.0	20	65	2.87	6.206	251.0	9667.99	3370.37	564.34	0.9816	1040.4	>2000
LP02650425	2	65	0.4	25	65	2.02	6.086	223.0	8258.59	4092.59	359.35	0.9226	60.6	>2000
LP02650625	2	65	0.6	25	65	2.85	7.064	237.0	11827.53	4152.78	533.93	0.9716	58.4	>2000
LP02650825	2	65	0.8	25	65	3.59	7.758	251.0	15106.24	4212.96	705.43	1.0499	55.4	>2000
LP02651025	2	65	1.0	25	65	4.25	8.274	265.0	18140.72	4273.15	874.05	1.1441	52.5	>2000
LP02650430	2	65	0.4	30	65	2.83	7.827	228.6	14003.15	4940.00	515.48	0.7853	29.4	>2000
LP02650630	2	65	0.6	30	65	3.96	9.007	245.4	19910.02	5026.67	764.62	0.9579	60.5	>2000
LP02650830	2	65	0.8	30	65	4.94	9.818	262.2	25274.14	5113.33	1008.66	1.0315	56.2	>2000
LP02651030	2	65	1.0	30	65	5.81	10.403	279.0	30193.55	5200.00	1248.00	1.1243	53.5	>2000
LP02800405	2	80	0.4	5	80	0.07	0.541	245.6	72.01	982.59	14.62	0.1044	374.2	>2000
LP02800605	2	80	0.6	5	80	0.11	0.657	248.4	107.07	985.00	21.89	0.1162	407.3	>2000
LP02800805	2	80	0.8	5	80	0.14	0.751	251.2	141.51	987.41	29.14	0.1257	433.3	>2000
LP02801005	2	80	1.0	5	80	0.18	0.831	254.0	175.36	989.81	36.36	0.1337	455.2	>2000
LP02800410	2	80	0.4	10	80	0.29	1.501	251.2	566.03	1974.81	58.27	0.2146	601.3	>2000
LP02800610	2	80	0.6	10	80	0.42	1.803	256.8	834.58	1984.44	87.12	0.2455	661.7	>2000
LP02800810	2	80	0.8	10	80	0.55	2.042	262.4	1094.31	1994.07	115.78	0.2684	707.6	>2000
LP02801010	2	80	1.0	10	80	0.67	2.241	268.0	1345.77	2003.70	144.27	0.2860	745.3	>2000
LP02800415	2	80	0.4	15	80	0.63	2.704	256.8	1877.80	2976.67	130.69	0.3519	808.1	>2000
LP02800615	2	80	0.6	15	80	0.92	3.219	265.2	2747.34	2998.33	195.07	0.4006	888.8	>2000
LP02800815	2	80	0.8	15	80	1.18	3.615	273.6	3576.32	3020.00	258.86	0.4338	949.4	>2000
LP02801015	2	80	1.0	15	80	1.44	3.936	282.0	4368.35	3041.67	322.06	0.4574	1000.7	>2000
LP02800420	2	80	0.4	20	80	1.10	4.084	262.4	4377.24	3988.15	231.57	0.5045	998.8	>2000
LP02800620	2	80	0.6	20	80	1.58	4.821	273.6	6357.89	4026.67	345.14	0.5678	1094.9	>2000
LP02800820	2	80	0.8	20	80	2.02	5.373	284.8	8221.72	4065.19	457.33	0.6074	1170.3	>2000
LP02801020	2	80	1.0	20	80	2.43	5.807	296.0	9981.98	4103.70	568.21	0.6328	1226.4	>2000
LP02800425	2	80	0.4	25	80	1.68	5.602	268.0	8411.07	5009.26	360.67	0.6145	74.3	>2000
LP02800625	2	80	0.6	25	80	2.39	6.560	282.0	12134.31	5069.44	536.76	0.6506	71.1	>2000
LP02800825	2	80	0.8	25	80	3.04	7.259	296.0	15596.85	5129.63	710.26	0.7064	67.3	>2000
LP02801025	2	80	1.0	25	80	3.63	7.795	310.0	18834.01	5189.81	881.29	0.7791	63.5	>2000
LP02800430	2	80	0.4	30	80	2.37	7.231	273.6	14305.26	6040.00	517.71	0.6099	74.5	>2000
LP02800630	2	80	0.6	30	80	3.35	8.403	290.4	20506.61	6126.67	769.40	0.6427	71.7	>2000
LP02800830	2	80	0.8	30	80	4.22	9.237	307.2	26212.50	6213.33	1016.73	0.6951	68.1	>2000
LP02801030	2	80	1.0	30	80	5.00	9.860	324.0	31500.00	6300.00	1260.00	0.7576	64.3	>2000

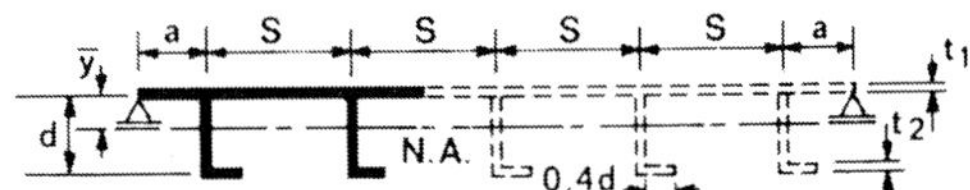

L-PLATE n angle stiffeners
edges pinned
free in-plane

CODE	n	DIMENSIONS					SECTION PROPERTIES					BUCKLING PROPERTIES		
		S	t_2	d	a	$\bar{y}$	rad. gyr.	A	I_{NA}	Z_{plate}	$Z_{stiff.}$	σ_{cr}/E	L_1	L_{euler}
To scale multiply by	–	t_1	t_1	t_1	t_1	t_1	t_1	t_1^2	t_1^4	t_1^3	t_1^3	10^{-3}	t_1	t_1
LP03350405	3	35	0.4	5	35	0.18	0.842	148.4	105.09	577.59	21.81	0.3809	259.6	>2000
LP03350605	3	35	0.6	5	35	0.27	1.005	152.6	154.25	581.20	32.58	0.4240	284.2	>2000
LP03350805	3	35	0.8	5	35	0.34	1.133	156.8	201.40	584.81	43.26	0.4555	302.5	>2000
LP03351005	3	35	1.0	5	35	0.42	1.238	161.0	246.70	588.43	53.86	0.4800	317.7	>2000
LP03350410	3	35	0.4	10	35	0.69	2.267	156.8	805.61	1169.63	86.52	0.8113	424.1	>2000
LP03350610	3	35	0.6	10	35	0.98	2.651	165.2	1161.14	1184.07	128.74	0.8971	464.2	>2000
LP03350810	3	35	0.8	10	35	1.24	2.931	173.6	1491.24	1198.52	170.32	0.9481	494.3	>2000
LP03351010	3	35	1.0	10	35	1.48	3.144	182.0	1799.45	1212.96	211.29	0.9794	518.3	>2000
LP03350415	3	35	0.4	15	35	1.47	3.977	165.2	2612.56	1776.11	193.11	1.2958	565.9	>2000
LP03350615	3	35	0.6	15	35	2.05	4.567	177.8	3707.75	1808.61	286.31	1.3974	619.4	>2000
LP03350815	3	35	0.8	15	35	2.55	4.968	190.4	4699.47	1841.11	377.54	1.4413	658.3	>2000
LP03351015	3	35	1.0	15	35	2.99	5.256	203.0	5606.99	1873.61	466.96	1.4561	689.0	>2000
LP03350420	3	35	0.4	20	35	2.49	5.862	173.6	5964.98	2397.04	340.63	1.7831	692.4	>2000
LP03350620	3	35	0.6	20	35	3.40	6.624	190.4	8354.62	2454.81	503.39	1.8767	762.0	>2000
LP03350820	3	35	0.8	20	35	4.17	7.111	207.2	10477.22	2512.59	661.85	1.8932	804.7	>2000
LP03351020	3	35	1.0	20	35	4.82	7.438	224.0	12392.86	2570.37	816.47	1.8767	843.8	>2000
LP03350425	3	35	0.4	25	35	3.71	7.861	182.0	11246.57	3032.41	528.23	1.1298	24.4	>2000
LP03350625	3	35	0.6	25	35	4.99	8.759	203.0	15574.97	3122.69	778.27	2.3099	883.2	>2000
LP03350825	3	35	0.8	25	35	6.03	9.298	224.0	19363.84	3212.96	1020.59	2.2876	937.3	>2000
LP03351025	3	35	1.0	25	35	6.89	9.637	245.0	22751.91	3303.24	1256.16	2.2321	978.2	>2000
LP03350430	3	35	0.4	30	35	5.11	9.936	190.4	18797.90	3682.22	755.09	0.7845	29.3	>2000
LP03350630	3	35	0.6	30	35	6.76	10.935	215.6	25780.24	3812.22	1109.43	1.7350	30.1	>2000
LP03350830	3	35	0.8	30	35	8.07	11.496	240.8	31825.91	3942.22	1451.45	2.6145	1057.8	>2000
LP03351030	3	35	1.0	30	35	9.14	11.826	266.0	37201.13	4072.22	1782.97	2.5196	1104.7	>2000
LP03500405	3	50	0.4	5	50	0.13	0.715	208.4	106.50	822.04	21.87	0.1698	343.9	>2000
LP03500605	3	50	0.6	5	50	0.19	0.860	212.6	157.28	825.65	32.70	0.1896	375.6	>2000
LP03500805	3	50	0.8	5	50	0.25	0.976	216.8	206.55	829.26	43.48	0.2049	400.2	>2000
LP03501005	3	50	1.0	5	50	0.31	1.073	221.0	254.38	832.87	54.19	0.2173	421.1	>2000
LP03500410	3	50	0.4	10	50	0.50	1.760	216.8	826.20	1658.52	86.95	0.3593	559.2	>2000
LP03500610	3	50	0.6	10	50	0.72	2.312	225.2	1203.46	1672.96	129.67	0.4047	613.9	>2000
LP03500810	3	50	0.8	10	50	0.92	2.584	233.6	1560.27	1687.41	171.92	0.4353	654.4	>2000
LP03501010	3	50	1.0	10	50	1.12	2.801	242.0	1898.76	1701.85	213.72	0.4569	687.7	>2000
LP03500415	3	50	0.4	15	50	1.08	3.468	225.2	2707.79	2509.44	194.51	0.5840	753.5	>2000
LP03500615	3	50	0.6	15	50	1.53	4.048	237.8	3896.29	2541.94	289.32	0.6471	822.1	>2000
LP03500815	3	50	0.8	15	50	1.94	4.467	250.4	4996.73	2574.44	382.62	0.6836	875.3	>2000
LP03501015	3	50	1.0	15	50	2.31	4.785	263.0	6021.74	2606.94	474.52	0.7051	920.4	>2000
LP03500420	3	50	0.4	20	50	1.85	5.169	233.6	6241.10	3374.81	343.85	0.8232	924.9	>2000
LP03500620	3	50	0.6	20	50	2.59	5.956	250.4	8883.07	3432.59	510.17	0.8946	1011.9	>2000
LP03500820	3	50	0.8	20	50	3.23	6.499	267.2	11286.23	3490.37	673.14	0.9281	1072.8	>2000
LP03501020	3	50	1.0	20	50	3.80	6.893	284.0	13492.96	3548.15	833.04	0.9417	1122.7	>2000
LP03500425	3	50	0.4	25	50	2.79	7.003	242.0	11867.25	4254.63	534.30	1.0657	1082.3	>2000
LP03500625	3	50	0.6	25	50	3.85	7.975	263.0	16727.07	4344.91	790.87	1.1366	1181.6	>2000
LP03500825	3	50	0.8	25	50	4.75	8.616	284.0	21082.75	4435.19	1041.30	1.1595	1256.2	>2000
LP03501025	3	50	1.0	25	50	5.53	9.061	305.0	25038.42	4525.46	1286.18	1.1595	1310.7	>2000
LP03500430	3	50	0.4	30	50	3.88	8.934	250.4	19986.90	5148.89	765.25	0.7849	29.2	>2000
LP03500630	3	50	0.6	30	50	5.29	10.066	275.6	27926.78	5278.89	1130.19	1.3664	1343.3	>2000
LP03500830	3	50	0.8	30	50	6.46	10.780	300.8	34956.38	5408.89	1485.15	1.3731	1424.5	>2000
LP03501030	3	50	1.0	30	50	7.45	11.254	326.0	41286.81	5538.89	1831.22	1.3558	1485.6	>2000
LP03650405	3	65	0.4	5	65	0.10	0.632	268.4	107.28	1056.48	21.90	0.0938	423.9	>2000
LP03650605	3	65	0.6	5	65	0.15	0.764	272.6	158.98	1070.09	32.77	0.1046	461.6	>2000
LP03650805	3	65	0.8	5	65	0.20	0.870	276.8	209.47	1073.70	43.59	0.1133	491.1	>2000
LP03651005	3	65	1.0	5	65	0.24	0.960	281.0	258.79	1077.31	54.37	0.1205	516.2	>2000
LP03650410	3	65	0.4	10	65	0.39	1.740	276.8	837.86	2147.41	87.19	0.1957	684.4	>2000
LP03650610	3	65	0.6	10	65	0.57	2.075	285.2	1227.98	2161.85	130.19	0.2226	756.2	>2000
LP03650810	3	65	0.8	10	65	0.74	2.335	293.6	1601.09	2176.30	172.82	0.2418	803.5	>2000
LP03651010	3	65	1.0	10	65	0.89	2.547	302.0	1958.61	2190.74	215.09	0.2562	845.0	>2000
LP03650415	3	65	0.4	15	65	0.85	3.113	285.2	2762.96	3242.78	195.29	0.3202	921.9	>2000
LP03650615	3	65	0.6	15	65	1.22	3.669	297.8	4008.86	3275.28	291.00	0.3605	1011.4	>2000
LP03650615	3	65	0.8	15	65	1.57	4.085	310.4	5179.06	3307.78	385.51	0.3864	1075.8	>2000
LP03651015	3	65	1.0	15	65	1.88	4.410	323.0	6282.41	3340.28	478.87	0.4038	1129.1	>2000
LP03650420	3	65	0.4	20	65	1.47	4.670	293.6	6404.36	4352.59	345.65	0.4563	1132.0	>2000
LP03650620	3	65	0.6	20	65	2.09	5.446	310.4	9207.22	4410.37	514.01	0.5060	1241.1	>2000
LP03650820	3	65	0.8	20	65	2.64	6.005	327.2	11798.53	4468.15	679.66	0.5344	1321.3	>2000
LP03651020	3	65	1.0	20	65	3.14	6.427	344.0	14209.30	4525.93	842.76	0.5506	1384.8	>2000

L-PLATE n angle stiffeners edges pinned free in-plane

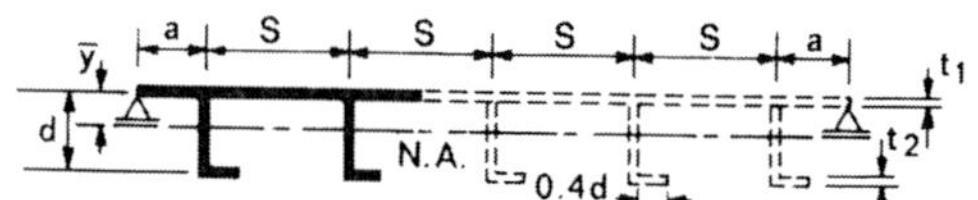

CODE	n	S	t_2	d	a	$\bar{y}$	rad. gyr.	A	I_{NA}	Z_{plate}	$Z_{stiff.}$	σ_{cr}/E	L_1	L_{euler}
To scale multiply by	$-$	t_1	t_1	t_1	t_1	t_1	t_1	t_1^2	t_1^4	t_1^3	t_1^3	10^{-3}	t_1	t_1
LP03650425	3	65	0.4	25	65	2.24	6.367	302.0	12241.31	5476.85	537.73	0.5983	1331.5	>2000
LP03650625	3	65	0.6	25	65	3.13	7.350	323.0	17451.14	5567.13	798.12	0.6532	1455.7	>2000
LP03650825	3	65	0.8	25	65	3.92	8.034	344.0	22202.03	5657.41	1053.45	0.6798	1547.5	>2000
LP03651025	3	65	1.0	25	65	4.62	8.532	365.0	26573.20	5747.69	1304.10	0.6915	1619.9	>2000
LP03650430	3	65	0.4	30	65	3.13	8.169	310.4	20716.24	6615.56	771.02	0.7424	1514.3	>2000
LP03650630	3	65	0.6	30	65	4.34	9.345	335.6	29305.78	6745.56	1142.28	0.7983	1658.1	>2000
LP03650830	3	65	0.8	30	65	5.39	10.133	360.8	37045.68	6875.56	1505.19	0.8197	1760.5	>2000
LP03651030	3	65	1.0	30	65	6.30	10.689	386.0	44102.33	7005.56	1860.49	0.8241	1840.7	>2000
LP03800405	3	80	0.4	5	80	0.08	0.573	328.4	107.78	1310.93	21.92	0.0587	499.2	>2000
LP03800605	3	80	0.6	5	80	0.12	0.694	332.6	160.07	1314.54	32.81	0.0653	543.2	>2000
LP03800805	3	80	0.8	5	80	0.16	0.792	336.8	211.34	1318.15	43.67	0.0707	577.3	>2000
LP03801005	3	80	1.0	5	80	0.20	0.876	341.0	261.64	1321.76	54.48	0.0753	607.3	>2000
LP03800410	3	80	0.4	10	80	0.32	1.584	336.8	845.37	2636.30	87.34	0.1206	803.2	>2000
LP03800610	3	80	0.6	10	80	0.47	1.898	345.2	1243.97	2650.74	130.52	0.1380	882.9	>2000
LP03800810	3	80	0.8	10	80	0.61	2.146	353.6	1628.05	2665.19	173.40	0.1508	943.8	>2000
LP03801010	3	80	1.0	10	80	0.75	2.350	362.0	1998.62	2679.63	215.97	0.1607	992.7	>2000
LP03800415	3	80	0.4	15	80	0.70	2.847	345.2	2798.94	3976.11	195.78	0.1978	1080.6	>2000
LP03800615	3	80	0.6	15	80	1.02	3.378	357.8	4083.67	4008.61	292.08	0.2249	1184.9	>2000
LP03800815	3	80	0.8	15	80	1.31	3.784	370.4	5302.32	4041.11	387.37	0.2435	1264.9	>2000
LP03801015	3	80	1.0	15	80	1.59	4.107	383.0	6461.41	4073.61	481.70	0.2567	1331.0	>2000
LP03800420	3	80	0.4	20	80	1.22	4.291	353.6	6512.22	5330.37	346.80	0.2835	1331.7	>2000
LP03800620	3	80	0.6	20	80	1.75	5.045	370.4	9426.35	5388.15	516.50	0.3187	1459.6	>2000
LP03800820	3	80	0.8	20	80	2.23	5.602	387.2	12152.07	5445.93	683.91	0.3407	1555.8	>2000
LP03801020	3	80	1.0	20	80	2.67	6.035	404.0	14712.87	5503.70	849.14	0.3549	1632.3	>2000
LP03800425	3	80	0.4	25	80	1.86	5.874	362.0	12491.37	6699.07	539.93	0.3744	1563.8	>2000
LP03800625	3	80	0.6	25	80	2.64	6.846	383.0	17948.35	6789.35	802.83	0.4154	1714.1	>2000
LP03800825	3	80	0.8	25	80	3.34	7.543	404.0	22988.86	6879.63	1061.43	0.4386	1826.0	>2000
LP03801025	3	80	1.0	25	80	3.97	8.069	425.0	27674.63	6969.91	1316.00	0.4516	1913.6	>2000
LP03800430	3	80	0.4	30	80	2.62	7.567	370.4	21209.29	8082.22	774.75	0.4684	1783.6	>2000
LP03800630	3	80	0.6	30	80	3.69	8.747	395.6	30266.48	8212.22	1150.18	0.6478	71.3	1357
LP03800830	3	80	0.8	30	80	4.62	9.570	420.8	38539.16	8342.22	1518.47	0.7056	67.4	1394
LP03801030	3	80	1.0	30	80	5.45	10.173	446.0	46160.31	8472.22	1880.14	0.7724	63.6	1383
LP05350405	5	35	0.4	5	35	0.20	0.882	224.0	174.29	867.59	36.32	0.1695	390.9	>2000
LP05350605	5	35	0.6	5	35	0.29	1.051	231.0	255.28	873.61	54.22	0.1887	426.6	>2000
LP05350805	5	35	0.8	5	35	0.38	1.182	238.0	332.63	879.63	71.97	0.2028	454.3	>2000
LP05351005	5	35	1.0	5	35	0.46	1.288	245.0	406.68	885.65	89.56	0.2140	477.4	>2000
LP05350410	5	35	0.4	10	35	0.76	2.364	238.0	1330.53	1759.26	143.94	0.3612	636.6	>2000
LP05350610	5	35	0.6	10	35	1.07	2.754	252.0	1910.71	1783.33	214.00	0.3993	697.5	>2000
LP05350810	5	35	0.8	10	35	1.35	3.032	266.0	2446.12	1807.41	282.90	0.4221	742.8	>2000
LP05351010	5	35	1.0	10	35	1.61	3.242	280.0	2943.45	1831.48	350.71	0.4365	778.5	>2000
LP05350415	5	35	0.4	15	35	1.61	4.130	252.0	4299.11	2675.00	321.00	0.5784	853.1	>2000
LP05350615	5	35	0.6	15	35	2.23	4.717	273.0	6073.15	2729.17	475.40	0.6232	931.5	>2000
LP05350815	5	35	0.8	15	35	2.76	5.107	294.0	7668.37	2783.33	626.25	0.6431	988.7	>2000
LP05351015	5	35	1.0	15	35	3.21	5.381	315.0	9120.54	2837.50	773.86	0.6506	1033.8	>2000
LP05350420	5	35	0.4	20	35	2.71	6.065	266.0	9784.46	3614.81	565.80	0.8007	1046.8	>2000
LP05350620	5	35	0.6	20	35	3.67	6.810	294.0	13632.65	3711.11	835.00	0.8409	1140.6	>2000
LP05350820	5	35	0.8	20	35	4.47	7.272	322.0	17026.92	3807.41	1096.53	0.8489	1208.2	>2000
LP05351020	5	35	1.0	20	35	5.14	7.574	350.0	20076.19	3903.70	1351.28	0.8430	1263.3	>2000
LP05350425	5	35	0.4	25	35	4.02	8.106	280.0	18396.58	4578.70	876.77	1.0179	1224.8	>2000
LP05350625	5	35	0.6	25	35	5.36	8.968	315.0	25334.82	4729.17	1289.77	1.0444	1334.6	>2000
LP05350825	5	35	0.8	25	35	6.43	9.467	350.0	31369.05	4879.63	1669.10	1.0346	1411.4	>2000
LP05351025	5	35	1.0	25	35	7.31	9.770	385.0	36745.81	5030.09	2076.64	1.0116	1469.6	>2000
LP05350430	5	35	0.4	30	35	5.51	10.214	294.0	30673.47	5566.67	1252.50	0.7844	29.3	>2000
LP05350630	5	35	0.6	30	35	7.23	11.157	336.0	41825.89	5783.33	1837.06	1.2295	1512.6	>2000
LP05350830	5	35	0.8	30	35	8.57	11.664	378.0	51428.57	6000.00	2400.00	1.1984	1596.9	>2000
LP05351030	5	35	1.0	30	35	9.64	11.947	420.0	59946.43	6216.67	2944.74	1.1568	1665.3	>2000
LP05500405	5	50	0.4	5	50	0.14	0.751	314.0	176.88	1234.26	36.42	0.0755	516.3	>2000
LP05500605	5	50	0.6	5	50	0.21	0.901	321.0	260.81	1240.28	54.45	0.0843	564.8	>2000
LP05500805	5	50	0.8	5	50	0.27	1.021	328.0	341.97	1246.30	72.37	0.0912	601.1	>2000
LP05501005	5	50	1.0	5	50	0.34	1.120	335.0	420.55	1252.31	90.17	0.0968	628.6	>2000
LP05500410	5	50	0.4	10	50	0.55	2.042	328.0	1367.89	2492.59	144.73	0.1598	843.0	>2000
LP05500610	5	50	0.6	10	50	0.79	2.410	342.0	1986.84	2516.67	215.71	0.1799	924.2	>2000
LP05500810	5	50	0.8	10	50	1.01	2.686	356.0	2569.29	2540.74	285.83	0.1935	982.3	>2000
LP05501010	5	50	1.0	10	50	1.22	2.904	370.0	3119.37	2564.81	355.13	0.2033	1031.3	>2000

		DIMENSIONS				SECTION PROPERTIES						BUCKLING PROPERTIES		
CODE	n	S	t_2	d	a	$\bar{y}$	rad. gyr.	A	I_{NA}	Z_{plate}	$Z_{stiff.}$	σ_{cr}/E	L_1	L_{euler}
To scale multiply by	−	t_1	t_1	t_1	t_1	t_1	t_1	t_1^2	t_1^4	t_1^3	t_1^3	10^{-3}	t_1	t_1
LP05500415	5	50	0.4	15	50	1.18	3.615	342.0	4470.39	3775.00	323.57	0.2598	1126.0	>2000
LP05500615	5	50	0.6	15	50	1.67	4.202	363.0	6408.32	3829.17	480.87	0.2877	1233.1	>2000
LP05500815	5	50	0.8	15	50	2.11	4.619	384.0	8191.41	3883.33	635.45	0.3040	1312.3	>2000
LP05501015	5	50	1.0	15	50	2.50	4.930	405.0	9843.75	3937.50	787.50	0.3138	1374.8	>2000
LP05500420	5	50	0.4	20	50	2.02	5.373	356.0	10277.15	5081.48	571.67	0.3667	1385.8	>2000
LP05500620	5	50	0.6	20	50	2.81	6.158	384.0	14562.50	5177.78	847.27	0.3983	1514.6	>2000
LP05500820	5	50	0.8	20	50	3.50	6.689	412.0	18433.66	5274.07	1116.86	0.4133	1608.6	>2000
LP05501020	5	50	1.0	20	50	4.09	7.066	440.0	21969.79	5370.37	1380.95	0.4198	1682.8	>2000
LP05500425	5	50	0.4	25	50	3.04	7.259	370.0	19496.06	6412.04	887.82	0.4762	1626.4	>2000
LP05500625	5	50	0.6	25	50	4.17	8.217	405.0	27343.75	6562.50	1312.50	0.5073	1773.9	>2000
LP05500825	5	50	0.8	25	50	5.11	8.833	440.0	34327.65	6712.96	1726.19	0.5178	1881.1	>2000
LP05501025	5	50	1.0	25	50	5.92	9.250	475.0	40632.71	6863.43	2130.03	1.9443	40.5	701
LP05500430	5	50	0.4	30	50	4.22	9.237	384.0	32765.63	7766.67	1270.91	0.5853	1851.6	>2000
LP05500630	5	50	0.6	30	50	5.70	10.339	426.0	45538.73	7983.33	1874.35	1.5435	44.9	893
LP05500830	5	50	0.8	30	50	6.92	11.014	468.0	56769.23	8200.00	2460.00	1.7153	43.7	897
LP05501030	5	50	1.0	30	50	7.94	11.448	510.0	66838.24	8416.67	3030.00	1.8999	41.4	880
LP05650405	5	65	0.4	5	65	0.11	0.664	404.0	178.32	1600.93	36.48	0.0417	635.2	>2000
LP05650605	5	65	0.6	5	65	0.16	0.801	411.0	263.91	1606.94	54.58	0.0465	692.4	>2000
LP05650805	5	65	0.8	5	65	0.22	0.912	418.0	347.29	1612.96	72.58	0.0504	738.5	>2000
LP05651005	5	65	1.0	5	65	0.26	1.004	425.0	428.55	1618.98	90.50	0.0536	767.7	>2000
LP05650410	5	65	0.4	10	65	0.43	1.823	418.0	1389.15	3225.93	145.17	0.0870	1027.6	>2000
LP05650610	5	65	0.6	10	65	0.63	2.168	432.0	2031.25	3250.00	216.67	0.0989	1128.2	>2000
LP05650810	5	65	0.8	10	65	0.81	2.434	446.0	2642.75	3274.07	287.48	0.1075	1204.5	>2000
LP05651010	5	65	1.0	10	65	0.98	2.648	460.0	3226.45	3298.15	357.63	0.1139	1267.1	>2000
LP05650415	5	65	0.4	15	65	0.94	3.253	432.0	4570.31	4875.00	325.00	0.1423	1379.0	>2000
LP05650615	5	65	0.6	15	65	1.34	3.820	453.0	6610.31	4929.17	483.95	0.1601	1512.8	>2000
LP05650815	5	65	0.8	15	65	1.71	4.239	474.0	8515.82	4983.33	640.71	0.1717	1612.7	>2000
LP05651015	5	65	1.0	15	65	2.05	4.562	495.0	10303.98	5037.50	795.39	0.1795	1692.5	>2000
LP06650420	5	65	0.4	20	65	1.61	4.860	446.0	10571.00	6540.15	574.96	0.2029	1699.8	>2000
LP05650620	5	65	0.6	20	65	2.28	5.651	474.0	15139.24	6644.44	854.29	0.2249	1861.0	>2000
LP05650820	5	65	0.8	20	65	2.87	6.206	502.0	19335.99	6740.74	1128.68	1.1024	53.5	631
LP05651020	5	65	1.0	20	65	3.40	6.619	530.0	23220.13	6837.04	1398.48	1.2023	50.8	642
LP05650425	5	65	0.4	25	65	2.45	6.621	460.0	20165.31	8245.37	894.08	0.9260	60.3	737
LP05650625	5	65	0.6	25	65	3.41	7.604	495.0	28622.16	8395.83	1325.66	0.9864	57.7	824
LP05650825	5	65	0.8	25	65	4.25	8.274	530.0	36281.45	8546.30	1748.11	1.0796	54.3	854
LP05651025	5	65	1.0	25	65	4.98	8.753	565.0	43291.39	8696.76	2162.18	1.1813	51.4	860
LP05650430	5	65	0.4	30	65	3.42	8.477	474.0	34063.29	9966.67	1281.43	0.7851	29.4	1044
LP05650630	5	65	0.6	30	65	4.71	9.640	516.0	47956.40	10183.33	1896.21	0.9685	58.3	1061
LP05650830	5	65	0.8	30	65	5.81	10.403	558.0	60387.10	10400.00	2496.00	1.0580	55.2	1091
LP05651030	5	65	1.0	30	65	6.75	10.929	600.0	71662.50	10616.67	3082.26	1.1611	52.2	1087
LP05800405	5	80	0.4	5	80	0.09	0.602	494.0	179.23	1967.59	36.51	0.0261	753.6	>2000
LP05800605	5	80	0.6	5	80	0.13	0.729	501.0	265.91	1973.61	54.65	0.0290	822.8	>2000
LP05800805	5	80	0.8	5	80	0.18	0.831	508.0	350.72	1979.63	72.72	0.0315	877.9	>2000
LP05801005	5	80	1.0	5	80	0.22	0.918	515.0	433.76	1985.65	90.71	0.0335	914.0	>2000
LP05800410	5	80	0.4	10	80	0.35	1.662	508.0	1402.89	3959.26	145.44	0.0536	1204.8	>2000
LP05800610	5	80	0.6	10	80	0.52	1.987	522.0	2060.34	3983.33	217.27	0.0613	1324.2	>2000
LP05800810	5	80	0.8	10	80	0.67	2.241	536.0	2691.54	4007.41	288.53	0.0570	1416.5	>2000
LP05801010	5	80	1.0	10	80	0.82	2.449	550.0	3298.48	4031.48	359.24	0.0715	1487.0	>2000
LP05800415	5	80	0.4	15	80	0.78	2.980	522.0	4635.78	5975.00	325.91	0.0879	1617.9	>2000
LP05800615	5	80	0.6	15	80	1.12	3.525	543.0	6745.34	6029.17	485.93	0.0999	1776.2	>2000
LP05800815	5	80	0.8	15	80	1.44	3.936	564.0	8736.70	6083.33	644.12	0.1081	1895.6	>2000
LP05801015	5	80	1.0	15	80	1.73	4.261	585.0	10622.60	6137.50	800.54	0.8025	65.6	507
LP05800420	5	80	0.4	20	80	1.34	4.482	536.0	10766.17	8014.81	577.07	0.6235	73.5	610
LP05800620	5	80	0.6	20	80	1.91	5.248	564.0	15531.91	8111.11	858.82	0.6731	69.4	689
LP05800820	5	80	0.8	20	80	2.43	5.807	592.0	19963.96	8207.41	1136.41	0.7414	65.2	724
LP05801020	5	80	1.0	20	80	2.90	6.236	620.0	24107.53	8303.70	1410.06	0.8066	63.8	743
LP05800425	5	80	0.4	25	80	2.05	6.122	550.0	20615.53	10078.70	898.10	0.6181	73.9	841
LP05800625	5	80	0.6	25	80	2.88	7.102	585.0	29507.21	10229.17	1334.24	0.6625	70.1	944
LP05800825	5	80	0.8	25	80	3.63	7.795	620.0	37668.01	10379.63	1762.58	0.7278	65.8	985
LP05801025	5	80	1.0	25	80	4.29	8.308	655.0	45215.09	10530.09	2183.66	0.7944	62.4	1001
LP05800430	5	80	0.4	30	80	2.87	7.872	564.0	34946.81	12156.67	1288.24	0.6124	74.2	1093
LP05800630	5	80	0.6	30	80	4.01	9.052	606.0	49655.94	12383.33	1910.57	0.6528	70.8	1220
LP05800830	5	80	0.8	30	80	5.00	9.860	648.0	63000.00	12600.00	2520.00	0.7150	66.9	1264
LP05801030	5	80	1.0	30	80	5.87	10.442	690.0	75228.26	12816.67	3117.57	0.7822	63.1	1273

L-PLATE n angle stiffeners
edges pinned
free in-plane

CODE	n	S	t_2	d	a	$\bar{y}$	rad. gyr.	A	I_{NA}	Z_{plate}	$Z_{stiff.}$	σ_{cr}/E	L_1	L_{euler}
To scale multiply by	—	t_1	t_1	t_1	t_1	t_1	t_1	t_1^2	t_1^4	t_1^3	t_1^3	10^{-3}	t_1	t_1
LQ02350405	2	35	0.4	5	7	0.33	1.111	54.6	67.40	204.44	14.43	2.6073	75.5	903
LQ02350605	2	35	0.6	5	7	0.47	1.302	57.4	97.30	206.85	21.48	2.9144	82.8	862
LQ02350805	2	35	0.8	5	7	0.60	1.442	60.2	125.14	209.26	28.43	3.1791	88.2	832
LQ02351005	2	35	1.0	5	7	0.71	1.549	63.0	151.17	211.67	35.28	3.4367	92.7	806
LQ02350410	2	35	0.4	10	7	1.20	2.884	60.2	500.55	418.52	56.86	4.4213	89.0	703
LQ02350610	2	35	0.6	10	7	1.64	3.268	65.8	702.74	428.15	84.07	5.2758	124.7	651
LQ02350810	2	35	0.8	10	7	2.02	3.516	71.4	882.91	437.78	110.60	5.6755	136.8	634
LQ02351010	2	35	1.0	10	7	2.34	3.686	77.0	1045.89	447.41	136.50	5.9695	145.1	624
LQ02350415	2	35	0.4	15	7	2.46	4.902	65.8	1581.16	642.22	126.11	4.0992	134.0	738
LQ02350615	2	35	0.6	15	7	3.27	5.413	74.2	2174.19	663.89	185.43	5.2315	126.2	662
LQ02350815	2	35	0.8	15	7	3.92	5.706	82.6	2689.10	685.56	242.75	6.1052	133.6	618
LQ02351015	2	35	1.0	15	7	4.45	5.881	91.0	3147.53	707.22	298.36	6.5243	174.5	603
LQ02350420	2	35	0.4	20	7	4.03	7.033	71.4	3531.65	875.56	221.19	1.7652	19.5	1149
LQ02350620	2	35	0.6	20	7	5.23	7.608	82.6	4780.63	914.07	323.67	4.3977	185.6	727
LQ02350820	2	35	0.8	20	7	6.14	7.897	93.8	5849.61	952.59	422.07	5.0560	179.8	685
LQ02351020	2	35	1.0	20	7	6.86	8.045	105.0	6796.19	991.11	517.10	5.5082	188.1	661
LQ02350425	2	35	0.4	25	7	5.84	9.214	77.0	6536.80	1118.52	341.24	1.1306	24.3	1456
LQ02350625	2	35	0.6	25	7	7.42	9.802	91.0	8743.13	1178.70	497.27	2.5172	24.5	982
LQ02350825	2	35	0.8	25	7	8.57	10.057	105.0	10619.05	1238.89	646.38	4.2711	246.7	747
LQ02351025	2	35	1.0	25	7	9.45	10.159	119.0	12281.16	1299.07	789.98	4.6051	249.5	725
LQ02350430	2	35	0.4	30	7	7.85	11.412	82.6	10756.42	1371.11	485.51	0.7855	29.2	1769
LQ02350630	2	35	0.6	30	7	9.78	11.975	99.4	14255.13	1457.78	704.96	1.7525	29.3	1194
LQ02350830	2	35	0.8	30	7	11.15	12.175	116.2	17225.47	1544.44	913.97	3.0558	29.5	896
LQ02351030	2	35	1.0	30	7	12.18	12.222	133.0	19867.67	1631.11	1114.94	3.9504	322.6	781
LQ02500405	2	50	0.4	5	10	0.24	0.956	75.6	69.05	290.00	14.50	1.1924	103.4	1900
LQ02500605	2	50	0.6	5	10	0.34	1.133	78.4	100.70	292.41	21.63	1.3286	111.7	1813
LQ02500805	2	50	0.8	5	10	0.44	1.269	81.2	130.71	294.81	28.68	1.4492	118.6	1748
LQ02501005	2	50	1.0	5	10	0.54	1.377	84.0	159.23	297.22	35.67	1.5664	124.4	1691
LQ02500410	2	50	0.4	10	10	0.89	2.537	81.2	522.82	589.63	57.37	2.2774	155.8	1392
LQ02500610	2	50	0.6	10	10	1.24	2.931	86.8	745.62	599.26	85.16	2.5998	174.9	1318
LQ02500810	2	50	0.8	10	10	1.56	3.205	92.4	948.92	608.89	112.41	2.8318	187.5	1275
LQ02501010	2	50	1.0	10	10	1.84	3.405	98.0	1136.05	618.52	139.17	3.0265	197.3	1244
LQ02500415	2	50	0.4	15	10	1.87	4.396	86.8	1677.65	898.89	127.74	3.0052	153.7	1224
LQ02500615	2	50	0.6	15	10	2.55	4.968	95.2	2349.74	920.56	188.77	3.2693	28.2	1190
LQ02500815	2	50	0.8	15	10	3.13	5.333	103.6	2946.72	942.22	248.20	3.5431	27.0	1156
LQ02501015	2	50	1.0	15	10	3.62	5.579	112.0	3485.49	963.89	306.18	4.1265	256.7	1080
LQ02500420	2	50	0.4	20	10	3.12	6.409	92.4	3795.67	1217.78	224.82	1.7651	19.5	1618
LQ02500620	2	50	0.6	20	10	4.17	7.111	103.6	5238.61	1256.30	330.93	3.1785	27.9	1220
LQ02500820	2	50	0.8	20	10	5.02	7.523	114.8	6496.63	1294.81	433.61	3.4495	27.3	1185
LQ02501020	2	50	1.0	20	10	5.71	7.776	126.0	7619.05	1333.33	533.33	3.8245	26.1	1135
LQ02500425	2	50	0.4	25	10	4.59	8.512	98.0	7100.34	1546.30	347.92	1.1302	24.3	>2000
LQ02500625	2	50	0.6	25	10	6.03	9.298	112.0	9681.92	1606.48	510.29	2.4876	25.6	1394
LQ02500825	2	50	0.8	25	10	7.14	9.720	126.0	11904.76	1666.67	666.67	3.2906	28.0	1223
LQ02501025	2	50	1.0	25	10	8.04	9.956	140.0	13876.49	1726.85	817.98	3.6860	27.0	1165
LQ02500430	2	50	0.4	30	10	6.25	10.666	103.6	11786.87	1884.44	496.39	0.7852	29.2	>2000
LQ02500630	2	50	0.6	30	10	8.07	11.496	120.4	15912.96	1971.11	725.73	1.7442	29.7	1685
LQ02500830	2	50	0.8	30	10	9.45	11.903	137.2	19437.90	2057.78	945.70	2.8577	30.3	1322
LQ02501030	2	50	1.0	30	10	10.52	12.103	154.0	22558.44	2144.44	1158.00	3.4870	28.6	1201
LQ02650405	2	65	0.4	5	13	0.19	0.851	96.6	69.98	375.56	14.54	0.6677	128.3	>2000
LQ02650605	2	65	0.6	5	13	0.27	1.016	99.4	102.67	377.96	21.71	0.7406	138.7	>2000
LQ02650805	2	65	0.8	5	13	0.35	1.145	102.2	133.99	380.37	28.83	0.8058	147.0	>2000
LQ02651005	2	65	1.0	5	13	0.43	1.250	105.0	164.05	382.78	35.89	0.8692	154.1	>2000
LQ02650410	2	65	0.4	10	13	0.70	2.290	102.2	535.94	760.74	57.66	1.2777	196.9	>2000
LQ02650610	2	65	0.6	10	13	1.00	2.676	107.8	771.80	770.37	85.77	1.4642	217.9	>2000
LQ02650810	2	65	0.8	10	13	1.27	2.955	113.4	990.48	780.00	113.45	1.6087	232.9	>2000
LQ02651010	2	65	1.0	10	13	1.51	3.168	119.0	1194.40	789.63	140.73	1.7337	245.0	>2000
LQ02650415	2	65	0.4	15	13	1.50	4.014	107.8	1736.55	1155.56	128.66	1.8620	37.6	>2000
LQ02650615	2	65	0.6	15	13	2.09	4.603	116.2	2461.83	1177.22	190.71	1.9776	36.0	1977
LQ02650815	2	65	0.8	15	13	2.60	5.002	124.6	3117.50	1198.89	251.42	2.1572	34.3	1913
LQ02651015	2	65	1.0	15	13	3.05	5.286	133.0	3716.73	1220.56	310.90	2.4184	32.3	1822
LQ02650420	2	65	0.4	20	13	2.54	5.911	113.4	3961.90	1560.00	226.91	1.7656	19.5	>2000
LQ02650620	2	65	0.6	20	13	3.47	6.669	124.6	5542.22	1598.52	335.22	1.9576	35.9	>2000
LQ02650820	2	65	0.8	20	13	4.24	7.151	135.8	6943.54	1637.04	440.62	2.1141	34.6	1956
LQ02651020	2	65	1.0	20	13	4.90	7.472	147.0	8206.80	1675.56	543.42	2.3508	32.9	1872

L-PLATE — n angle stiffeners / edges pinned / free in-plane

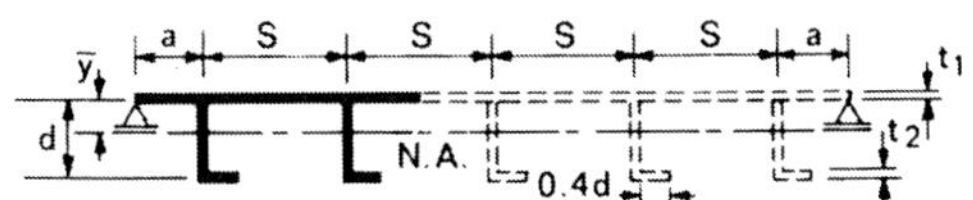

CODE	n	S	t_2	d	a	$\bar{y}$	rad. gyr.	A	I_{NA}	Z_{plate}	$Z_{stiff.}$	σ_{cr}/E	L_1	L_{euler}
To scale multiply by	–	t_1	t_1	t_1	t_1	t_1	t_1	t_1^2	t_1^4	t_1^3	t_1^3	10^{-3}	t_1	t_1
LG02650425	2	65	0.4	25	13	3.78	7.920	119.0	7464.99	1974.07	351.82	1.1303	24.3	>2000
LG02650625	2	65	0.6	25	13	5.08	8.811	133.0	10324.25	2034.26	518.16	1.9140	36.0	>2000
LG02650825	2	65	0.8	25	13	6.12	9.340	147.0	12823.13	2094.44	679.28	2.0669	35.0	1998
LG02651025	2	65	1.0	25	13	6.99	9.670	161.0	15055.64	2154.63	835.85	2.2896	33.6	1916
LG02650430	2	65	0.4	30	13	5.20	10.004	124.6	12469.98	2397.78	502.83	0.7851	29.2	>2000
LG02650630	2	65	0.6	30	13	6.87	10.990	141.4	17078.36	2484.44	738.50	1.7022	33.2	>2000
LG02650830	2	65	0.8	30	13	8.19	11.539	158.2	21062.96	2571.11	965.84	2.0045	35.6	>2000
LG02651030	2	65	1.0	30	13	9.26	11.857	175.0	24603.43	2657.78	1186.12	2.2273	34.4	1958
LG02800405	2	80	0.4	5	16	0.15	0.775	117.6	70.58	461.11	14.56	0.4221	152.7	>2000
LG02800605	2	80	0.6	5	16	0.22	0.929	120.4	103.95	463.52	21.77	0.4659	164.5	>2000
LG02800805	2	80	0.8	5	16	0.29	1.051	123.2	136.15	465.93	28.92	0.5055	174.2	>2000
LG02801005	2	80	1.0	5	16	0.36	1.152	126.0	167.26	468.33	36.03	0.5439	182.3	>2000
LG02800410	2	80	0.4	10	16	0.58	2.102	123.2	544.59	931.85	57.84	0.7976	234.1	>2000
LG02800610	2	80	0.6	10	16	0.84	2.476	128.8	789.44	941.48	86.17	0.9164	257.9	>2000
LG02800810	2	80	0.8	10	16	1.07	2.754	134.4	1019.05	951.11	114.13	1.0113	275.5	>2000
LG02801010	2	80	1.0	10	16	1.29	2.970	140.0	1235.24	960.74	141.75	1.0946	289.8	>2000
LG02800415	2	80	0.4	15	16	1.26	3.714	128.8	1776.24	1412.22	129.25	1.2529	305.6	>2000
LG02800615	2	80	0.6	15	16	1.77	4.302	137.2	2539.61	1433.89	191.98	1.4402	341.0	>2000
LG02800815	2	80	0.8	15	16	2.23	4.717	145.6	3239.01	1455.56	253.55	1.4536	41.8	>2000
LG02801015	2	80	1.0	15	16	2.63	5.023	154.0	3884.90	1477.22	314.06	1.6796	383.6	>2000
LG02800420	2	80	0.4	20	16	2.14	5.507	134.4	4076.19	1902.22	228.27	1.2468	45.2	>2000
LG02800620	2	80	0.6	20	16	2.97	6.289	145.6	5758.24	1940.74	338.06	1.3143	43.8	>2000
LG02800820	2	80	0.8	20	16	3.67	6.810	156.8	7270.75	1979.26	445.33	1.4269	41.9	>2000
LG02801020	2	80	1.0	20	16	4.29	7.175	168.0	8647.62	2017.78	550.30	1.5954	39.6	>2000
LG02800425	2	80	0.4	25	16	3.21	7.426	140.0	7720.24	2401.85	354.37	1.1305	24.4	>2000
LG02800625	2	80	0.6	25	16	4.38	8.371	154.0	10791.40	2462.04	523.43	1.2991	43.9	>2000
LG02800825	2	80	0.8	25	16	5.36	8.968	168.0	13511.90	2522.22	687.88	1.4005	42.4	>2000
LG02801025	2	80	1.0	25	16	6.18	9.365	182.0	15962.68	2582.41	848.24	1.5559	40.3	>2000
LG02800430	2	80	0.4	30	16	4.45	9.433	145.6	12956.04	2911.11	507.10	0.7851	29.3	>2000
LG02800630	2	80	0.6	30	16	5.99	10.511	162.4	17942.36	2997.78	747.14	1.2751	44.1	>2000
LG02800830	2	80	0.8	30	16	7.23	11.157	179.2	22307.14	3084.44	979.76	1.3740	42.9	>2000
LG02801030	2	80	1.0	30	16	8.27	11.564	196.0	26210.20	3171.11	1205.92	1.5215	41.1	>2000
LG03350405	3	35	0.4	5	7	0.29	1.051	92.4	102.11	349.44	21.69	1.0374	152.5	>2000
LG03350605	3	35	0.6	5	7	0.42	1.238	96.6	148.02	353.06	32.31	1.1589	166.8	>2000
LG03350805	3	35	0.8	5	7	0.54	1.377	100.8	191.07	356.67	42.80	1.2540	177.8	>2000
LG03351005	3	35	1.0	5	7	0.64	1.485	105.0	231.61	360.28	53.16	1.3360	187.1	>2000
LG03350410	3	35	0.4	10	7	1.07	2.754	100.8	764.29	713.33	85.60	2.1631	246.2	1693
LG03350610	3	35	0.6	10	7	1.48	3.144	109.2	1079.67	727.78	126.77	2.4000	271.1	1624
LG03350810	3	35	0.8	10	7	1.84	3.405	117.6	1363.27	742.22	167.00	2.5498	289.3	1589
LG03351010	3	35	1.0	10	7	2.14	3.587	126.0	1621.43	756.67	206.36	2.6568	303.9	1568
LG03350415	3	35	0.4	15	7	2.23	4.717	109.2	2429.26	1091.67	190.16	3.2847	319.9	1384
LG03350615	3	35	0.6	15	7	2.99	5.256	121.8	3364.19	1124.17	280.18	3.5739	359.0	1344
LG03350815	3	35	0.8	15	7	3.62	5.579	134.4	4182.59	1156.67	367.41	3.6959	383.9	1334
LG03351015	3	35	1.0	15	7	4.13	5.782	147.0	4914.41	1189.17	452.22	3.7529	403.3	1334
LG03350420	3	35	0.4	20	7	3.67	6.810	117.6	5453.06	1484.44	334.00	1.7649	19.5	1913
LG03350620	3	35	0.6	20	7	4.82	7.438	134.4	7435.71	1542.22	489.88	3.8270	22.5	1307
LG03350820	3	35	0.8	20	7	5.71	7.776	151.2	9142.86	1600.00	640.00	4.4410	462.0	1224
LG03351020	3	35	1.0	20	7	6.43	7.965	168.0	10657.14	1657.78	795.26	4.4274	488.0	1235
LG03350425	3	35	0.4	25	7	5.36	8.968	126.0	10133.93	1891.67	515.91	1.1298	24.4	>2000
LG03350625	3	35	0.6	25	7	6.89	9.637	147.0	13651.15	1981.94	753.70	2.4985	25.0	1638
LG03350825	3	35	0.8	25	7	8.04	9.956	168.0	16651.79	2072.22	981.58	3.9699	27.1	1302
LG03351025	3	35	1.0	25	7	8.93	10.107	189.0	19308.04	2162.50	1201.39	4.6156	554.0	1211
LG03350430	3	35	0.4	30	7	7.23	11.157	134.4	16730.36	2313.33	734.82	0.7846	29.2	>2000
LG03350630	3	35	0.6	30	7	9.14	11.826	159.6	22320.68	2443.33	1069.78	1.7558	32.7	1973
LG03350830	3	35	0.8	30	7	10.52	12.103	184.8	27070.13	2573.33	1389.60	2.9625	30.5	1521
LG03351030	3	35	1.0	30	7	11.57	12.205	210.0	31281.43	2703.33	1697.44	4.1446	30.1	1280
LG03500405	3	50	0.4	5	10	0.21	0.901	128.4	104.32	496.11	21.78	0.4643	202.7	>2000
LG03500605	3	50	0.6	5	10	0.31	1.073	132.6	152.63	499.72	32.51	0.5197	221.3	>2000
LG03500805	3	50	0.8	5	10	0.39	1.205	136.8	198.68	503.33	43.14	0.5647	235.7	>2000
LG03501005	3	50	1.0	5	10	0.48	1.312	141.0	242.69	506.94	53.68	0.6041	247.6	>2000
LG03500410	3	50	0.4	10	10	0.79	2.410	136.8	794.74	1006.67	86.29	0.9708	327.4	>2000
LG03500610	3	50	0.6	10	10	1.12	2.801	145.2	1139.26	1021.11	128.23	1.0967	360.1	>2000
LG03500810	3	50	0.8	10	10	1.41	3.079	153.6	1456.25	1035.56	169.45	1.1954	384.4	>2000
LG03501010	3	50	1.0	10	10	1.67	3.287	162.0	1750.00	1050.00	210.00	1.2539	404.0	>2000

L-PLATE n angle stiffeners / edges pinned / free in-plane

CODE	n	S	t_2	d	a	$\bar{y}$	rad. gyr.	A	I_{NA}	Z_{plate}	$Z_{stiff.}$	σ_{cr}/E	L_1	L_{euler}
To scale multiply by	—	t_1	t_1	t_1	t_1	t_1	t_1	t_1^2	t_1^4	t_1^3	t_1^3	10^{-3}	t_1	t_1
LQ03500415	3	50	0.4	15	10	1.67	4.202	145.2	2563.33	1531.67	192.35	1.5541	436.5	>2000
LQ03500615	3	50	0.6	15	10	2.31	4.785	157.8	3613.05	1564.17	284.71	1.7259	480.9	>2000
LQ03500815	3	50	0.8	15	10	2.85	5.170	170.4	4553.87	1596.67	374.87	1.8291	513.1	>2000
LQC3501015	3	50	1.0	15	10	3.32	5.436	183.0	5408.30	1629.17	463.03	1.8974	538.7	>2000
LQ03500420	3	50	0.4	20	10	2.81	6.158	153.6	5825.00	2071.11	338.91	1.7652	19.5	>2000
LQ03500620	3	50	0.6	20	10	3.80	6.893	170.4	8095.77	2128.89	499.83	2.2220	36.3	>2000
LQ03500820	3	50	0.8	20	10	4.62	7.342	187.2	10092.31	2186.67	656.00	2.4069	627.7	>2000
LQ03501020	3	50	1.0	20	10	5.29	7.632	204.0	11882.35	2244.44	808.00	2.4502	659.1	>2000
LQ03500425	3	50	0.4	25	10	4.17	8.217	162.0	10937.50	2625.00	525.00	1.1301	24.3	>2000
LQ03500625	3	50	0.6	25	10	5.53	9.061	183.0	15023.05	2715.28	771.71	2.1271	36.3	>2000
LQ03500825	3	50	0.8	25	10	6.62	9.540	204.0	18566.18	2805.56	1010.00	2.3587	35.1	>2000
LQ03501025	3	50	1.0	25	10	7.50	9.825	225.0	21718.75	2895.83	1241.07	2.6738	33.2	>2000
LQ03500430	3	50	0.4	30	10	5.70	10.339	170.4	18215.49	3193.33	749.74	0.7850	29.2	>2000
LQ03500630	3	50	0.6	30	10	7.45	11.254	195.6	24772.09	3323.33	1098.73	1.7205	31.5	>2000
LQ03500830	3	50	0.8	30	10	8.80	11.735	220.8	30404.35	3453.33	1434.46	2.2361	36.2	>2000
LQ03501030	3	50	1.0	30	10	9.88	11.995	246.0	35396.34	3583.33	1759.09	2.5612	34.5	>2000
LQ03650405	3	65	0.4	5	13	0.16	0.801	164.4	105.57	642.78	21.83	0.2568	249.9	>2000
LQ03650605	3	65	0.6	5	13	0.24	0.960	168.6	155.27	646.39	32.62	0.2870	272.3	>2000
LQ03650805	3	65	0.8	5	13	0.31	1.084	172.8	203.13	650.00	43.33	0.3121	289.8	>2000
LQ03651005	3	65	1.0	5	13	0.38	1.187	177.0	249.26	653.61	53.97	0.3343	304.4	>2000
LQ03650410	3	65	0.4	10	13	0.63	2.168	172.8	812.50	1300.00	86.67	0.5310	402.2	>2000
LQ03650610	3	65	0.6	10	13	0.89	2.547	181.2	1175.17	1314.44	129.05	0.6055	442.1	>2000
LQ03650810	3	65	0.8	10	13	1.14	2.826	189.6	1513.92	1328.89	170.86	0.6606	472.3	>2000
LQ03651010	3	65	1.0	10	13	1.36	3.042	198.0	1831.82	1343.33	212.11	0.7047	496.7	>2000
LQ03650415	3	65	0.4	15	13	1.34	3.820	181.2	2644.12	1971.67	193.58	0.8621	538.2	>2000
LQ03650615	3	65	0.6	15	13	1.88	4.410	193.8	3769.45	2004.17	287.32	0.9726	592.0	>2000
LQ03650815	3	65	0.8	15	13	2.35	4.820	206.4	4795.64	2036.67	379.24	1.0463	631.8	>2000
LQ03651015	3	65	1.0	15	13	2.77	5.119	219.0	5739.81	2069.17	469.47	1.0996	664.0	>2000
LQ03650420	3	65	0.4	20	13	2.28	5.651	189.6	6055.70	2657.78	341.71	1.2133	660.2	>2000
LQ03650620	3	65	0.6	20	13	3.14	6.427	206.4	8525.58	2715.56	505.66	1.3475	727.9	>2000
LQ03650820	3	65	0.8	20	13	3.87	6.935	223.2	10735.48	2773.33	665.60	1.4258	776.6	>2000
LQ03651020	3	65	1.0	20	13	4.50	7.286	240.0	12740.00	2831.11	821.94	1.4754	815.1	>2000
LQC3650425	3	65	0.4	25	13	3.41	7.604	198.0	11448.86	3358.33	530.26	1.1303	24.4	>2000
LQ03650625	3	65	0.6	25	13	4.62	8.532	219.0	15943.92	3448.61	782.46	1.3264	46.8	>2000
LQ03650825	3	65	0.8	25	13	5.63	9.107	240.0	19906.25	3538.89	1027.42	1.4596	44.3	>2000
LQ03651025	3	65	1.0	25	13	6.47	9.482	261.0	23464.44	3629.17	1265.99	1.6589	41.3	>2000
LQ03650430	3	65	0.4	30	13	4.71	9.640	206.4	19182.56	4073.33	758.48	0.7851	29.3	>2000
LQ03650630	3	65	0.6	30	13	6.30	10.689	231.6	26461.40	4203.33	1116.30	1.2919	47.2	>2000
LQ03650830	3	65	0.8	30	13	7.57	11.302	256.8	32803.74	4333.33	1462.50	1.4208	45.1	>2000
LQ03651030	3	65	1.0	30	13	8.62	11.678	282.0	38460.64	4463.33	1798.66	1.6100	42.5	>2000
LQ03800405	3	80	0.4	5	16	0.13	0.729	200.4	106.36	789.44	21.86	0.1609	295.4	>2000
LQ03800605	3	80	0.6	5	16	0.20	0.876	204.6	156.98	793.06	32.69	0.1793	321.2	>2000
LQ03800805	3	80	0.8	5	16	0.26	0.993	208.8	206.03	796.67	43.45	0.1949	341.4	>2000
LQ03801005	3	80	1.0	5	16	0.32	1.091	213.0	253.61	800.28	54.15	0.2087	358.5	>2000
LQ03800410	3	80	0.4	10	16	0.52	1.987	208.8	824.14	1593.33	86.91	0.3279	473.0	>2000
LQ03800610	3	80	0.6	10	16	0.75	2.350	217.2	1199.17	1607.78	129.58	0.3758	519.5	>2000
LQ03800810	3	80	0.8	10	16	0.96	2.624	225.6	1553.19	1622.22	171.76	0.4123	555.7	>2000
LQ03801010	3	80	1.0	10	16	1.15	2.841	234.0	1888.46	1636.67	213.48	0.4421	584.7	>2000
LQ03800415	3	80	0.4	15	16	1.12	3.525	217.2	2698.14	2411.67	194.37	0.5348	633.1	>2000
LQ03800615	3	80	0.6	15	16	1.59	4.107	229.8	3876.84	2444.17	289.02	0.6096	696.8	>2000
LQ03800815	3	80	0.8	15	16	2.00	4.526	242.4	4965.59	2476.67	382.11	0.6621	744.7	>2000
LQ03801015	3	80	1.0	15	16	2.38	4.842	255.0	5977.72	2509.17	473.76	0.7018	782.8	>2000
LQ03800420	3	80	0.4	20	16	1.91	5.248	225.6	6212.77	3244.44	343.53	0.7616	779.9	>2000
LQ03800620	3	80	0.6	20	16	2.67	6.035	242.4	8827.72	3302.22	509.49	0.8574	857.8	>2000
LQ03800820	3	80	0.8	20	16	3.33	6.573	259.2	11200.00	3360.00	672.00	0.9190	918.0	>2000
LQ03801020	3	80	1.0	20	16	3.91	6.961	276.0	13373.91	3417.78	831.35	0.9615	961.0	>2000
LQ03800425	3	80	0.4	25	16	2.88	7.102	234.0	11802.88	4091.67	533.70	0.8385	59.5	>2000
LQC3800625	3	80	0.6	25	16	3.97	8.069	255.0	16604.78	4181.94	789.60	0.8940	57.0	>2000
LQ03800825	3	80	0.8	25	16	4.89	8.701	276.0	20896.74	4272.22	1039.19	0.9506	53.4	>2000
LQ03801025	3	80	1.0	25	16	5.68	9.136	297.0	24786.93	4362.50	1283.09	1.1324	49.3	>2000
LQ03800430	3	80	0.4	30	16	4.01	9.052	242.4	19862.38	4953.33	764.23	0.7852	29.3	>2000
LQ03800630	3	80	0.6	30	16	5.45	10.173	267.6	27696.19	5083.33	1128.08	0.8800	57.4	>2000
LQ03800830	3	80	0.8	30	16	6.64	10.873	292.8	34613.11	5213.33	1481.68	0.9686	54.3	>2000
LQ03801030	3	80	1.0	30	16	7.64	11.331	318.0	40831.13	5343.33	1826.20	1.1015	50.5	>2000

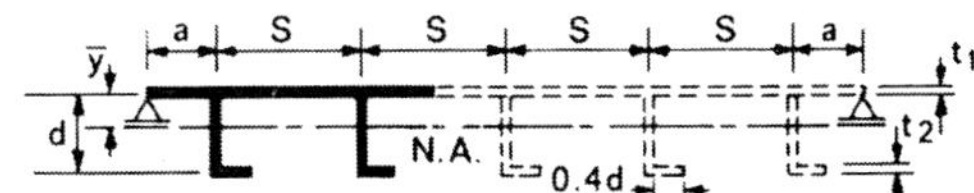

L-PLATE — n angle stiffeners, edges pinned, free in-plane

CODE	n	S	t_2	d	a	$\bar{y}$	rad. gyr.	A	I_{NA}	Z_{plate}	$Z_{stiff.}$	σ_{cr}/E	L_1	L_{euler}
To scale multiply by	–	t_1	t_1	t_1	t_1	t_1	t_1	t_1^2	t_1^4	t_1^3	t_1^3	10^{-3}	t_1	t_1
LQ05350405	5	35	0.4	5	7	0.27	1.010	168.0	171.28	639.44	36.19	0.3143	285.1	>2000
LQ05350605	5	35	0.6	5	7	0.39	1.193	175.0	248.96	645.46	53.96	0.3507	311.9	>2000
LQ05350805	5	35	0.8	5	7	0.49	1.330	182.0	322.16	651.49	71.50	0.3786	332.0	>2000
LQ05351005	5	35	1.0	5	7	0.60	1.439	189.0	391.37	657.50	88.85	0.4019	348.6	>2000
LQ05350410	5	35	0.4	10	7	0.99	2.661	182.0	1288.64	1302.96	143.01	0.6677	464.8	>2000
LQ05350610	5	35	0.6	10	7	1.38	3.054	196.0	1828.06	1327.04	212.01	0.7396	509.4	>2000
LQ05350810	5	35	0.8	10	7	1.71	3.321	210.0	2316.19	1351.11	279.54	0.7851	542.6	>2000
LQ05351010	5	35	1.0	10	7	2.01	3.512	224.0	2762.65	1375.19	345.72	0.8169	569.1	>2000
LQ05350415	5	35	0.4	15	7	2.07	4.581	196.0	4113.14	1990.56	318.02	1.0628	621.8	>2000
LQ05350615	5	35	0.6	15	7	2.80	5.136	217.0	5724.28	2044.72	469.19	1.1473	680.8	>2000
LQ05350815	5	35	0.8	15	7	3.40	5.478	238.0	7143.28	2098.89	615.98	1.1885	724.5	>2000
LQ05351015	5	35	1.0	15	7	3.91	5.701	259.0	8416.87	2153.06	758.91	1.2094	762.3	>2000
LQ05350420	5	35	0.4	20	7	3.43	6.642	210.0	9264.76	2702.22	559.08	1.4549	763.1	>2000
LQ05350620	5	35	0.6	20	7	4.54	7.305	238.0	12699.16	2798.52	821.30	1.5306	838.1	>2000
LQ05350820	5	35	0.8	20	7	5.41	7.676	266.0	15671.18	2894.81	1074.36	1.5503	887.0	>2000
LQ05351020	5	35	1.0	20	7	6.12	7.892	294.0	18312.93	2991.11	1319.61	1.5478	928.7	>2000
LQ05350425	5	35	0.4	25	7	5.02	8.780	224.0	17266.56	3437.96	864.29	1.1298	24.4	>2000
LQ05350625	5	35	0.6	25	7	6.52	9.501	259.0	23380.19	3588.43	1264.85	1.8654	975.1	>2000
LQ05350825	5	35	0.8	25	7	7.65	9.865	294.0	28613.95	3738.89	1649.51	1.8536	1036.5	>2000
LQ05351025	5	35	1.0	25	7	8.55	10.053	329.0	33248.64	3889.35	2021.03	1.8215	1087.3	>2000
LQ05350430	5	35	0.4	30	7	6.81	10.957	238.0	28573.11	4197.78	1231.96	0.7845	29.3	>2000
LQ05350630	5	35	0.6	30	7	8.68	11.697	280.0	38311.07	4414.44	1796.83	1.7356	30.0	>2000
LQ05350830	5	35	0.8	30	7	10.06	12.030	322.0	46598.76	4631.11	2337.20	2.0895	1176.9	>2000
LQ05351030	5	35	1.0	30	7	11.13	12.173	364.0	53938.19	4847.78	2857.86	2.0274	1230.9	>2000
LQ05500405	5	50	0.4	5	10	0.19	0.864	234.0	174.68	908.33	36.33	0.1401	377.7	>2000
LQ05500605	5	50	0.6	5	10	0.28	1.031	241.0	256.09	914.35	54.26	0.1568	412.4	>2000
LQ05500805	5	50	0.8	5	10	0.36	1.161	248.0	334.01	920.37	72.03	0.1700	435.2	>2000
LQ05501005	5	50	1.0	5	10	0.44	1.266	255.0	408.70	926.39	89.65	0.1814	461.3	>2000
LQ05500410	5	50	0.4	10	10	0.73	2.321	248.0	1336.02	1840.74	144.06	0.2959	613.6	>2000
LQ05500610	5	50	0.6	10	10	1.03	2.708	262.0	1921.76	1864.81	214.26	0.3338	674.7	>2000
LQ05500810	5	50	0.8	10	10	1.30	2.988	276.0	2463.77	1888.89	283.33	0.3602	718.6	>2000
LQ05501010	5	50	1.0	10	10	1.55	3.199	290.0	2968.39	1912.96	351.36	0.3801	758.4	>2000
LQ05500415	5	50	0.4	15	10	1.55	4.062	262.0	4323.95	2797.22	321.38	0.4800	823.1	>2000
LQ05500615	5	50	0.6	15	10	2.15	4.651	283.0	6120.91	2851.39	476.21	0.5326	902.0	>2000
LQ05500815	5	50	0.8	15	10	2.66	5.046	304.0	7741.78	2905.56	627.60	0.5645	960.5	>2000
LQ05501015	5	50	1.0	15	10	3.12	5.326	325.0	9220.67	2959.72	775.85	0.5854	1010.1	>2000
LQ05500420	5	50	0.4	20	10	2.61	5.976	276.0	9855.07	3777.78	566.67	0.6752	1012.6	>2000
LQ05500620	5	50	0.6	20	10	3.55	6.729	304.0	13763.16	3874.07	836.80	0.7344	1108.4	>2000
LQ05500820	5	50	0.8	20	10	4.34	7.202	332.0	17220.88	3970.37	1099.45	0.7644	1179.0	>2000
LQ05501020	5	50	1.0	20	10	5.00	7.515	360.0	20333.33	4066.67	1355.56	0.7800	1235.2	>2000
LQ05500425	5	50	0.4	25	10	3.88	7.998	290.0	18552.44	4782.41	878.40	0.8714	1187.4	>2000
LQ05500625	5	50	0.6	25	10	5.19	8.877	325.0	25612.98	4932.87	1293.08	0.9294	1298.8	>2000
LQ05500825	5	50	0.8	25	10	6.25	9.394	360.0	31770.83	5083.33	1694.44	0.9514	1380.4	>2000
LQ05501025	5	50	1.0	25	10	7.12	9.713	395.0	37265.95	5233.80	2084.26	0.9570	1445.5	>2000
LQ05500430	5	50	0.4	30	10	5.33	10.093	304.0	30967.11	5811.11	1255.20	0.7848	29.2	>2000
LQ05500630	5	50	0.6	30	10	7.02	11.061	346.0	42333.82	6027.78	1842.45	1.1114	1477.1	>2000
LQ05500830	5	50	0.8	30	10	8.35	11.593	388.0	52144.33	6244.44	2408.57	1.1204	1569.4	>2000
LQ05501030	5	50	1.0	30	10	9.42	11.896	430.0	60854.65	6461.11	2956.78	1.1129	1642.7	>2000
LQ05650405	5	65	0.4	5	13	0.15	0.767	300.0	176.58	1177.22	36.41	0.0774	464.7	>2000
LQ05650605	5	65	0.6	5	13	0.22	0.921	307.0	260.16	1183.24	54.43	0.0865	507.2	>2000
LQ05650805	5	65	0.8	5	13	0.29	1.042	314.0	340.87	1189.26	72.32	0.0939	539.3	>2000
LQ05651005	5	65	1.0	5	13	0.35	1.142	321.0	418.91	1195.28	90.10	0.1004	566.7	>2000
LQ05650410	5	65	0.4	10	13	0.57	2.084	314.0	1363.48	2378.52	144.64	0.1612	755.2	>2000
LQ05650610	5	65	0.6	10	13	0.82	2.456	328.0	1977.74	2402.59	215.51	0.1836	825.5	>2000
LQ05650810	5	65	0.8	10	13	1.05	2.733	342.0	2554.39	2426.67	285.49	0.2000	881.2	>2000
LQ05651010	5	65	1.0	10	13	1.26	2.950	356.0	3097.85	2450.74	354.61	0.2128	929.3	>2000
LQ05650415	5	65	0.4	15	13	1.23	3.683	328.0	4449.92	3603.89	323.27	0.2634	1010.8	>2000
LQ05650615	5	65	0.6	15	13	1.74	4.271	349.0	6367.53	3658.06	480.23	0.2968	1106.6	>2000
LQ05650815	5	65	0.8	15	13	2.19	4.687	370.0	8126.76	3712.22	634.37	0.3190	1180.0	>2000
LQ05651015	5	65	1.0	15	13	2.59	4.994	391.0	9753.12	3766.39	785.88	0.3349	1238.9	>2000
LQ05650420	5	65	0.4	20	13	2.11	5.466	342.0	10217.54	4853.33	570.98	0.3749	1242.4	>2000
LQ05650620	5	65	0.6	20	13	2.92	6.249	370.0	14447.57	4949.63	845.82	0.4161	1361.7	>2000
LQ05650820	5	65	0.8	20	13	3.62	6.773	398.0	18256.62	5045.93	1114.44	0.4476	1450.2	>2000
LQ05651020	5	65	1.0	20	13	4.23	7.142	426.0	21727.70	5142.22	1377.38	0.4561	1521.0	>2000

161

L-PLATE   n angle stiffeners
edges pinned
free in-plane

CODE	n	S	t_2	d	a	$\bar{y}$	rad. gyr.	A	I_{NA}	Z_{plate}	$Z_{stiff.}$	σ_{cr}/E	L_1	L_{euler}
To scale multiply by	$-$	t_1	t_1	t_1	t_1	t_1	t_1	t_1^2	t_1^4	t_1^3	t_1^3	10^{-3}	t_1	t_1
LQ05650425	5	65	0.4	25	13	3.16	7.375	356.0	19361.54	6125.85	886.52	0.4907	1459.6	>2000
LQ05650625	5	65	0.6	25	13	4.32	8.324	391.0	27091.99	6277.31	1309.80	0.5361	1598.1	>2000
LQ05650825	5	65	0.8	25	13	5.28	8.927	426.0	33949.53	6427.78	1721.73	0.5596	1700.3	>2000
LQ05651025	5	65	1.0	25	13	6.10	9.330	461.0	40132.98	6578.24	2123.54	0.5719	1781.8	>2000
LQ05650430	5	65	0.4	30	13	4.38	9.373	370.0	32507.03	7424.44	1268.73	0.6076	1665.3	>2000
LQ05650630	5	65	0.6	30	13	5.90	10.459	412.0	45067.72	7641.11	1869.88	0.6534	1820.1	>2000
LQ05650830	5	65	0.8	30	13	7.14	11.114	454.0	56077.53	7857.78	2452.72	1.2010	50.5	1051
LQ05651030	5	65	1.0	30	13	8.17	11.529	496.0	65930.44	8074.44	3019.53	1.3830	46.7	1015
LQ05800405	5	80	0.4	5	16	0.12	0.697	366.0	177.80	1446.11	36.46	0.0484	548.2	>2000
LQ05800605	5	80	0.6	5	16	0.18	0.839	373.0	252.78	1452.13	54.53	0.0540	599.3	>2000
LQ05800805	5	80	0.8	5	16	0.24	0.953	380.0	345.35	1458.15	72.50	0.0586	634.6	>2000
LQ05801005	5	80	1.0	5	16	0.29	1.049	387.0	425.63	1464.17	90.38	0.0627	664.5	>2000
LQ05800410	5	80	0.4	10	16	0.47	1.907	380.0	1381.40	2916.30	145.01	0.0994	881.4	>2000
LQ05800610	5	80	0.6	10	16	0.69	2.261	394.0	2014.97	2940.37	216.32	0.1138	968.6	>2000
LQ05800810	5	80	0.8	10	16	0.88	2.532	408.0	2615.69	2964.44	286.88	0.1246	1034.9	>2000
LQ05801010	5	80	1.0	10	16	1.07	2.748	422.0	3186.81	2988.52	356.72	0.1334	1091.0	>2000
LQ05800415	5	80	0.4	15	16	1.03	3.392	394.0	4533.69	4410.56	324.48	0.1627	1183.0	>2000
LQ05800615	5	80	0.6	15	16	1.46	3.968	415.0	6535.71	4464.72	482.83	0.1852	1299.3	>2000
LQ05800815	5	80	0.8	15	16	1.86	4.388	436.0	8395.18	4518.89	638.80	0.2010	1387.0	>2000
LQ05801015	5	80	1.0	15	16	2.22	4.709	457.0	10131.77	4573.06	792.51	0.2127	1457.7	>2000
LQ05800420	5	80	0.4	20	16	1.76	5.064	408.0	10462.75	5928.89	573.76	0.2331	1458.9	>2000
LQ05800620	5	80	0.6	20	16	2.48	5.851	436.0	14924.77	6025.19	851.73	0.2622	1600.7	>2000
LQ05800820	5	80	0.8	20	16	3.10	6.399	464.0	18997.70	6121.48	1124.35	0.2809	1706.6	>2000
LQ05801020	5	80	1.0	20	16	3.66	6.800	492.0	22747.97	6217.78	1392.04	0.2937	1791.3	>2000
LQ05800425	5	80	0.4	25	16	2.67	6.870	422.0	19917.56	7471.30	891.80	0.3075	1715.7	>2000
LQ05800625	5	80	0.6	25	16	3.69	7.848	457.0	28143.80	7621.76	1320.84	0.3414	1880.5	>2000
LQ05800825	5	80	0.8	25	16	4.57	8.500	492.0	35543.70	7772.22	1740.05	0.8411	59.4	929
LQ05801025	5	80	1.0	25	16	5.34	8.957	527.0	42281.88	7922.69	2150.31	0.9788	54.0	910
LQ05800430	5	80	0.4	30	16	3.72	8.776	436.0	33580.73	9037.78	1277.59	0.6811	68.4	1064
LQ05800630	5	80	0.6	30	16	5.08	9.921	478.0	47046.65	9254.44	1888.19	0.7315	65.3	1186
LQ05800830	5	80	0.8	30	16	6.23	10.653	520.0	59012.31	9471.11	2492.72	0.8197	60.6	1210
LQ05801030	5	80	1.0	30	16	7.21	11.146	562.0	69814.06	9687.78	3062.88	0.9487	55.5	1176

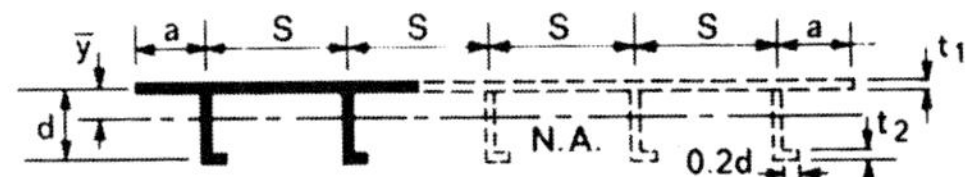

CODE	n	S	t_2	d	a	$\bar{y}$	rad. gyr.	A	I_{NA}	Z_{plate}	$Z_{stiff.}$	σ_{cr}/E	L_1	L_{euler}
To scale multiply by	–	t_1	t_1	t_1	t_1	t_1	t_1	t_1^2	t_1^4	t_1^3	t_1^3	10^{-3}	t_1	t_1
JF02350405	2	35	0.4	5	7	0.26	0.961	53.8	49.69	190.95	10.48	FULER	–	–
JF02350605	2	35	0.6	5	7	0.37	1.133	56.2	72.15	193.10	15.60	EULER	–	–
JF02350805	2	35	0.8	5	7	0.48	1.262	58.6	93.29	195.24	20.63	FULER	–	–
JF02351005	2	35	1.0	5	7	0.57	1.363	61.0	113.25	197.38	25.59	EULER	–	–
JF02350410	2	35	0.4	10	7	0.96	2.523	58.6	373.15	390.48	41.26	3.4101	48.7	126
JF02350610	2	35	0.6	10	7	1.32	2.688	63.4	528.71	399.05	60.95	3.6470	45.8	143
JF02350810	2	35	0.8	10	7	1.64	3.133	68.2	669.40	407.62	80.09	3.9009	46.6	151
JF02351010	2	35	1.0	10	7	1.92	3.307	73.0	798.17	416.19	98.76	4.2998	49.4	151
JF02350415	2	35	0.4	15	7	1.99	4.332	63.4	1189.59	598.57	91.42	2.6483	88.0	257
JF02350615	2	35	0.6	15	7	2.68	4.840	70.6	1654.04	617.86	134.22	3.2364	78.3	258
JF02350815	2	35	0.8	15	7	3.24	5.150	77.8	2063.75	637.14	175.48	3.4943	76.4	258
JF02351015	2	35	1.0	15	7	3.71	5.350	85.0	2432.65	656.43	215.39	3.7460	77.7	251
JF02350420	2	35	0.4	20	7	3.28	6.266	68.2	2677.61	815.24	160.19	1.9939	135.1	413
JF02350620	2	35	0.6	20	7	4.32	6.867	77.8	3668.89	849.52	233.97	2.6084	117.8	357
JF02350820	2	35	0.8	20	7	5.13	7.200	87.4	4530.28	883.81	304.57	2.8813	112.8	333
JF02351020	2	35	1.0	20	7	5.77	7.392	97.0	5300.34	918.10	372.56	3.0588	113.0	317
JF02350425	2	35	0.4	25	7	4.79	8.267	73.0	4988.58	1040.48	246.89	1.5771	189.4	521
JF02350625	2	35	0.6	25	7	6.18	8.916	85.0	6757.35	1094.05	358.98	2.1201	162.6	422
JF02350825	2	35	0.8	25	7	7.22	9.240	97.0	8281.79	1147.62	465.70	2.3940	153.4	383
JF02351025	2	35	1.0	25	7	8.03	9.406	109.0	9642.58	1201.19	568.13	2.5300	151.9	362
JF02350430	2	35	0.4	30	7	6.48	10.301	77.8	8255.01	1274.29	350.95	1.2934	251.7	588
JF02350630	2	35	0.6	30	7	8.20	10.963	92.2	11081.13	1351.43	508.30	1.7718	214.0	461
JF02350830	2	35	0.8	30	7	9.46	11.257	106.6	13508.44	1428.57	657.53	2.0334	199.0	411
JF02351030	2	35	1.0	30	7	10.41	11.383	121.0	15679.34	1505.71	800.51	2.1617	194.7	387
JF02500405	2	50	0.4	5	10	0.19	0.823	74.8	50.71	270.95	10.54	FULER	–	–
JF02500605	2	50	0.6	5	10	0.27	0.981	77.2	74.29	273.10	15.71	EULER	–	–
JF02500805	2	50	0.8	5	10	0.35	1.103	79.6	96.82	275.24	20.83	FULER	–	–
JF02501005	2	50	1.0	5	10	0.43	1.202	82.0	118.39	277.38	25.89	EULER	–	–
JF02500410	2	50	0.4	10	10	0.70	2.206	79.6	387.27	550.48	41.66	1.8237	54.2	149
JF02500610	2	50	0.6	10	10	1.00	2.568	84.4	556.40	559.05	61.79	1.9025	54.7	175
JF02500810	2	50	0.8	10	10	1.26	2.827	89.2	712.71	567.62	81.50	2.0360	55.7	187
JF02501010	2	50	1.0	10	10	1.49	3.021	94.0	858.16	576.19	100.83	2.2496	56.8	190
JF02500415	2	50	0.4	15	10	1.49	3.851	84.4	1251.90	838.57	92.68	1.9241	48.7	266
JF02500615	2	50	0.6	15	10	2.06	4.396	91.6	1770.03	857.86	136.82	2.0637	71.7	297
JF02500815	2	50	0.8	15	10	2.55	4.759	98.8	2237.25	877.14	179.71	2.1906	77.3	313
JF02501015	2	50	1.0	15	10	2.97	5.013	106.0	2663.92	896.43	221.47	2.3855	81.5	316
JF02500420	2	50	0.4	20	10	2.51	5.653	89.2	2850.82	1135.24	163.01	1.8582	134.5	403
JF02500620	2	50	0.6	20	10	3.40	6.345	98.8	3977.33	1169.52	239.61	2.1064	117.9	425
JF02500820	2	50	0.8	20	10	4.13	6.775	108.4	4975.15	1203.81	313.55	2.1980	116.6	438
JF02501020	2	50	1.0	20	10	4.75	7.056	118.0	5875.71	1238.10	385.19	2.3139	119.2	434
JF02500425	2	50	0.4	25	10	3.72	7.554	94.0	5363.48	1440.48	252.08	1.5428	188.6	586
JF02500625	2	50	0.6	25	10	4.95	8.355	106.0	7399.76	1494.05	369.12	1.8993	165.1	551
JF02500825	2	50	0.8	25	10	5.93	8.821	118.0	9180.79	1547.62	481.48	2.0206	159.3	535
JF02501025	2	50	1.0	25	10	6.73	9.105	130.0	10777.24	1601.19	589.91	2.0988	160.5	520
JF02500430	2	50	0.4	30	10	5.10	9.517	98.8	8948.99	1754.29	359.41	1.2928	247.9	748
JF02500630	2	50	0.6	30	10	6.68	10.395	113.2	12231.10	1831.43	524.45	1.6614	215.0	649
JF02500830	2	50	0.8	30	10	7.90	10.870	127.6	15077.12	1908.57	682.21	1.8054	205.0	610
JF02501030	2	50	1.0	30	10	8.87	11.139	142.0	17619.72	1985.71	834.00	1.8714	204.6	587
JF02650405	2	65	0.4	5	13	0.15	0.732	95.8	51.29	350.95	10.57	EULER	–	–
JF02650605	2	65	0.6	5	13	0.21	0.877	98.2	75.51	353.10	15.78	EULER	–	–
JF02650805	2	65	0.8	5	13	0.28	0.991	100.6	98.87	355.24	20.94	EULER	–	–
JF02651005	2	65	1.0	5	13	0.34	1.086	103.0	121.44	357.38	26.06	EULER	–	–
JF02650410	2	65	0.4	10	13	0.56	1.983	100.6	395.49	710.48	41.88	1.0869	68.3	171
JF02650610	2	65	0.6	10	13	0.80	2.332	105.4	573.06	719.05	62.27	1.1313	68.1	204
JF02650810	2	65	0.8	10	13	1.02	2.590	110.2	739.50	727.62	82.32	1.2043	68.6	222
JF02651010	2	65	1.0	10	13	1.22	2.792	115.0	896.23	736.19	102.05	1.3201	69.3	229
JF02650415	2	65	0.4	15	13	1.20	3.498	105.4	1289.37	1078.57	93.40	1.1454	65.3	315
JF02650615	2	65	0.6	15	13	1.68	4.045	112.6	1842.76	1097.86	138.33	1.2239	74.1	355
JF02650815	2	65	0.8	15	13	2.10	4.429	119.8	2349.92	1117.14	182.21	1.3253	81.4	375
JF02651015	2	65	1.0	15	13	2.48	4.711	127.0	2818.70	1136.43	225.14	1.4686	86.5	379
JF02650420	2	65	0.4	20	13	2.03	5.181	110.2	2958.02	1455.24	164.63	1.1540	60.3	472
JF02650620	2	65	0.6	20	13	2.80	5.905	119.8	4177.63	1489.52	242.95	1.2791	56.1	513
JF02650820	2	65	0.8	20	13	3.46	6.385	129.4	5275.63	1523.81	319.00	1.4841	113.3	514
JF02651020	2	65	1.0	20	13	4.03	6.720	139.0	6277.22	1558.10	393.03	1.6017	121.8	520

J-PLATE — n angle stiffeners, edges free out-of-plane, free in-plane

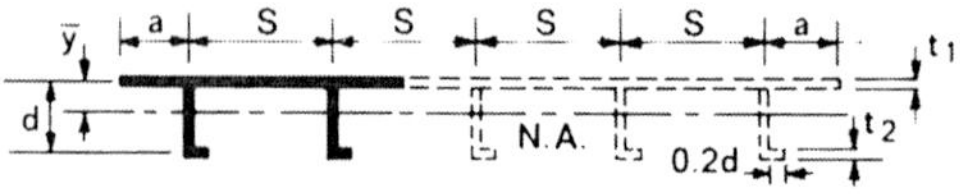

CODE	n	S	t_2	d	a	$\bar{y}$	rad. gyr.	A	I_{NA}	Z_{plate}	$Z_{stiff.}$	σ_{cr}/E	L_1	L_{euler}
To scale multiply by	–	t_1	t_1	t_1	t_1	t_1	t_1	t_1^2	t_1^4	t_1^3	t_1^3	10^{-3}	t_1	t_1
JF02650425	2	65	0.4	25	13	3.04	6.979	115.0	5601.45	1840.48	255.12	1.1382	59.9	643
JF02650625	2	65	0.6	25	13	4.13	7.852	127.0	7829.72	1894.05	375.24	1.2541	55.2	689
JF02650825	2	65	0.8	25	13	5.04	8.400	139.0	9808.15	1947.62	491.29	1.4496	49.6	678
JF02651025	2	65	1.0	25	13	5.79	8.763	151.0	11596.30	2001.19	603.81	1.6234	165.9	652
JF02650430	2	65	0.4	30	13	4.21	8.858	119.8	9399.67	2234.29	364.43	1.1106	59.7	822
JF02650630	2	65	0.6	30	13	5.63	9.850	134.2	13021.16	2311.43	534.39	1.2174	55.8	850
JF02650830	2	65	0.8	30	13	6.78	10.442	148.6	16202.42	2388.57	697.88	1.3998	50.5	799
JF02651030	2	65	1.0	30	13	7.73	10.814	163.0	19060.12	2465.71	855.87	1.5544	213.0	752
JF02800405	2	80	0.4	5	16	0.12	0.665	116.8	51.66	430.95	10.58	EULER	–	–
JF02800605	2	80	0.6	5	16	0.18	0.800	119.2	76.30	433.10	15.82	EULER	–	–
JF02800805	2	80	0.8	5	16	0.23	0.908	121.6	100.22	435.24	21.01	EULER	–	–
JF02801005	2	80	1.0	5	16	0.28	0.998	124.0	123.45	437.38	26.17	EULER	–	–
JF02800410	2	80	0.4	10	16	0.46	1.816	121.6	400.88	870.48	42.02	0.7168	83.3	190
JF02800610	2	80	0.6	10	16	0.66	2.150	126.4	584.18	879.05	62.58	0.7442	82.5	230
JF02800810	2	80	0.8	10	16	0.85	2.403	131.2	757.72	887.62	82.84	0.7873	82.6	253
JF02801010	2	80	1.0	10	16	1.03	2.605	136.0	922.55	896.19	102.84	0.8549	83.0	264
JF02800415	2	80	0.4	15	16	1.00	3.225	126.4	1314.40	1318.57	93.86	0.7533	81.9	357
JF02800615	2	80	0.6	15	16	1.41	3.764	133.6	1892.63	1337.86	139.31	0.7969	87.6	410
JF02800815	2	80	0.8	15	16	1.79	4.153	140.8	2428.98	1357.14	183.87	0.8606	91.7	436
JF02801015	2	80	1.0	15	16	2.13	4.449	148.0	2929.56	1376.43	227.60	0.9539	95.4	444
JF02800420	2	80	0.4	20	16	1.71	4.806	131.2	3030.89	1775.24	165.69	0.7701	74.4	536
JF02800620	2	80	0.6	20	16	2.39	5.538	140.8	4318.18	1809.52	245.16	0.8571	74.2	588
JF02800820	2	80	0.8	20	16	2.98	6.043	150.4	5492.20	1843.81	322.67	0.9789	110.4	600
JF02801020	2	80	1.0	20	16	3.50	6.410	160.0	6573.33	1878.10	398.38	1.0841	124.4	605
JF02800425	2	80	0.4	25	16	2.57	6.511	136.0	5765.93	2240.48	257.10	0.7651	73.3	734
JF02800625	2	80	0.6	25	16	3.55	7.415	148.0	8137.67	2294.05	379.33	0.8513	67.2	794
JF02800825	2	80	0.8	25	16	4.38	8.012	160.0	10270.83	2347.62	497.98	0.9916	59.9	794
JF02801025	2	80	1.0	25	16	5.09	8.427	172.0	12215.36	2401.19	613.44	1.1902	168.0	759
JF02800430	2	80	0.4	30	16	3.58	8.307	140.8	9715.91	2714.29	367.74	0.7552	73.4	945
JF02800630	2	80	0.6	30	16	4.87	9.360	155.2	13597.42	2791.43	541.11	0.8341	67.4	1012
JF02800830	2	80	0.8	30	16	5.94	10.026	169.6	17049.06	2868.57	708.71	0.9666	60.2	999
JF02801030	2	80	1.0	30	16	6.85	10.470	184.0	20171.74	2945.71	871.27	1.1522	53.6	934
JF03350405	3	35	0.4	5	7	0.23	0.908	91.2	75.16	326.43	15.76	EULER	–	–
JF03350605	3	35	0.6	5	7	0.33	1.075	94.8	109.53	329.64	23.47	EULER	–	–
JF03350805	3	35	0.8	5	7	0.43	1.202	98.4	142.07	332.86	31.07	EULER	–	–
JF03351005	3	35	1.0	5	7	0.51	1.302	102.0	172.98	336.07	38.57	EULER	–	–
JF03350410	3	35	0.4	10	7	0.85	2.403	98.4	568.29	665.71	62.13	3.3424	43.0	117
JF03350610	3	35	0.6	10	7	1.19	2.769	105.6	809.66	678.57	91.94	3.4957	41.1	136
JF03350810	3	35	0.8	10	7	1.49	3.021	112.8	1029.79	691.43	121.00	3.7075	41.9	146
JF03351010	3	35	1.0	10	7	1.75	3.205	120.0	1232.50	704.29	149.39	4.0549	43.1	149
JF03350415	3	35	0.4	15	7	1.79	4.153	105.6	1821.73	1017.86	137.90	2.7006	88.4	241
JF03350615	3	35	0.6	15	7	2.44	4.680	116.4	2549.52	1046.79	202.92	3.3516	75.7	244
JF03350815	3	35	0.8	15	7	2.97	5.013	127.2	3196.70	1075.71	265.76	3.6116	72.1	253
JF03351015	3	35	1.0	15	7	3.42	5.235	138.0	3782.20	1104.54	326.73	3.8593	72.6	256
JF03350420	3	35	0.4	20	7	2.98	6.043	112.8	4119.15	1382.36	242.00	2.0143	134.8	414
JF03350620	3	35	0.6	20	7	3.96	6.684	127.2	5683.02	1434.29	354.35	2.7048	116.5	390
JF03350820	3	35	0.8	20	7	4.75	7.056	141.6	7050.85	1485.71	462.22	3.0467	109.1	378
JF03351020	3	35	1.0	20	7	5.38	7.284	156.0	8276.92	1537.14	566.32	3.2494	108.1	369
JF03350425	3	35	0.4	25	7	4.38	8.012	120.0	7703.13	1760.71	373.48	1.5935	188.2	606
JF03350625	3	35	0.6	25	7	5.71	8.725	138.0	10506.11	1841.07	544.54	2.1898	161.0	527
JF03350825	3	35	0.8	25	7	6.73	9.105	156.0	12932.69	1921.43	707.89	2.5319	149.8	488
JF03351025	3	35	1.0	25	7	7.54	9.316	174.0	15099.68	2001.79	864.97	2.7159	146.7	466
JF03350430	3	35	0.4	30	7	5.94	10.026	127.2	12786.79	2151.43	531.53	1.3142	247.7	774
JF03350630	3	35	0.6	30	7	7.62	10.776	148.8	17277.82	2267.14	772.05	1.8243	211.0	644
JF03350830	3	35	0.8	30	7	8.87	11.139	170.4	21143.66	2382.86	1000.60	2.1434	194.7	580
JF03351030	3	35	1.0	30	7	9.84	11.318	192.0	24595.31	2498.57	1220.23	2.3153	188.9	545
JF03500405	3	50	0.4	5	10	0.17	0.776	127.2	76.53	463.57	15.83	EULER	–	–
JF03500605	3	50	0.6	5	10	0.24	0.927	130.8	112.41	466.79	23.62	EULER	–	–
JF03500805	3	50	0.8	5	10	0.31	1.045	134.4	146.88	470.00	31.33	EULER	–	–
JF03501005	3	50	1.0	5	10	0.38	1.142	138.0	180.03	473.21	38.97	EULER	–	–
JF03500410	3	50	0.4	10	10	0.63	2.091	134.4	587.50	940.00	62.67	1.6883	50.5	145
JF03500610	3	50	0.6	10	10	0.89	2.447	141.6	847.88	952.86	93.07	1.7498	51.1	171
JF03500810	3	50	0.8	10	10	1.13	2.707	148.8	1090.32	965.71	122.91	1.8594	51.8	185
JF03501010	3	50	1.0	10	10	1.35	2.906	156.0	1317.31	978.57	152.22	2.0351	52.6	191

J-PLATE — n angle stiffeners, edges free out-of-plane, free in-plane

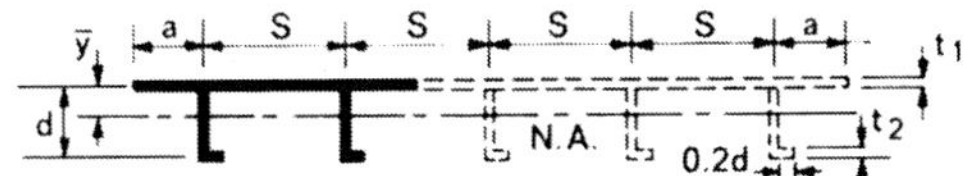

CODE	n	S	t_2	d	à	$\bar{y}$	rad. gyr.	A	I_{NA}	Z_{plate}	$Z_{stiff.}$	σ_{cr}/E	L_1	L_{euler}
To scale multiply by	–	t_1	t_1	t_1	t_1	t_1	t_1	t_1^2	t_1^4	t_1^3	t_1^3	10^{-3}	t_1	t_1
JF03500415	3	50	0.4	15	10	1.33	3.671	141.6	1907.73	1429.29	139.60	1.7449	46.8	260
JF03500615	3	50	0.6	15	10	1.86	4.219	152.4	2712.62	1458.21	206.44	1.8816	46.7	252
JF03500815	3	50	0.8	15	10	2.32	4.594	163.2	3444.49	1487.14	271.57	2.0791	60.2	304
JF03501015	3	50	1.0	15	10	2.72	4.864	174.0	4116.92	1516.07	335.13	2.2768	71.3	309
JF03500420	3	50	0.4	20	10	2.26	5.414	148.8	4361.29	1931.43	245.82	1.7275	46.0	395
JF03500620	3	50	0.6	20	10	3.09	6.125	163.2	6123.53	1982.86	362.09	1.8699	43.6	434
JF03500820	3	50	0.8	20	10	3.78	6.583	177.6	7697.30	2034.29	474.67	2.1150	40.3	439
JF03501020	3	50	1.0	20	10	4.38	6.894	192.0	9125.00	2085.71	584.00	2.3718	110.3	434
JF03500425	3	50	0.4	25	10	3.37	7.265	156.0	8233.17	2446.43	380.56	1.5688	188.2	564
JF03500625	3	50	0.6	25	10	4.53	8.107	174.0	11435.88	2526.79	558.55	1.8147	43.9	587
JF03500825	3	50	0.8	25	10	5.47	8.617	192.0	14257.81	2607.14	730.00	2.0451	40.7	583
JF03501025	3	50	1.0	25	10	6.25	8.943	210.0	16796.88	2687.50	895.83	2.2251	152.3	571
JF03500430	3	50	0.4	30	10	4.63	9.188	163.2	13777.94	2974.29	543.13	1.3072	247.4	784
JF03500630	3	50	0.6	30	10	6.14	10.129	184.8	18961.36	3090.00	794.57	1.7342	212.4	732
JF03500830	3	50	0.8	30	10	7.33	10.667	206.4	23483.72	3205.71	1035.69	1.9231	199.1	709
JF03501030	3	50	1.0	30	10	8.29	10.989	228.0	27532.89	3321.43	1268.18	2.0082	196.2	696
JF03650405	3	65	0.4	5	13	0.13	0.688	163.2	77.30	600.71	15.87	EULER	–	–
JF03650605	3	65	0.6	5	13	0.19	0.827	166.8	114.05	603.93	23.71	EULER	–	–
JF03650805	3	65	0.8	5	13	0.25	0.937	170.4	149.65	607.14	31.48	EULER	–	–
JF03651005	3	65	1.0	5	13	0.30	1.029	174.0	184.16	610.36	39.20	EULER	–	–
JF03650410	3	65	0.4	10	13	0.49	1.874	170.4	598.59	1214.29	62.96	1.0005	64.2	170
JF03650610	3	65	0.6	10	13	0.71	2.214	177.6	870.61	1227.14	93.71	1.0333	64.4	201
JF03650810	3	65	0.8	10	13	0.91	2.470	184.8	1127.27	1240.00	124.00	1.0905	64.7	220
JF03651010	3	65	1.0	10	13	1.09	2.672	192.0	1370.31	1252.86	153.86	1.1819	64.9	230
JF03650415	3	65	0.4	15	13	1.06	3.321	177.6	1958.87	1840.71	140.56	1.0382	61.9	306
JF03650615	3	65	0.6	15	13	1.50	3.864	188.4	2813.40	1859.64	208.47	1.1084	64.8	349
JF03650815	3	65	0.8	15	13	1.90	4.253	199.2	3602.71	1898.57	274.97	1.2056	70.9	370
JF03651015	3	65	1.0	15	13	2.25	4.544	210.0	4336.88	1927.50	340.15	1.3355	75.8	377
JF03650420	3	65	0.4	20	13	1.82	4.940	184.8	4509.09	2480.00	248.00	1.0405	59.6	465
JF03650620	3	65	0.6	20	13	2.53	5.670	199.2	6404.82	2531.43	366.62	1.1357	56.3	515
JF03650820	3	65	0.8	20	13	3.15	6.168	213.6	8125.84	2582.86	482.13	1.2924	52.2	526
JF03651020	3	65	1.0	20	13	3.68	6.524	228.0	9705.26	2634.29	594.84	1.5115	48.0	514
JF03650425	3	65	0.4	25	13	2.73	6.679	192.0	8564.45	3132.14	384.65	1.0288	59.5	641
JF03650625	3	65	0.6	25	13	3.75	7.574	210.0	12046.88	3212.50	566.91	1.1175	55.8	702
JF03650825	3	65	0.8	25	13	4.61	8.155	228.0	15164.47	3292.86	743.55	1.2695	51.1	710
JF03651025	3	65	1.0	25	13	5.34	8.553	246.0	17997.33	3373.21	915.21	1.4858	46.4	687
JF03650430	3	65	0.4	30	13	3.80	8.506	199.2	14410.84	3797.14	549.93	1.0103	59.5	830
JF03650630	3	65	0.6	30	13	5.14	9.540	220.8	20095.92	3912.86	808.23	1.0919	56.3	898
JF03650830	3	65	0.8	30	13	6.24	10.182	242.4	25128.71	4028.57	1057.50	1.2344	51.9	899
JF03651030	3	65	1.0	30	13	7.16	10.601	264.0	29669.32	4144.29	1298.96	1.4423	47.3	858
JF03800405	3	80	0.4	5	16	0.11	0.625	199.2	77.79	737.86	15.89	EULER	–	–
JF03800605	3	80	0.6	5	16	0.16	0.753	202.8	115.11	741.07	23.76	EULER	–	–
JF03800805	3	80	0.8	5	16	0.20	0.857	206.4	151.45	744.29	31.58	EULER	–	–
JF03801005	3	80	1.0	5	16	0.25	0.943	210.0	186.88	747.50	39.34	EULER	–	–
JF03800410	3	80	0.4	10	16	0.41	1.713	206.4	605.81	1488.57	63.15	0.6595	78.8	191
JF03800610	3	80	0.6	10	16	0.59	2.036	213.6	885.67	1501.43	94.12	0.6787	78.3	228
JF03800810	3	80	0.8	10	16	0.76	2.284	220.8	1152.17	1514.29	124.71	0.7117	78.7	252
JF03801010	3	80	1.0	10	16	0.92	2.484	228.0	1406.58	1527.14	154.93	0.7641	78.7	265
JF03800415	3	80	0.4	15	16	0.88	3.054	213.6	1992.77	2252.14	141.18	0.6937	77.5	346
JF03800615	3	80	0.6	15	16	1.26	3.584	224.4	2881.83	2281.07	209.79	0.7208	80.8	401
JF03800815	3	80	0.8	15	16	1.61	3.973	235.2	3712.50	2310.00	277.20	0.7750	84.0	431
JF03801015	3	80	1.0	15	16	1.92	4.273	246.0	4492.45	2338.93	343.48	0.8530	86.6	443
JF03800420	3	80	0.4	20	16	1.52	4.569	220.8	4608.70	3028.57	249.41	0.6928	73.3	527
JF03800620	3	80	0.6	20	16	2.14	5.297	235.2	6600.00	3080.00	369.60	0.7585	70.5	588
JF03800820	3	80	0.8	20	16	2.69	5.812	249.6	8430.77	3131.43	487.11	0.8620	68.4	606
JF03801020	3	80	1.0	20	16	3.18	6.194	264.0	10127.27	3182.86	602.16	0.9942	100.6	601
JF03800425	3	80	0.4	25	16	2.30	6.209	228.0	8791.12	3817.86	387.32	0.6888	74.1	729
JF03800625	3	80	0.6	25	16	3.20	7.122	246.0	12479.04	3898.21	572.47	0.7543	67.8	803
JF03800825	3	80	0.8	25	16	3.98	7.742	264.0	15823.86	3978.57	752.70	0.8629	62.0	817
JF03801025	3	80	1.0	25	16	4.65	8.185	282.0	18891.29	4058.93	928.51	1.0137	56.1	796
JF03800430	3	80	0.4	30	16	3.21	7.946	235.2	14850.00	4620.00	554.40	0.6818	74.3	945
JF03800630	3	80	0.6	30	16	4.42	9.024	256.8	20912.38	4735.71	817.40	0.7419	68.1	1033
JF03800830	3	80	0.8	30	16	5.43	9.728	278.4	26349.28	4851.43	1072.42	0.8456	63.7	1043
JF03801030	3	80	1.0	30	16	6.30	10.213	300.0	31293.00	4967.14	1320.38	0.9908	56.2	1010

n angle stiffeners
edges free out-of-plane
free in-plane

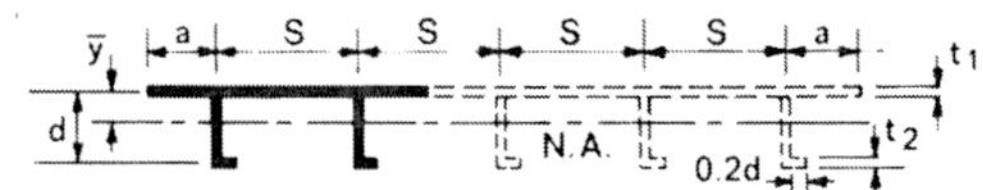

CODE	n	S	t_2	d	a	$\bar{y}$	rad. gyr.	A	I_{NA}	Z_{plate}	$Z_{stiff.}$	σ_{cr}/E	L_1	L_{euler}
To scale multiply by	–	t_1	t_1	t_1	t_1	t_1	t_1	t_1^2	t_1^4	t_1^3	t_1^3	10^{-3}	t_1	t_1
JF05350405	5	35	0.4	5	7	0.21	0.871	166.0	125.95	597.38	26.30	EULER	–	–
JF05350605	5	35	0.6	5	7	0.31	1.034	172.0	183.98	602.74	39.19	EULER	–	–
JF05350805	5	35	0.8	5	7	0.39	1.159	178.0	239.14	608.10	51.91	EULER	–	–
JF05351005	5	35	1.0	5	7	0.48	1.259	184.0	291.72	613.45	64.48	3.6585	34.0	61
JF05350410	5	35	0.4	10	7	0.79	2.318	178.0	956.55	1216.19	103.82	3.2562	37.8	117
JF05350610	5	35	0.6	10	7	1.11	2.683	190.0	1367.89	1237.62	153.79	3.3638	38.3	136
JF05350810	5	35	0.8	10	7	1.39	2.939	202.0	1745.21	1259.05	202.61	3.5671	39.5	146
JF05351010	5	35	1.0	10	7	1.64	3.128	214.0	2094.24	1280.48	250.37	3.8971	40.4	150
JF05350415	5	35	0.4	15	7	1.66	4.025	190.0	3077.76	1856.43	230.68	2.7056	87.6	230
JF05350615	5	35	0.6	15	7	2.27	4.561	208.0	4326.65	1904.64	339.92	3.5230	31.7	230
JF05350815	5	35	0.8	15	7	2.79	4.908	226.0	5443.81	1952.86	445.76	3.6698	69.4	245
JF05351015	5	35	1.0	15	7	3.23	5.145	244.0	6458.38	2001.07	548.60	3.9119	69.7	251
JF05350420	5	35	0.4	20	7	2.77	5.879	202.0	6980.86	2518.10	405.21	2.0166	134.6	401
JF05350620	5	35	0.6	20	7	3.72	6.544	226.0	9677.88	2603.31	594.35	2.7262	116.5	384
JF05350820	5	35	0.8	20	7	4.48	6.942	250.0	12049.07	2689.52	776.36	3.1087	108.4	383
JF05351020	5	35	1.0	20	7	5.11	7.194	274.0	14180.05	2775.24	952.29	3.3339	106.6	383
JF05350425	5	35	0.4	25	7	4.09	7.821	214.0	13088.98	3201.19	625.93	1.5985	187.8	606
JF05350625	5	35	0.6	25	7	5.38	8.575	244.0	17939.93	3335.12	914.33	2.1989	160.8	562
JF05350825	5	35	0.8	25	7	6.39	8.992	274.0	22156.33	3469.05	1190.36	2.5635	149.3	539
JF05351025	5	35	1.0	25	7	7.20	9.235	304.0	25926.02	3602.98	1456.17	2.7709	145.6	525
JF05350430	5	35	0.4	30	7	5.58	9.816	226.0	21775.22	3905.71	891.52	1.3215	246.8	832
JF05350630	5	35	0.6	30	7	7.21	10.623	262.0	29566.03	4098.57	1297.54	1.8335	210.3	743
JF05350830	5	35	0.8	30	7	8.46	11.035	298.0	36289.93	4291.43	1684.49	2.1637	194.0	692
JF05351030	5	35	1.0	30	7	9.43	11.253	334.0	42291.92	4464.29	2056.11	2.3528	187.7	664
JF05500405	5	50	0.4	5	10	0.15	0.743	232.0	128.05	848.81	26.41	EULER	–	–
JF05500605	5	50	0.6	5	10	0.22	0.890	238.0	188.42	854.17	39.42	EULER	–	–
JF05500805	5	50	0.8	5	10	0.29	1.005	244.0	246.58	859.52	52.32	EULER	–	–
JF05501005	5	50	1.0	5	10	0.35	1.100	250.0	302.71	864.88	65.10	EULER	–	–
JF05500410	5	50	0.4	10	10	0.57	2.011	244.0	986.34	1719.05	104.64	1.6130	49.2	147
JF05500610	5	50	0.6	10	10	0.82	2.362	256.0	1427.73	1740.48	155.53	1.6649	49.9	172
JF05500810	5	50	0.8	10	10	1.04	2.621	268.0	1840.80	1761.90	205.56	1.7608	50.4	187
JF05501010	5	50	1.0	10	10	1.25	2.822	280.0	2229.17	1783.33	254.76	1.9150	50.8	194
JF05500415	5	50	0.4	15	10	1.23	3.542	256.0	3212.40	2610.71	233.30	1.6523	46.5	259
JF05500615	5	50	0.6	15	10	1.72	4.091	274.0	4585.20	2658.93	345.39	1.7700	46.1	293
JF05500815	5	50	0.8	15	10	2.16	4.472	292.0	5840.75	2707.14	454.80	1.9576	48.0	306
JF05501015	5	50	1.0	15	10	2.54	4.752	310.0	6999.50	2755.36	561.77	2.1797	63.3	310
JF05500420	5	50	0.4	20	10	2.09	5.242	268.0	7363.18	3523.81	411.11	1.6362	46.0	390
JF05500620	5	50	0.6	20	10	2.88	5.963	292.0	10383.56	3609.52	606.40	1.7644	46.6	433
JF05500820	5	50	0.8	20	10	3.54	6.438	316.0	13097.05	3695.24	795.90	1.9724	41.1	444
JF05501020	5	50	1.0	20	10	4.12	6.767	340.0	15568.63	3780.95	980.25	2.2834	37.9	435
JF05500425	5	50	0.4	25	10	3.13	7.054	280.0	13932.29	4458.33	636.90	1.5705	188.2	542
JF05500625	5	50	0.6	25	10	4.23	7.920	310.0	19443.04	4592.26	936.29	1.7133	44.2	589
JF05500825	5	50	0.8	25	10	5.15	8.459	340.0	24325.98	4726.19	1225.31	1.9147	41.4	597
JF05501025	5	50	1.0	25	10	5.91	8.812	370.0	28733.81	4860.12	1505.35	2.2151	38.2	579
JF05500430	5	50	0.4	30	10	4.32	8.945	292.0	23363.01	5414.29	909.60	1.3095	247.1	762
JF05500630	5	50	0.6	30	10	5.76	9.925	328.0	32309.45	5607.14	1333.02	1.7456	212.4	733
JF05500830	5	50	0.8	30	10	6.92	10.503	364.0	40153.85	5800.00	1740.00	1.9649	198.1	732
JF05501030	5	50	1.0	30	10	7.88	10.862	400.0	47193.75	5992.86	2133.05	2.0660	193.8	735
JF05650405	5	65	0.4	5	13	0.12	0.659	298.0	129.22	1100.24	26.47	EULER	–	–
JF05650605	5	65	0.6	5	13	0.17	0.793	304.0	190.93	1105.60	39.55	EULER	–	–
JF05650805	5	65	0.8	5	13	0.23	0.900	310.0	250.86	1110.95	52.55	EULER	–	–
JF05651005	5	65	1.0	5	13	0.28	0.989	316.0	309.10	1116.31	65.45	1.0155	63.8	93
JF05650410	5	65	0.4	10	13	0.45	1.799	310.0	1003.44	2221.90	105.09	0.9542	63.0	173
JF05650610	5	65	0.6	10	13	0.65	2.132	322.0	1463.04	2243.33	156.51	0.9810	63.3	203
JF05650810	5	65	0.8	10	13	0.84	2.384	334.0	1858.60	2264.76	207.23	1.0301	63.6	223
JF05651010	5	65	1.0	10	13	1.01	2.585	346.0	2312.62	2286.19	257.29	1.1090	63.7	234
JF05650415	5	65	0.4	15	13	0.98	3.197	322.0	3291.85	3365.00	234.77	0.9824	61.3	305
JF05650615	5	65	0.6	15	13	1.39	3.735	340.0	4743.36	3413.21	348.51	1.0429	62.9	349
JF05650815	5	65	0.8	15	13	1.76	4.125	358.0	6091.34	3461.43	460.06	1.1327	67.0	372
JF05651015	5	65	1.0	15	13	2.09	4.421	376.0	7350.65	3509.64	569.57	1.2521	71.0	381
JF05650420	5	65	0.4	20	13	1.68	4.768	334.0	7594.41	4529.52	414.47	0.9828	59.6	459
JF05650620	5	65	0.6	20	13	2.35	5.500	358.0	10829.05	4615.24	613.42	1.0645	56.8	514
JF05650820	5	65	0.8	20	13	2.93	6.007	382.0	13782.90	4700.95	807.53	1.2007	53.4	531
JF05651020	5	65	1.0	20	13	3.45	6.376	406.0	16505.79	4786.67	997.22	1.3932	49.6	525

J-PLATE — n angle stiffeners, edges free out-of-plane, free in-plane

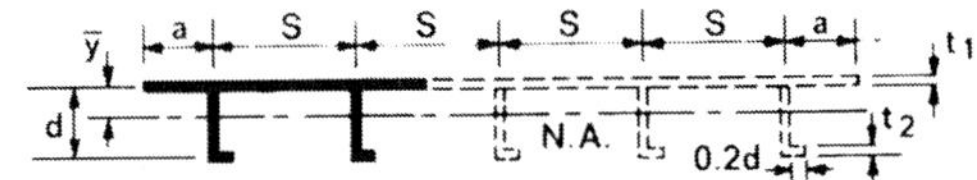

CODE	n	S	t_2	d	a	$\bar{y}$	rad. gyr.	A	I_{NA}	Z_{plate}	$Z_{stiff.}$	σ_{cr}/E	L_1	L_{euler}
To scale multiply by	−	t_1	t_1	t_1	t_1	t_1	t_1	t_1^2	t_1^4	t_1^3	t_1^3	10^{-3}	t_1	t_1
JF05650425	5	65	0.4	25	13	2.53	6.463	346.0	14453.88	5715.48	643.22	0.9729	59.6	632
JF05650625	5	65	0.6	25	13	3.49	7.369	376.0	20418.47	5849.40	949.29	1.0493	56.4	700
JF05650825	5	65	0.8	25	13	4.31	7.970	406.0	25790.23	5983.33	1246.53	1.1822	52.2	716
JF05651025	5	65	1.0	25	13	5.02	8.390	436.0	30691.54	6117.26	1535.90	1.3736	47.8	701
JF05650430	5	65	0.4	30	13	3.52	8.250	358.0	24365.36	6922.86	920.13	0.9581	59.7	819
JF05650630	5	65	0.6	30	13	4.80	9.308	394.0	34133.76	7115.71	1354.35	1.0282	56.9	899
JF05650830	5	65	0.8	30	13	5.86	9.980	430.0	42831.63	7308.57	1774.34	1.1530	53.3	913
JF05651030	5	65	1.0	30	13	6.76	10.431	466.0	50707.08	7501.43	2181.86	1.3392	50.6	886
JF05800405	5	80	0.4	5	16	0.10	0.598	364.0	129.97	1351.67	26.50	EULER	–	–
JF05800605	5	80	0.6	5	16	0.14	0.721	370.0	192.55	1357.02	39.63	EULER	–	–
JF05800805	5	80	0.8	5	16	0.19	0.821	376.0	253.63	1362.38	52.69	EULER	–	–
JF05801005	5	80	1.0	5	16	0.23	0.906	382.0	313.29	1367.74	65.67	EULER	–	–
JF05800410	5	80	0.4	10	16	0.37	1.643	376.0	1014.54	2724.76	105.38	0.6289	77.0	195
JF05800610	5	80	0.6	10	16	0.54	1.957	388.0	1486.34	2746.19	157.14	0.6443	77.1	231
JF05800810	5	80	0.8	10	16	0.70	2.201	400.0	1937.33	2767.62	208.32	0.6722	77.2	255
JF05801010	5	80	1.0	10	16	0.85	2.398	412.0	2369.34	2789.05	258.93	0.7170	77.2	270
JF05800415	5	80	0.4	15	16	0.81	2.936	388.0	3344.27	4119.29	235.71	0.6473	76.4	345
JF05800615	5	80	0.6	15	16	1.16	3.456	406.0	4850.11	4167.50	350.54	0.6796	78.9	400
JF05800815	5	80	0.8	15	16	1.49	3.844	424.0	6263.92	4215.71	463.51	0.7276	81.4	432
JF05801015	5	80	1.0	15	16	1.78	4.146	442.0	7596.93	4263.93	574.73	0.7965	83.5	447
JF05800420	5	80	0.4	20	16	1.40	4.402	400.0	7749.33	5535.24	416.63	0.6537	73.4	520
JF05800620	5	80	0.6	20	16	1.98	5.125	424.0	11135.85	5620.95	618.01	0.7099	71.0	586
JF05800820	5	80	0.8	20	16	2.50	5.643	448.0	14266.67	5706.67	815.24	0.7991	69.2	611
JF05801020	5	80	1.0	20	16	2.97	6.033	472.0	17180.79	5792.38	1008.62	0.9215	71.6	609
JF05800425	5	80	0.4	25	16	2.12	5.995	412.0	14808.35	6972.62	647.33	0.6501	74.3	717
JF05800625	5	80	0.6	25	16	2.97	6.910	442.0	21102.59	7106.55	957.88	0.7065	68.7	798
JF05800825	5	80	0.8	25	16	3.71	7.542	472.0	26844.99	7240.48	1260.78	0.8014	63.5	821
JF05801025	5	80	1.0	25	16	4.36	8.001	502.0	32134.48	7374.40	1556.72	0.9348	58.0	809
JF05800430	5	80	0.4	30	16	2.97	7.687	424.0	25055.66	8431.43	927.02	0.6442	74.5	930
JF05800630	5	80	0.6	30	16	4.11	8.777	460.0	35434.57	8624.29	1368.59	0.6961	69.0	1029
JF05800830	5	80	0.8	30	16	5.08	9.503	496.0	44796.77	8817.14	1797.67	0.7863	63.6	1051
JF05801030	5	80	1.0	30	16	5.92	10.014	532.0	53348.68	9010.00	2215.57	0.9156	58.0	1028

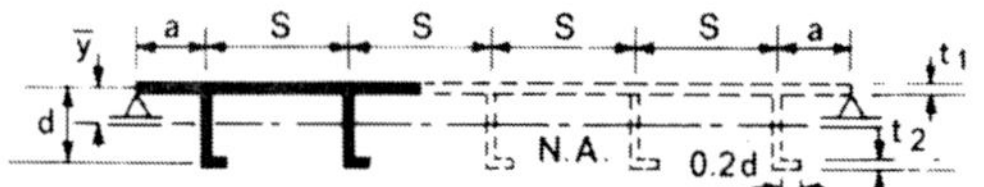

J-PLATE n angle stiffeners · edges pinned · free in-plane

CODE	n	S	t2	d	a	ȳ	rad. gyr.	A	I_{NA}	Z_{plate}	$Z_{stiff.}$	$\sigma_{cr/E}$	L_1	L_{euler}
To scale multiply by	—	t_1	t_1	t_1	t_1	t_1	t_1	t_1^2	t_1^4	t_1^3	t_1^3	10^{-3}	t_1	t_1
JP01350405	1	35	0.4	5	35	0.10	0.599	72.4	25.99	268.81	5.30	1.3847	115.9	1695
JP01350605	1	35	0.6	5	35	0.14	0.723	73.6	38.50	269.88	7.93	1.5385	131.1	1594
JP01350805	1	35	0.8	5	35	0.19	0.823	74.8	50.71	270.95	10.54	1.6520	139.8	1526
JP01351005	1	35	1.0	5	35	0.23	0.908	76.0	62.64	272.02	13.13	1.7388	147.0	1475
JP01350410	1	35	0.4	10	35	0.37	1.647	74.8	202.85	541.90	21.07	2.9036	193.8	1148
JP01350610	1	35	0.6	10	35	0.54	1.962	77.2	297.15	546.19	31.42	3.1974	34.7	1077
JP01350810	1	35	0.8	10	35	0.70	2.206	79.6	387.27	550.48	41.66	3.2997	35.0	1043
JP01351010	1	35	1.0	10	35	0.85	2.403	82.0	473.58	554.76	51.78	3.4561	34.8	1004
JP01350415	1	35	0.4	15	35	0.82	2.943	77.2	668.59	819.29	47.14	2.8432	84.7	1142
JP01350615	1	35	0.6	15	35	1.17	3.464	80.8	969.48	828.93	70.10	3.2508	32.6	1043
JP01350815	1	35	0.8	15	35	1.49	3.851	84.4	1251.90	838.57	92.68	3.4174	32.1	995
JP01351015	1	35	1.0	15	35	1.79	4.153	88.0	1518.11	848.21	114.92	3.6364	31.3	944
JP01350420	1	35	0.4	20	35	1.41	4.411	79.6	1549.08	1100.95	83.32	2.1419	130.6	1296
JP01350620	1	35	0.6	20	35	1.99	5.135	84.4	2225.59	1118.10	123.58	3.0457	109.0	1054
JP01350820	1	35	0.8	20	35	2.51	5.653	89.2	2850.82	1135.24	163.01	3.3700	31.7	973
JP01351020	1	35	1.0	20	35	2.98	6.043	94.0	3432.62	1152.38	201.67	3.6049	30.6	915
JP01350425	1	35	0.4	25	35	2.13	6.008	82.0	2959.86	1386.90	129.44	1.7140	181.0	1428
JP01350625	1	35	0.6	25	35	2.98	6.922	88.0	4216.97	1413.69	191.53	2.4660	151.6	1146
JP01350825	1	35	0.8	25	35	3.72	7.554	94.0	5363.48	1440.48	252.08	3.0234	136.6	999
JP01351025	1	35	1.0	25	35	4.38	8.012	100.0	6419.27	1467.26	311.24	3.3783	129.6	914
JP01350430	1	35	0.4	30	35	2.99	7.703	84.4	5007.58	1677.14	185.37	1.4314	237.0	1539
JP01350630	1	35	0.6	30	35	4.13	8.792	91.6	7080.13	1715.71	273.65	2.0638	198.6	1227
JP01350830	1	35	0.8	30	35	5.10	9.517	98.8	8948.99	1754.29	359.41	2.5612	178.0	1057
JP01351030	1	35	1.0	30	35	5.94	10.026	106.0	10655.66	1792.86	442.94	2.8990	167.4	956
JP01500405	1	50	0.4	5	50	0.07	0.506	102.4	26.19	383.10	5.31	0.6213	159.7	>2000
JP01500605	1	50	0.6	5	50	0.10	0.613	103.6	38.94	384.17	7.95	0.6902	174.1	>2000
JP01500805	1	50	0.8	5	50	0.13	0.701	104.8	51.46	385.24	10.58	0.7437	185.7	>2000
JP01501005	1	50	1.0	5	50	0.17	0.776	106.0	63.78	386.31	13.19	0.7866	194.7	>2000
JP01500410	1	50	0.4	10	50	0.27	1.402	104.8	205.85	770.48	21.15	1.2836	257.0	>2000
JP01500610	1	50	0.6	10	50	0.39	1.683	107.2	303.54	774.76	31.59	1.4564	283.4	>2000
JP01500810	1	50	0.8	10	50	0.51	1.906	109.6	398.05	779.05	41.95	1.5767	303.0	>2000
JP01501010	1	50	1.0	10	50	0.63	2.091	112.0	489.58	783.33	52.22	1.6636	319.2	>2000
JP01500415	1	50	0.4	15	50	0.59	2.524	107.2	682.98	1162.14	47.39	1.5585	47.2	>2000
JP01500615	1	50	0.6	15	50	0.85	3.003	110.8	999.40	1171.79	70.64	1.6092	46.9	>2000
JP01500815	1	50	0.8	15	50	1.10	3.373	114.4	1301.22	1181.43	93.62	1.6861	46.8	>2000
JP01501015	1	50	1.0	15	50	1.33	3.671	118.0	1589.78	1191.07	116.34	1.7828	46.5	1996
JP01500420	1	50	0.4	20	50	1.02	3.811	109.6	1592.21	1558.10	83.90	1.5512	47.0	>2000
JP01500620	1	50	0.6	20	50	1.47	4.497	114.4	2313.29	1575.24	124.83	1.6059	46.0	>2000
JP01500820	1	50	0.8	20	50	1.88	5.010	119.2	2992.39	1592.38	165.14	1.6963	44.7	>2000
JP01501020	1	50	1.0	20	50	2.26	5.414	124.0	3634.41	1609.52	204.85	1.8112	43.2	1931
JP01500425	1	50	0.4	25	50	1.56	5.227	112.0	3059.90	1958.33	130.56	1.5350	46.9	>2000
JP01500625	1	50	0.6	25	50	2.22	6.118	118.0	4416.05	1985.12	193.90	1.5877	46.1	>2000
JP01500825	1	50	0.8	25	50	2.82	6.767	124.0	5678.76	2011.90	256.06	1.6761	44.7	>2000
JP01501025	1	50	1.0	25	50	3.37	7.265	130.0	6860.98	2038.69	317.13	1.7932	43.0	1894
JP01500430	1	50	0.4	30	50	2.20	6.745	114.4	5204.90	2362.86	187.25	1.3955	238.8	>2000
JP01500630	1	50	0.6	30	50	3.11	7.835	121.6	7464.97	2401.43	277.60	1.5612	46.3	>2000
JP01500830	1	50	0.8	30	50	3.91	8.610	128.8	9547.63	2440.00	366.00	1.6489	45.1	1985
JP01501030	1	50	1.0	30	50	4.63	9.188	136.0	11481.62	2478.57	452.61	1.7668	43.5	1865
JP01650405	1	65	0.4	5	65	0.05	0.446	132.4	26.30	497.38	5.32	0.3448	197.3	>2000
JP01650605	1	65	0.6	5	65	0.08	0.542	133.6	39.17	498.45	7.96	0.3821	214.7	>2000
JP01650805	1	65	0.8	5	65	0.10	0.620	134.8	51.88	499.52	10.60	0.4119	228.3	>2000
JP01651005	1	65	1.0	5	65	0.13	0.688	136.0	64.41	500.60	13.22	0.4364	239.6	>2000
JP01650410	1	65	0.4	10	65	0.21	1.241	134.8	207.52	999.05	21.19	0.6989	315.2	>2000
JP01650610	1	65	0.6	10	65	0.31	1.496	137.2	307.14	1003.33	31.68	0.7983	347.3	>2000
JP01650810	1	65	0.8	10	65	0.40	1.702	139.6	404.20	1007.62	42.11	0.8709	371.8	>2000
JP01651010	1	65	1.0	10	65	0.49	1.874	142.0	498.83	1011.90	52.47	0.9261	392.4	>2000
JP01650415	1	65	0.4	15	65	0.46	2.244	137.2	691.07	1505.00	47.53	0.9240	61.7	>2000
JP01650615	1	65	0.6	15	65	0.67	2.687	140.8	1016.57	1514.64	70.95	0.9510	63.6	>2000
JP01650815	1	65	0.8	15	65	0.87	3.035	144.4	1330.06	1524.29	94.15	0.9883	62.6	>2000
JP01651015	1	65	1.0	15	65	1.06	3.321	148.0	1632.39	1533.93	117.14	1.0368	62.5	>2000
JP01650420	1	65	0.4	20	65	0.80	3.403	139.6	1616.81	2015.24	84.22	0.9240	61.0	>2000
JP01650620	1	65	0.6	20	65	1.16	4.047	144.4	2364.54	2032.38	125.53	0.9596	59.8	>2000
JP01650820	1	65	0.8	20	65	1.50	4.541	149.2	3077.03	2049.52	166.34	1.0155	60.5	>2000
JP01651020	1	65	1.0	20	65	1.82	4.940	154.0	3757.58	2066.67	206.67	1.0807	56.9	>2000

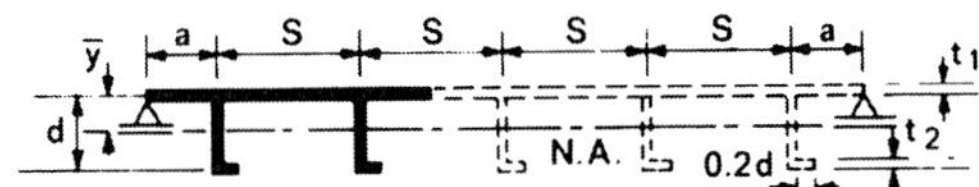

CODE	n	S	t_2	d	a	$\bar{y}$	rad. gyr.	A	I_{NA}	Z_{plate}	$Z_{stiff.}$	σ_{cr}/E	L_1	L_{euler}
To scale multiply by	$-$	t_1	t_1	t_1	t_1	t_1	t_1	t_1^2	t_1^4	t_1^3	t_1^3	10^{-3}	t_1	t_1
JP01650425	1	65	0.4	25	65	1.23	4.686	142.0	3117.66	2529.76	131.17	0.9196	61.0	>2000
JP01650625	1	65	0.6	25	65	1.77	5.535	148.0	4534.42	2556.55	195.23	0.9539	59.5	>2000
JP01650825	1	65	0.8	25	65	2.27	6.175	154.0	5871.21	2583.33	258.33	1.0101	57.5	>2000
JP01651025	1	65	1.0	25	65	2.73	6.679	160.0	7137.04	2610.12	320.54	1.0806	55.2	>2000
JP01650430	1	65	0.4	30	65	1.75	6.070	144.4	5320.22	3048.57	188.29	0.9133	61.1	>2000
JP01650630	1	65	0.6	30	65	2.49	7.126	151.6	7697.49	3087.14	279.84	0.9453	59.7	>2000
JP01650830	1	65	0.8	30	65	3.17	7.904	158.8	9920.40	3125.71	369.80	0.9995	57.7	>2000
JP01651030	1	65	1.0	30	65	3.80	8.506	166.0	12009.04	3164.29	458.28	1.0699	55.3	>2000
JP01800405	1	80	0.4	5	80	0.04	0.403	162.4	26.36	611.67	5.32	0.2168	233.4	>2000
JP01800605	1	80	0.6	5	80	0.06	0.490	163.6	39.33	612.74	7.97	0.2394	253.3	>2000
JP01800805	1	80	0.8	5	80	0.08	0.563	164.8	52.14	613.81	10.61	0.2578	268.9	>2000
JP01801005	1	80	1.0	5	80	0.11	0.625	166.0	64.82	614.88	13.24	0.2733	282.1	>2000
JP01800410	1	80	0.4	10	80	0.17	1.125	164.8	208.58	1227.62	21.22	0.4308	370.4	>2000
JP01800610	1	80	0.6	10	80	0.25	1.360	167.2	309.45	1231.90	31.74	0.4939	407.8	>2000
JP01800810	1	80	0.8	10	80	0.33	1.551	169.6	408.18	1236.19	42.21	0.5413	436.5	>2000
JP01801010	1	80	1.0	10	80	0.41	1.713	172.0	504.84	1240.48	52.63	0.5784	460.0	>2000
JP01800415	1	80	0.4	15	80	0.38	2.041	167.2	696.26	1847.86	47.61	0.6094	76.3	>2000
JP01800615	1	80	0.6	15	80	0.55	2.453	170.8	1027.72	1857.50	71.14	0.6234	77.0	>2000
JP01800815	1	80	0.8	15	80	0.72	2.781	174.4	1348.97	1867.14	94.48	0.6439	77.6	>2000
JP01801015	1	80	1.0	15	80	0.88	3.054	178.0	1660.64	1876.79	117.65	0.6713	77.5	>2000
JP01800420	1	80	0.4	20	80	0.66	3.103	169.6	1632.70	2472.38	84.42	0.6120	75.1	>2000
JP01800620	1	80	0.6	20	80	0.96	3.708	174.4	2398.17	2489.52	125.98	0.6358	74.2	>2000
JP01800820	1	80	0.8	20	80	1.25	4.182	179.2	3133.33	2506.67	167.11	0.6701	73.4	>2000
JP01801020	1	80	1.0	20	80	1.52	4.569	184.0	3840.58	2523.81	207.84	0.7109	74.3	>2000
JP01800425	1	80	0.4	25	80	1.02	4.283	172.0	3155.28	3101.19	131.57	0.6103	74.9	>2000
JP01800625	1	80	0.6	25	80	1.47	5.091	178.0	4612.89	3127.98	196.08	0.6355	74.4	>2000
JP01800825	1	80	0.8	25	80	1.90	5.711	184.0	6000.91	3154.76	259.80	0.6741	70.6	>2000
JP01801025	1	80	1.0	25	80	2.30	6.209	190.0	7325.93	3181.55	322.77	0.7208	68.0	>2000
JP01800430	1	80	0.4	30	80	1.44	5.562	174.4	5395.87	3734.29	188.96	0.6078	75.0	>2000
JP01800630	1	80	0.6	30	80	2.08	6.576	181.6	7853.19	3772.86	281.29	0.6314	74.3	>2000
JP01800830	1	80	0.8	30	80	2.67	7.341	188.8	10174.58	3811.43	372.28	0.6693	70.3	>2000
JP01801030	1	80	1.0	30	80	3.21	7.946	196.0	12375.00	3850.00	462.00	0.7169	67.6	>2000
JP02350405	2	35	0.4	5	35	0.13	0.685	109.8	51.55	404.29	10.58	0.6156	181.4	>2000
JP02350605	2	35	0.6	5	35	0.19	0.823	112.2	76.07	406.43	15.81	0.6838	198.5	>2000
JP02350805	2	35	0.8	5	35	0.24	0.933	114.6	99.83	408.57	20.99	0.7348	211.1	>2000
JP02351005	2	35	1.0	5	35	0.30	1.025	117.0	122.86	410.71	26.14	0.7749	221.8	>2000
JP02350410	2	35	0.4	10	35	0.49	1.867	114.6	399.30	817.14	41.98	1.2892	294.5	>2000
JP02350610	2	35	0.6	10	35	0.70	2.206	119.4	580.90	825.71	62.49	1.4377	323.5	>2000
JP02350810	2	35	0.8	10	35	0.90	2.461	124.2	752.33	834.29	82.69	1.5316	345.0	>2000
JP02351010	2	35	1.0	10	35	1.09	2.663	129.0	914.73	842.86	102.61	1.5933	362.5	>2000
JP02350415	2	35	0.4	15	35	1.06	3.309	119.4	1307.04	1238.57	93.73	2.0685	394.8	>2000
JP02350615	2	35	0.6	15	35	1.49	3.851	126.6	1877.84	1257.86	139.03	2.2610	432.4	1887
JP02350815	2	35	0.8	15	35	1.88	4.240	133.8	2405.38	1277.14	183.38	2.3593	460.5	1815
JP02351015	2	35	1.0	15	35	2.23	4.532	141.0	2896.28	1296.43	226.87	2.4065	482.8	1767
JP02350420	2	35	0.4	20	35	1.80	4.922	124.2	3009.34	1668.57	165.38	2.1425	130.6	1950
JP02350620	2	35	0.6	20	35	2.51	5.653	133.8	4276.23	1702.86	244.51	2.9928	109.6	1608
JP02350820	2	35	0.8	20	35	3.12	6.152	143.4	5427.06	1737.14	321.59	3.1395	563.7	1536
JP02351020	2	35	1.0	20	35	3.66	6.510	153.0	6483.66	1771.43	396.80	3.1458	590.6	1504
JP02350425	2	35	0.4	25	35	2.71	6.657	129.0	5717.05	2107.14	256.52	1.7090	181.4	>2000
JP02350625	2	35	0.6	25	35	3.72	7.554	141.0	8045.21	2160.71	378.13	2.4423	152.1	1752
JP02350825	2	35	0.8	25	35	4.58	8.137	153.0	10130.72	2214.29	496.00	2.9435	138.0	1554
JP02351025	2	35	1.0	25	35	5.30	8.537	165.0	12026.52	2267.86	610.58	3.2488	131.5	1444
JP02350430	2	35	0.4	30	35	3.77	8.480	133.8	9621.52	2554.29	366.77	1.4194	238.1	>2000
JP02350630	2	35	0.6	30	35	5.10	9.517	148.2	13423.48	2631.43	539.12	2.0557	198.8	1881
JP02350830	2	35	0.8	30	35	6.20	10.162	162.6	16791.14	2708.57	705.49	2.5110	179.3	1652
JP02351030	2	35	1.0	30	35	7.12	10.585	177.0	19830.51	2785.71	866.67	2.8051	169.8	1521
JP02500405	2	50	0.4	5	50	0.09	0.580	154.8	52.07	575.71	10.61	0.2757	241.2	>2000
JP02500605	2	50	0.6	5	50	0.13	0.701	157.2	77.19	577.86	15.86	0.3064	262.5	>2000
JP02500805	2	50	0.8	5	50	0.18	0.798	159.6	101.75	580.00	21.09	0.3304	279.5	>2000
JP02501005	2	50	1.0	5	50	0.22	0.881	162.0	125.77	582.14	26.29	0.3501	293.5	>2000
JP02500410	2	50	0.4	10	50	0.35	1.597	159.6	407.02	1160.00	42.18	0.5689	388.6	>2000
JP02500610	2	50	0.6	10	50	0.51	1.906	164.4	597.08	1168.57	62.92	0.6440	427.2	>2000
JP02500810	2	50	0.8	10	50	0.66	2.146	169.2	779.20	1177.14	83.44	0.6965	456.3	>2000
JP02501010	2	50	1.0	10	50	0.80	2.342	174.0	954.02	1185.71	103.75	0.7348	480.4	>2000

J-PLATE — n angle stiffeners, edges pinned, free in-plane

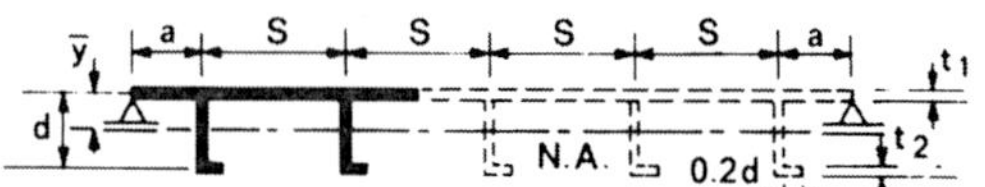

CODE	n	S	t_2	d	a	$\bar{y}$	rad. gyr.	A	I_{NA}	Z_{plate}	$Z_{stiff.}$	$\sigma_{cr/E}$	L_1	L_{euler}
To scale multiply by	–	t_1	t_1	t_1	t_1	t_1	t_1	t_1^2	t_1^4	t_1^3	t_1^3	10^{-3}	t_1	t_1
JP02500415	2	50	0.4	15	50	0.77	2.859	164.4	1343.43	1752.86	94.38	0.9244	521.2	>2000
JP02500615	2	50	0.6	15	50	1.10	3.373	171.6	1951.84	1772.14	140.43	1.0348	572.6	>2000
JP02500815	2	50	0.8	15	50	1.41	3.758	178.8	2524.83	1791.43	185.78	1.1034	611.3	>2000
JP02501015	2	50	1.0	15	50	1.69	4.060	186.0	3066.53	1810.71	230.45	1.1473	642.0	>2000
JP02500420	2	50	0.4	20	50	1.32	4.292	169.2	3116.78	2354.29	166.89	1.3095	642.4	>2000
JP02500620	2	50	0.6	20	50	1.88	5.010	178.8	4488.59	2388.57	247.70	1.4423	704.8	>2000
JP02500820	2	50	0.8	20	50	2.38	5.530	188.4	5761.36	2422.86	326.94	1.5138	754.7	>2000
JP02501020	2	50	1.0	20	50	2.83	5.924	198.0	6949.49	2457.14	404.71	1.5509	787.7	>2000
JP02500425	2	50	0.4	25	50	2.01	5.854	174.0	5962.64	2964.29	259.38	1.5349	46.8	>2000
JP02500625	2	50	0.6	25	50	2.82	6.767	186.0	8518.15	3017.86	384.09	1.6042	45.8	>2000
JP02500825	2	50	0.8	25	50	3.54	7.405	198.0	10858.59	3071.43	505.88	1.7201	43.9	>2000
JP02501025	2	50	1.0	25	50	4.17	7.874	210.0	13020.83	3125.00	625.00	1.9310	923.9	>2000
JP02500430	2	50	0.4	30	50	2.82	7.516	178.8	10099.33	3582.86	371.56	1.3958	238.8	>2000
JP02500630	2	50	0.6	30	50	3.91	8.610	193.2	14321.74	3660.00	549.00	1.5690	46.0	>2000
JP02500830	2	50	0.8	30	50	4.86	9.349	207.6	18145.66	3737.14	721.66	1.6878	46.6	>2000
JP02501030	2	50	1.0	30	50	5.68	9.875	222.0	21648.65	3814.29	890.00	1.8362	42.5	>2000
JP02650405	2	65	0.4	5	65	0.07	0.512	199.8	52.35	747.14	10.62	0.1529	297.5	>2000
JP02650605	2	65	0.6	5	65	0.10	0.620	202.2	77.82	749.29	15.89	0.1695	323.2	>2000
JP02650805	2	65	0.8	5	65	0.14	0.709	204.6	102.83	751.43	21.15	0.1828	343.5	>2000
JP02651005	2	65	1.0	5	65	0.17	0.785	207.0	127.42	753.57	26.38	0.1941	360.7	>2000
JP02650410	2	65	0.4	10	65	0.27	1.418	204.6	411.34	1502.86	42.29	0.3095	476.5	>2000
JP02650610	2	65	0.6	10	65	0.40	1.702	209.4	606.30	1511.43	63.16	0.3530	523.2	>2000
JP02650810	2	65	0.8	10	65	0.52	1.926	214.2	794.77	1520.00	83.86	0.3849	559.8	>2000
JP02651010	2	65	1.0	10	65	0.64	2.112	219.0	977.17	1528.57	104.39	0.4094	589.2	>2000
JP02650415	2	65	0.4	15	65	0.60	2.552	209.4	1364.18	2267.14	94.75	0.5047	639.0	>2000
JP02650615	2	65	0.6	15	65	0.87	3.035	216.6	1995.08	2286.43	141.22	0.5726	702.7	>2000
JP02650815	2	65	0.8	15	65	1.13	3.406	223.8	2596.25	2305.71	187.13	0.6185	754.5	>2000
JP02651015	2	65	1.0	15	65	1.36	3.705	231.0	3170.45	2325.00	232.50	0.6506	789.4	>2000
JP02650420	2	65	0.4	20	65	1.05	3.852	214.2	3179.08	3040.00	167.72	0.7211	788.5	>2000
JP02650620	2	65	0.6	20	65	1.50	4.541	223.8	4615.55	3074.29	249.51	0.8085	866.0	>2000
JP02650820	2	65	0.8	20	65	1.92	5.056	233.4	5966.75	3108.57	330.01	0.8623	926.5	>2000
JP02651020	2	65	1.0	20	65	2.30	5.459	243.0	7242.80	3142.86	409.30	0.8962	969.7	>2000
JP02650425	2	65	0.4	25	65	1.60	5.281	219.0	6107.31	3821.43	260.98	0.9233	60.8	>2000
JP02650625	2	65	0.6	25	65	2.27	6.175	231.0	8806.82	3875.00	387.50	0.9694	60.5	>2000
JP02650825	2	65	0.8	25	65	2.88	6.824	243.0	11316.87	3928.57	511.63	1.0420	56.2	>2000
JP02651025	2	65	1.0	25	65	3.43	7.320	255.0	13664.22	3982.14	633.52	1.1366	1136.2	>2000
JP02650430	2	65	0.4	30	65	2.25	6.812	223.8	10384.99	4611.43	374.26	0.9149	60.8	>2000
JP02650630	2	65	0.6	30	65	3.17	7.904	238.2	14880.60	4688.57	554.70	0.9571	59.1	>2000
JP02650830	2	65	0.8	30	65	3.99	8.677	252.6	19017.58	4765.71	731.18	1.0279	56.5	>2000
JP02651030	2	65	1.0	30	65	4.72	9.252	267.0	22853.93	4842.86	904.00	1.1181	53.8	>2000
JP02800405	2	80	0.4	5	80	0.06	0.463	244.8	52.53	918.57	10.63	0.0961	351.6	>2000
JP02800605	2	80	0.6	5	80	0.08	0.563	247.2	78.22	920.71	15.91	0.1062	381.2	>2000
JP02800805	2	80	0.8	5	80	0.11	0.644	249.6	103.53	922.86	21.18	0.1144	404.8	>2000
JP02801005	2	80	1.0	5	80	0.14	0.714	252.0	128.47	925.00	26.43	0.1214	425.2	>2000
JP02800410	2	80	0.4	10	80	0.22	1.288	249.6	414.10	1845.71	42.36	0.1907	558.8	>2000
JP02800610	2	80	0.6	10	80	0.33	1.551	254.4	612.26	1854.29	63.32	0.2184	614.5	>2000
JP02800810	2	80	0.8	10	80	0.43	1.762	259.2	804.94	1862.86	84.13	0.2393	656.6	>2000
JP02801010	2	80	1.0	10	80	0.53	1.939	264.0	992.42	1871.43	104.80	0.2558	691.7	>2000
JP02800415	2	80	0.4	15	80	0.50	2.327	254.4	1377.59	2781.43	94.98	0.3110	753.6	>2000
JP02800615	2	80	0.6	15	80	0.72	2.781	261.6	2023.45	2800.71	141.72	0.3558	825.3	>2000
JP02800815	2	80	0.8	15	80	0.94	3.136	268.8	2643.75	2820.00	188.00	0.3874	881.9	>2000
JP02801015	2	80	1.0	15	80	1.14	3.427	276.0	3240.49	2839.29	233.82	0.4108	928.1	>2000
JP02800420	2	80	0.4	20	80	0.86	3.524	259.2	3219.75	3725.71	168.26	0.4462	928.2	>2000
JP02800620	2	80	0.6	20	80	1.25	4.182	268.8	4700.00	3760.00	250.67	0.5062	1017.5	>2000
JP02800820	2	80	0.8	20	80	1.61	4.683	278.4	6105.75	3794.29	332.00	0.5457	1088.7	>2000
JP02801020	2	80	1.0	20	80	1.94	5.084	288.0	7444.44	3828.57	412.31	0.5728	1141.6	>2000
JP02800425	2	80	0.4	25	80	1.33	4.847	264.0	6202.65	4678.57	262.00	0.5913	1091.5	>2000
JP02800625	2	80	0.6	25	80	1.90	5.711	276.0	9001.36	4732.14	389.71	0.6635	1196.1	>2000
JP02800825	2	80	0.8	25	80	2.43	6.355	288.0	11631.94	4785.71	515.38	0.6976	69.0	>2000
JP02801025	2	80	1.0	25	80	2.92	6.859	300.0	14114.58	4839.29	639.15	0.7353	1341.3	>2000
JP02800430	2	80	0.4	30	80	1.88	6.272	268.8	10575.00	5640.00	376.00	0.6104	74.6	>2000
JP02800630	2	80	0.6	30	80	2.67	7.341	283.2	15261.86	5717.14	558.42	0.6413	72.1	>2000
JP02800830	2	80	0.8	30	80	3.39	8.121	297.6	19625.81	5794.29	737.45	0.6911	68.7	>2000
JP02801030	2	80	1.0	30	80	4.04	8.718	312.0	23711.54	5871.43	913.33	0.7515	65.3	>2000

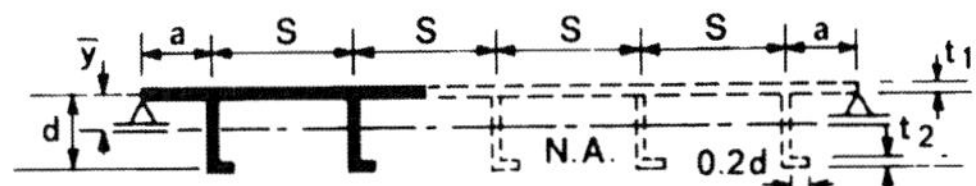

J-PLATE — n angle stiffeners, edges pinned, free in-plane

CODE	n	S	t_2	d	a	$\bar{y}$	rad. gyr.	A	I_{NA}	Z_{plate}	$Z_{stiff.}$	σ_{cr}/E	L_1	L_{euler}
To scale multiply by	–	t_1	t_1	t_1	t_1	t_1	t_1	t_1^2	t_1^4	t_1^3	t_1^3	10^{-3}	t_1	t_1
JP03350405	3	35	0.4	5	35	0.14	0.723	147.2	77.00	539.76	15.85	0.3465	242.7	>2000
JP03350605	3	35	0.6	5	35	0.21	0.867	150.8	113.42	542.98	23.67	0.3849	264.7	>2000
JP03350805	3	35	0.8	5	35	0.27	0.981	154.4	148.58	546.19	31.42	0.4139	282.1	>2000
JP03351005	3	35	1.0	5	35	0.33	1.075	158.0	182.56	549.40	39.11	0.4370	296.2	>2000
JP03350410	3	35	0.4	10	35	0.54	1.962	154.4	594.30	1092.38	62.85	0.7258	394.4	>2000
JP03350610	3	35	0.6	10	35	0.78	2.309	161.6	861.76	1105.24	93.46	0.8087	431.9	>2000
JP03350810	3	35	0.8	10	35	1.00	2.568	168.8	1112.80	1118.10	123.58	0.8614	460.4	>2000
JP03351010	3	35	1.0	10	35	1.19	2.769	176.0	1349.43	1130.95	153.23	0.8965	483.2	>2000
JP03350415	3	35	0.4	15	35	1.17	3.464	161.6	1938.95	1657.86	140.19	1.1653	527.3	>2000
JP03350615	3	35	0.6	15	35	1.64	4.011	172.4	2773.80	1686.79	207.69	1.2718	577.4	>2000
JP03350815	3	35	0.8	15	35	2.06	4.396	183.2	3540.07	1715.71	273.65	1.3265	614.7	>2000
JP03351015	3	35	1.0	15	35	2.44	4.680	194.0	4249.19	1744.64	338.19	1.3534	643.9	>2000
JP03350420	3	35	0.4	20	35	1.99	5.135	168.8	4451.18	2236.19	247.16	1.6218	648.1	>2000
JP03350620	3	35	0.6	20	35	2.75	5.861	183.2	6293.45	2287.62	364.86	1.7298	708.7	>2000
JP03350820	3	35	0.8	20	35	3.40	6.345	197.6	7954.66	2339.05	479.22	1.7677	756.5	>2000
JP03351020	3	35	1.0	20	35	3.96	6.684	212.0	9471.70	2390.48	590.59	1.7716	787.4	>2000
JP03350425	3	35	0.4	25	35	2.98	6.922	176.0	8433.95	2827.38	383.06	1.7092	181.3	>2000
JP03350625	3	35	0.6	25	35	4.06	7.800	194.0	11803.32	2907.74	563.65	2.1611	829.2	>2000
JP03350825	3	35	0.8	25	35	4.95	8.355	212.0	14799.53	2988.10	738.24	2.1681	878.9	>2000
JP03351025	3	35	1.0	25	35	5.71	8.725	230.0	17510.19	3068.45	907.57	2.1398	921.1	>2000
JP03350430	3	35	0.4	30	35	4.13	8.792	183.2	14160.26	3431.43	547.29	1.4182	238.1	>2000
JP03350630	3	35	0.6	30	35	5.54	9.793	204.8	19640.92	3547.14	802.89	2.0545	198.8	>2000
JP03350830	3	35	0.8	30	35	6.68	10.395	226.4	24462.19	3662.86	1048.91	2.4893	180.0	>2000
JP03351030	3	35	1.0	30	35	7.62	10.776	248.0	28796.37	3778.57	1286.76	2.4538	1039.5	>2000
JP03500405	3	50	0.4	5	50	0.10	0.613	207.2	77.87	768.33	15.90	0.1551	321.8	>2000
JP03500605	3	50	0.6	5	50	0.15	0.740	210.8	115.29	771.55	23.77	0.1724	350.4	>2000
JP03500805	3	50	0.8	5	50	0.20	0.841	214.4	151.77	774.76	31.59	0.1860	373.1	>2000
JP03501005	3	50	1.0	5	50	0.24	0.927	218.0	187.36	777.98	39.37	0.1973	392.4	>2000
JP03500410	3	50	0.4	10	50	0.39	1.683	214.4	607.09	1549.52	63.18	0.3199	518.9	>2000
JP03500610	3	50	0.6	10	50	0.57	2.002	221.6	888.36	1562.38	94.19	0.3620	569.9	>2000
JP03500810	3	50	0.8	10	50	0.73	2.248	228.8	1156.64	1575.24	124.83	0.3915	608.4	>2000
JP03501010	3	50	1.0	10	50	0.89	2.447	236.0	1413.14	1588.10	155.12	0.4132	640.0	>2000
JP03500415	3	50	0.4	15	50	0.85	3.003	221.6	1998.80	2343.57	141.29	0.5198	696.5	>2000
JP03500615	3	50	0.6	15	50	1.22	3.529	232.4	2894.16	2372.50	210.02	0.5813	764.8	>2000
JP03500815	3	50	0.8	15	50	1.55	3.918	243.2	3732.48	2401.43	277.60	0.6196	815.1	>2000
JP03501015	3	50	1.0	15	50	1.86	4.219	254.0	4521.04	2430.36	344.07	0.6443	855.5	>2000
JP03500420	3	50	0.4	20	50	1.47	4.497	228.8	4626.57	3150.48	249.66	0.7365	858.5	>2000
JP03500620	3	50	0.6	20	50	2.07	5.223	243.2	6635.53	3201.90	370.13	0.8101	940.3	>2000
JP03500820	3	50	0.8	20	50	2.61	5.740	257.6	8486.96	3253.33	488.00	0.8498	1003.2	>2000
JP03501020	3	50	1.0	20	50	3.09	6.125	272.0	10205.88	3304.76	603.48	0.8707	1048.3	>2000
JP03500425	3	50	0.4	25	50	2.22	6.118	236.0	8832.10	3970.24	387.79	0.9607	1010.7	>2000
JP03500625	3	50	0.6	25	50	3.10	7.032	254.0	12558.44	4050.60	573.46	1.0386	1102.5	>2000
JP03500825	3	50	0.8	25	50	3.86	7.657	272.0	15946.69	4130.95	754.35	1.0729	1173.7	>2000
JP03501025	3	50	1.0	25	50	4.53	8.107	290.0	19059.81	4211.31	930.92	1.0845	1225.9	>2000
JP03500430	3	50	0.4	30	50	3.11	7.835	243.2	14929.93	4802.86	555.19	1.1858	1148.4	>2000
JP03500630	3	50	0.6	30	50	4.28	8.919	264.8	21063.67	4918.57	819.04	1.2609	1256.4	>2000
JP03500830	3	50	0.8	30	50	5.28	9.633	286.4	26577.65	5034.29	1075.12	1.2840	1333.1	>2000
JP03501030	3	50	1.0	30	50	6.14	10.129	308.0	31602.27	5150.00	1324.29	1.2823	1391.5	>2000
JP03650405	3	65	0.4	5	65	0.08	0.542	267.2	78.35	996.90	15.92	0.0860	396.9	>2000
JP03650605	3	65	0.6	5	65	0.12	0.655	270.8	116.34	1000.12	23.82	0.0953	431.1	>2000
JP03650805	3	65	0.8	5	65	0.15	0.748	274.4	153.57	1003.33	31.68	0.1029	458.6	>2000
JP03651005	3	65	1.0	5	65	0.19	0.827	278.0	190.09	1006.55	39.51	0.1093	481.1	>2000
JP03650410	3	65	0.4	10	65	0.31	1.496	274.4	614.29	2006.67	63.37	0.1740	635.1	>2000
JP03650610	3	65	0.6	10	65	0.45	1.791	281.6	903.62	2019.52	94.59	0.1984	698.4	>2000
JP03650810	3	65	0.8	10	65	0.58	2.023	288.8	1182.27	2032.38	125.53	0.2163	750.2	>2000
JP03651010	3	65	1.0	10	65	0.71	2.214	296.0	1451.01	2045.24	156.18	0.2301	785.5	>2000
JP03650415	3	65	0.4	15	65	0.67	2.687	281.6	2033.15	3029.29	141.89	0.2836	853.2	>2000
JP03650615	3	65	0.6	15	65	0.97	3.184	292.4	2965.13	3058.21	211.34	0.3216	937.3	>2000
JP03650815	3	65	0.8	15	65	1.25	3.563	303.2	3848.75	3087.14	279.84	0.3472	1003.0	>2000
JP03651015	3	65	1.0	15	65	1.50	3.864	314.0	4688.99	3116.07	347.46	0.3653	1051.0	>2000
JP03650420	3	65	0.4	20	65	1.16	4.047	288.8	4729.09	4064.76	251.06	0.4052	1051.8	>2000
JP03650620	3	65	0.6	20	65	1.66	4.750	303.2	6842.22	4116.19	373.12	0.4538	1154.1	>2000
JP03650820	3	65	0.8	20	65	2.12	5.269	317.6	8818.14	4167.62	493.07	0.4838	1230.2	>2000
JP03651020	3	65	1.0	20	65	2.53	5.670	332.0	10674.70	4219.05	611.03	0.5028	1290.8	>2000

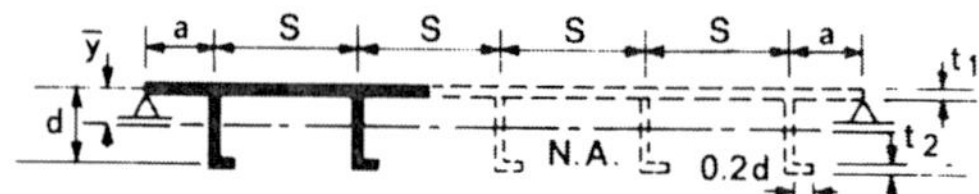

J-PLATE   n angle stiffeners / edges pinned / free in-plane

CODE	n	S	t_2	d	a	$\bar{y}$	rad. gyr.	A	I_{NA}	Z_{plate}	$Z_{stiff.}$	σ_{cr}/E	L_1	L_{euler}
To scale multiply by	−	t_1	t_1	t_1	t_1	t_1	t_1	t_1^2	t_1^4	t_1^3	t_1^3	10^{-3}	t_1	t_1
JP03650425	3	65	0.4	25	65	1.77	5.535	296.0	9068.83	5113.10	390.45	0.5338	1237.0	>2000
JP03650625	3	65	0.6	25	65	2.51	6.441	314.0	13024.98	5193.45	579.09	0.5897	1355.4	>2000
JP03650825	3	65	0.8	25	65	3.16	7.088	332.0	16679.22	5273.81	763.79	0.6206	1442.8	>2000
JP03651025	3	65	1.0	25	65	3.75	7.574	350.0	20078.13	5354.17	944.85	0.6375	1513.7	>2000
JP03650430	3	65	0.4	30	65	2.49	7.126	303.2	15394.99	6174.29	559.68	0.6662	1413.2	>2000
JP03650630	3	65	0.6	30	65	3.49	8.223	324.8	21960.78	6290.00	828.44	0.7259	1544.2	>2000
JP03650830	3	65	0.8	30	65	4.36	8.984	346.4	27960.28	6405.71	1090.70	0.7546	1641.8	>2000
JP03651030	3	65	1.0	30	65	5.14	9.540	368.0	33493.21	6521.43	1347.05	0.7667	1719.0	>2000
JP03800405	3	80	0.4	5	80	0.06	0.490	327.2	78.65	1225.48	15.93	0.0540	468.9	>2000
JP03800605	3	80	0.6	5	80	0.10	0.595	330.8	117.00	1228.69	23.85	0.0597	507.7	>2000
JP03800805	3	80	0.8	5	80	0.13	0.680	334.4	154.72	1231.90	31.74	0.0644	540.1	>2000
JP03801005	3	80	1.0	5	80	0.16	0.753	338.0	191.85	1235.12	39.60	0.0684	568.2	>2000
JP03800410	3	80	0.4	10	80	0.25	1.360	334.4	618.90	2463.81	63.48	0.1072	746.0	>2000
JP03800610	3	80	0.6	10	80	0.37	1.635	341.6	913.52	2476.67	94.85	0.1227	819.8	>2000
JP03800810	3	80	0.8	10	80	0.48	1.854	348.8	1199.08	2489.52	125.98	0.1345	876.3	>2000
JP03801010	3	80	1.0	10	80	0.59	2.036	356.0	1476.12	2502.38	156.87	0.1438	922.5	>2000
JP03800415	3	80	0.4	15	80	0.55	2.453	341.6	2055.43	3715.00	142.28	0.1747	1003.0	>2000
JP03800615	3	80	0.6	15	80	0.80	2.924	352.4	3011.93	3743.93	212.17	0.1998	1100.1	>2000
JP03800815	3	80	0.8	15	80	1.04	3.288	363.2	3926.60	3772.86	281.29	0.2175	1177.4	>2000
JP03801015	3	80	1.0	15	80	1.26	3.584	374.0	4803.06	3801.79	349.65	0.2307	1235.7	>2000
JP03800420	3	80	0.4	20	80	0.96	3.708	348.8	4796.33	4979.05	251.95	0.2506	1234.2	>2000
JP03800620	3	80	0.6	20	80	1.39	4.384	363.2	6980.62	5030.48	375.05	0.2841	1355.9	>2000
JP03800820	3	80	0.8	20	80	1.78	4.894	377.6	9044.07	5081.90	496.37	0.3061	1446.9	>2000
JP03801020	3	80	1.0	20	80	2.14	5.297	392.0	11000.00	5133.33	616.00	0.3213	1519.9	>2000
JP03800425	3	80	0.4	25	80	1.47	5.091	356.0	9225.77	6255.95	392.16	0.3321	1452.6	>2000
JP03800625	3	80	0.6	25	80	2.11	5.973	374.0	13341.83	6336.31	582.76	0.3722	1595.4	>2000
JP03800825	3	80	0.8	25	80	2.68	6.622	392.0	17187.50	6416.67	770.00	0.3968	1698.5	>2000
JP03801025	3	80	1.0	25	80	3.20	7.122	410.0	20798.40	6497.02	954.11	0.4122	1782.4	>2000
JP03800430	3	80	0.4	30	80	2.08	6.576	363.2	15706.39	7545.71	562.58	0.4171	1660.2	>2000
JP03800630	3	80	0.6	30	80	2.95	7.660	384.8	22578.12	7661.43	834.59	0.4621	1817.7	>2000
JP03800830	3	80	0.8	30	80	3.72	8.438	406.4	28934.65	7777.14	1101.03	0.7014	68.0	1218
JP03801030	3	80	1.0	30	80	4.42	9.024	428.0	34853.97	7892.86	1362.33	0.7661	64.5	1225
JP05350405	5	35	0.4	5	35	0.16	0.759	222.0	127.82	810.71	26.40	0.1541	365.1	>2000
JP05350605	5	35	0.6	5	35	0.23	0.908	228.0	187.91	816.07	39.40	0.1713	397.5	>2000
JP05350805	5	35	0.8	5	35	0.30	1.025	234.0	245.73	821.43	52.27	0.1843	423.9	>2000
JP05351005	5	35	1.0	5	35	0.36	1.121	240.0	301.43	826.79	65.03	0.1948	445.0	>2000
JP05350410	5	35	0.4	10	35	0.60	2.050	234.0	982.91	1642.86	104.55	0.3230	591.5	>2000
JP05350610	5	35	0.6	10	35	0.85	2.403	246.0	1420.73	1664.29	155.33	0.3598	648.4	>2000
JP05350810	5	35	0.8	10	35	1.09	2.663	258.0	1829.46	1685.71	205.22	0.3834	691.2	>2000
JP05351010	5	35	1.0	10	35	1.30	2.863	270.0	2212.96	1707.14	254.26	0.3995	725.6	>2000
JP05350415	5	35	0.4	15	35	1.28	3.605	246.0	3196.65	2496.43	233.00	0.5192	793.4	>2000
JP05350615	5	35	0.6	15	35	1.79	4.153	264.0	4554.33	2544.64	344.76	0.5665	867.7	>2000
JP05350815	5	35	0.8	15	35	2.23	4.532	282.0	5792.55	2592.86	453.75	0.5912	925.3	>2000
JP05351015	5	35	1.0	15	35	2.63	4.807	300.0	6932.81	2641.07	560.23	0.6039	965.6	>2000
JP05350420	5	35	0.4	20	35	2.17	5.326	258.0	7317.83	3371.43	410.43	0.7245	975.4	>2000
JP05350620	5	35	0.6	20	35	2.98	6.043	282.0	10297.87	3457.14	605.00	0.7727	1064.3	>2000
JP05350820	5	35	0.8	20	35	3.66	6.510	306.0	12967.32	3542.86	793.60	0.7904	1129.2	>2000
JP05351020	5	35	1.0	20	35	4.24	6.830	330.0	15393.94	3628.57	976.92	0.7935	1180.5	>2000
JP05350425	5	35	0.4	25	35	3.24	7.157	270.0	13831.02	4267.86	635.64	0.9298	1143.7	>2000
JP05350625	5	35	0.6	25	35	4.38	8.012	300.0	19257.81	4401.79	933.71	0.9705	1247.6	>2000
JP05350825	5	35	0.8	25	35	5.30	8.537	330.0	24053.03	4535.71	1221.15	0.9751	1319.5	>2000
JP05351025	5	35	1.0	25	35	6.08	8.878	360.0	28374.57	4669.64	1499.43	0.9644	1377.5	>2000
JP05350430	5	35	0.4	30	35	4.47	9.064	282.0	23170.21	5185.71	907.50	1.1294	1301.0	>2000
JP05350630	5	35	0.6	30	35	5.94	10.026	318.0	31966.98	5378.57	1328.82	1.1554	1416.6	>2000
JP05350830	5	35	0.8	30	35	7.12	10.585	354.0	39661.02	5571.43	1733.33	1.1426	1498.5	>2000
JP05351030	5	35	1.0	30	35	8.08	10.926	390.0	46557.69	5764.29	2123.68	1.1159	1561.3	>2000
JP05500405	5	50	0.4	5	50	0.11	0.644	312.0	129.41	1153.57	26.48	0.0690	482.9	>2000
JP05500605	5	50	0.6	5	50	0.17	0.776	318.0	191.33	1158.93	39.57	0.0767	525.5	>2000
JP05500805	5	50	0.8	5	50	0.22	0.881	324.0	251.54	1164.29	52.58	0.0828	559.8	>2000
JP05501005	5	50	1.0	5	50	0.27	0.969	330.0	310.13	1169.64	65.50	0.0879	587.8	>2000
JP05500410	5	50	0.4	10	50	0.43	1.762	324.0	1006.17	2329.57	105.16	0.1422	780.0	>2000
JP05500610	5	50	0.6	10	50	0.63	2.091	336.0	1468.75	2350.00	156.67	0.1609	856.2	>2000
JP05500810	5	50	0.8	10	50	0.80	2.342	348.0	1908.05	2371.43	207.50	0.1741	913.6	>2000
JP05501010	5	50	1.0	10	50	0.97	2.542	360.0	2326.39	2392.86	257.69	0.1838	959.0	>2000

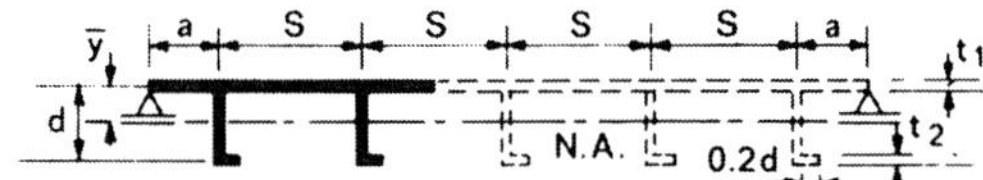

J-PLATE n angle stiffeners / edges pinned / free in-plane

CODE	n	S	t_2	d	a	$\bar{y}$	rad. gyr.	A	I_{NA}	Z_{plate}	$Z_{stiff.}$	σ_{cr}/E	L_1	L_{euler}
To scale multiply by	–	t_1	t_1	t_1	t_1	t_1	t_1	t_1^2	t_1^4	t_1^3	t_1^3	10^{-3}	t_1	t_1
JP05500415	5	50	0.4	15	50	0.94	3.136	336.0	3304.69	3525.00	235.00	0.2311	1045.6	>2000
JP05500615	5	50	0.6	15	50	1.33	3.671	354.0	4769.33	3573.21	349.01	0.2584	1146.7	>2000
JP05500815	5	50	0.8	15	50	1.69	4.060	372.0	6133.06	3621.43	460.91	0.2755	1221.8	>2000
JP05501015	5	50	1.0	15	50	2.02	4.359	390.0	7409.86	3669.64	570.83	0.2867	1281.7	>2000
JP05500420	5	50	0.4	20	50	1.61	4.683	348.0	7632.18	4742.86	415.00	0.3277	1288.3	>2000
JP05500620	5	50	0.6	20	50	2.26	5.414	372.0	10903.23	4828.57	614.55	0.3604	1411.9	>2000
JP05500820	5	50	0.8	20	50	2.83	5.924	396.0	13898.99	4914.29	809.41	0.3781	1501.6	>2000
JP05501020	5	50	1.0	20	50	3.33	6.299	420.0	16666.67	5000.00	1000.00	0.3878	1571.0	>2000
JP05500425	5	50	0.4	25	50	2.43	6.355	360.0	14539.93	5982.14	644.23	0.4280	1513.6	>2000
JP05500625	5	50	0.6	25	50	3.37	7.265	390.0	20582.93	6116.07	951.39	0.4627	1653.5	>2000
JP05500825	5	50	0.8	25	50	4.17	7.874	420.0	26041.67	6250.00	1250.00	0.4783	1757.7	>2000
JP05501025	5	50	1.0	25	50	4.86	8.304	450.0	31032.99	6383.93	1540.95	0.4841	1838.8	>2000
JP05500430	5	50	0.4	30	50	3.39	8.121	372.0	24532.26	7242.86	921.82	0.5296	1725.3	>2000
JP05500630	5	50	0.6	30	50	4.63	9.188	408.0	34444.85	7435.71	1357.83	0.5632	1882.2	>2000
JP05500830	5	50	0.8	30	50	5.68	9.875	444.0	43297.30	7628.57	1780.00	1.7161	43.8	811
JP05501030	5	50	1.0	30	50	6.56	10.341	480.0	51328.13	7821.43	2190.00	1.8932	41.7	803
JP05650405	5	65	0.4	5	65	0.09	0.569	402.0	130.29	1496.43	26.52	0.0382	595.5	>2000
JP05650605	5	65	0.6	5	65	0.13	0.688	408.0	193.24	1501.79	39.67	0.0424	649.4	>2000
JP05650805	5	65	0.8	5	65	0.17	0.785	414.0	254.83	1507.14	52.75	0.0458	682.3	>2000
JP05651005	5	65	1.0	5	65	0.21	0.866	420.0	315.10	1512.50	65.76	0.0487	721.0	>2000
JP05650410	5	65	0.4	10	65	0.34	1.569	414.0	1019.32	3014.29	105.50	0.0773	953.7	>2000
JP05650610	5	65	0.6	10	65	0.49	1.874	426.0	1496.48	3035.71	157.41	0.0882	1047.9	>2000
JP05650810	5	65	0.8	10	65	0.64	2.112	438.0	1954.34	3057.14	208.78	0.0961	1119.3	>2000
JP05651010	5	65	1.0	10	65	0.78	2.307	450.0	2394.07	3078.57	259.64	0.1024	1174.1	>2000
JP05650415	5	65	0.4	15	65	0.74	2.811	426.0	3367.08	4553.57	236.11	0.1261	1279.5	>2000
JP05650615	5	65	0.6	15	65	1.06	3.321	444.0	4897.11	4601.79	351.41	0.1429	1405.3	>2000
JP05650815	5	65	0.8	15	65	1.36	3.705	462.0	6340.91	4650.00	465.00	0.1543	1501.3	>2000
JP05651015	5	65	1.0	15	65	1.64	4.007	480.0	7708.01	4698.21	576.97	0.1624	1576.9	>2000
JP05650420	5	65	0.4	20	65	1.28	4.225	438.0	7817.35	6114.29	417.56	0.1801	1579.8	>2000
JP05650620	5	65	0.6	20	65	1.82	4.940	462.0	11272.73	6200.00	620.00	0.2016	1730.5	>2000
JP05650820	5	65	0.8	20	65	2.30	5.459	486.0	14485.60	6285.71	818.60	0.2150	1845.0	>2000
JP05651020	5	65	1.0	20	65	2.75	5.856	510.0	17490.20	6371.43	1013.64	1.1715	54.0	579
JP05650425	5	65	0.4	25	65	1.94	5.767	450.0	14965.28	7696.43	649.10	0.2373	1856.5	>2000
JP05650625	5	65	0.6	25	65	2.73	6.679	480.0	21411.13	7830.36	961.62	0.9828	58.2	729
JP05650825	5	65	0.8	25	65	3.43	7.320	510.0	27328.43	7964.29	1267.05	1.0709	55.2	763
JP05651025	5	65	1.0	25	65	4.05	7.794	540.0	32805.27	8098.21	1565.95	1.1693	52.5	774
JP05650430	5	65	0.4	30	65	2.73	7.409	462.0	25363.64	9300.00	930.00	0.9164	60.6	843
JP05650630	5	65	0.6	30	65	3.80	8.506	498.0	36027.11	9492.86	1374.83	0.9685	58.5	940
JP05650830	5	65	0.8	30	65	4.72	9.252	534.0	45707.87	9685.71	1808.00	1.0540	55.6	976
JP05651030	5	65	1.0	30	65	5.53	9.787	570.0	54592.11	9878.57	2230.65	1.1549	53.1	981
JP05800405	5	80	0.4	5	80	0.07	0.516	492.0	130.84	1839.29	26.55	0.0240	704.3	>2000
JP05800605	5	80	0.6	5	80	0.11	0.625	498.0	194.47	1844.64	39.73	0.0265	760.5	>2000
JP05800805	5	80	0.8	5	80	0.14	0.714	504.0	256.94	1850.00	52.86	0.0286	807.9	>2000
JP05801005	5	80	1.0	5	80	0.17	0.790	510.0	318.32	1855.36	65.93	0.0304	852.5	>2000
JP05800410	5	80	0.4	10	80	0.28	1.428	504.0	1027.78	3700.00	105.71	0.0476	1118.8	>2000
JP05800610	5	80	0.6	10	80	0.41	1.713	516.0	1514.53	3721.43	157.88	0.0545	1229.5	>2000
JP05800810	5	80	0.8	10	80	0.53	1.939	528.0	1984.85	3742.86	209.60	0.0598	1315.7	>2000
JP05801010	5	80	1.0	10	80	0.65	2.126	540.0	2439.81	3764.29	260.89	0.0639	1386.1	>2000
JP05800415	5	80	0.4	15	80	0.61	2.570	516.0	3407.70	5582.14	236.82	0.0776	1502.5	>2000
JP05800615	5	80	0.6	15	80	0.88	3.054	534.0	4981.92	5630.36	352.95	0.0887	1648.7	>2000
JP05800815	5	80	0.8	15	80	1.14	3.427	552.0	6480.98	5678.57	467.65	0.0966	1763.1	>2000
JP05801015	5	80	1.0	15	80	1.38	3.726	570.0	7912.01	5726.79	580.98	0.1025	1852.9	>2000
JP05800420	5	80	0.4	20	80	1.06	3.878	528.0	7939.39	7485.71	419.20	0.1114	1851.2	>2000
JP05800620	5	80	0.6	20	80	1.52	4.569	552.0	11521.74	7571.43	623.53	0.6595	72.9	608
JP05800820	5	80	0.8	20	80	1.94	5.084	576.0	14888.89	7657.14	824.62	0.7141	71.5	649
JP05801020	5	80	1.0	20	80	2.33	5.487	600.0	18066.67	7742.86	1022.64	0.7714	69.1	671
JP05800425	5	80	0.4	25	80	1.62	5.314	540.0	15248.84	9410.71	652.23	0.6173	74.2	735
JP05800625	5	80	0.6	25	80	2.30	6.209	570.0	21977.80	9544.64	968.30	0.6580	71.2	831
JP05800825	5	80	0.8	25	80	2.92	6.859	600.0	28229.17	9678.57	1278.30	0.7184	67.8	876
JP05801025	5	80	1.0	25	80	3.47	7.354	630.0	34071.18	9812.50	1582.66	0.7819	64.5	897
JP05800430	5	80	0.4	30	80	2.28	6.853	552.0	25923.91	11357.14	935.29	0.6130	74.3	956
JP05800630	5	80	0.6	30	80	3.21	7.946	588.0	37125.00	11550.00	1386.00	0.6511	71.3	1075
JP05800830	5	80	0.8	30	80	4.04	8.718	624.0	47443.08	11742.86	1826.67	0.7106	67.6	1124
JP05801030	5	80	1.0	30	80	4.77	9.290	660.0	56965.91	11935.71	2258.11	0.7762	64.0	1141

J-PLATE

n angle stiffeners
edges pinned
free in-plane

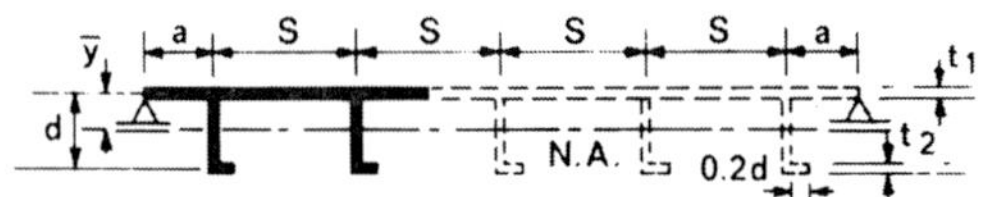

| CODE | n | DIMENSIONS | | | | $\bar{y}$ | rad. gyr. | A | I_{NA} | Z_{plate} | $Z_{stiff.}$ | σ_{cr}/E | L_1 | L_{euler} |
		S	t_2	d	a									
To scale multiply by	–	t_1	t_1	t_1	t_1	t_1	t_1	t_1^2	t_1^4	t_1^3	t_1^3	10^{-3}	t_1	t_1
JQ02350405	2	35	0.4	5	7	0.26	0.961	53.8	49.69	190.95	10.48	2.4162	71.8	936
JQ02350605	2	35	0.6	5	7	0.37	1.133	56.2	72.15	193.10	15.60	2.6838	78.2	895
JQ02350805	2	35	0.8	5	7	0.48	1.262	58.6	93.29	195.24	20.63	2.9223	83.0	864
JQ02351005	2	35	1.0	5	7	0.57	1.363	61.0	113.25	197.38	25.59	3.1585	87.0	836
JQ02350410	2	35	0.4	10	7	0.96	2.523	58.6	373.15	390.48	41.26	4.1540	48.2	722
JQ02350610	2	35	0.6	10	7	1.32	2.888	63.4	528.71	399.05	60.95	4.9406	120.7	670
JQ02350810	2	35	0.8	10	7	1.64	3.133	68.2	669.40	407.62	80.09	5.3229	130.1	652
JQ02351010	2	35	1.0	10	7	1.92	3.307	73.0	798.17	416.19	98.76	5.6381	136.7	638
JQ02350415	2	35	0.4	15	7	1.99	4.332	63.4	1189.59	598.57	91.42	2.8255	86.0	889
JQ02350615	2	35	0.6	15	7	2.68	4.840	70.6	1654.04	617.86	134.22	4.1205	73.1	745
JQ02350815	2	35	0.8	15	7	3.24	5.150	77.8	2063.75	637.14	175.48	5.2507	66.1	665
JQ02351015	2	35	1.0	15	7	3.71	5.350	85.0	2432.65	656.43	215.39	6.1456	62.1	618
JQ02350420	2	35	0.4	20	7	3.28	6.266	68.2	2677.61	815.24	160.19	2.1278	131.5	1036
JQ02350620	2	35	0.6	20	7	4.32	6.867	77.8	3668.89	849.52	233.97	3.0509	111.8	877
JQ02350820	2	35	0.8	20	7	5.13	7.200	87.4	4530.28	883.81	304.57	3.8975	102.3	783
JQ02351020	2	35	1.0	20	7	5.77	7.392	97.0	5300.34	918.10	372.56	4.6198	95.2	723
JQ02350425	2	35	0.4	25	7	4.79	8.267	73.0	4988.58	1040.48	246.89	1.7177	183.6	1162
JQ02350625	2	35	0.6	25	7	6.18	8.916	85.0	6757.35	1094.05	358.98	2.4198	155.6	994
JQ02350825	2	35	0.8	25	7	7.22	9.240	97.0	8281.79	1147.62	465.70	3.0609	141.2	892
JQ02351025	2	35	1.0	25	7	8.03	9.406	109.0	9642.58	1201.19	568.13	3.6190	133.2	825
JQ02350430	2	35	0.4	30	7	6.48	10.301	77.8	8255.01	1274.29	350.95	1.4463	242.8	1273
JQ02350630	2	35	0.6	30	7	8.20	10.963	92.2	11081.13	1351.43	508.30	2.0114	204.5	1097
JQ02350830	2	35	0.8	30	7	9.46	11.257	106.6	13508.44	1428.57	657.53	2.5161	186.4	990
JQ02351030	2	35	1.0	30	7	10.41	11.383	121.0	15679.34	1505.71	800.51	2.9537	176.1	918
JQ02500405	2	50	0.4	5	10	0.19	0.823	74.8	50.71	270.95	10.54	1.1064	97.5	1969
JQ02500605	2	50	0.6	5	10	0.27	0.981	77.2	74.29	273.10	15.71	1.2232	105.3	1884
JQ02500805	2	50	0.8	5	10	0.35	1.103	79.6	96.82	275.24	20.83	1.3290	111.5	1819
JQ02501005	2	50	1.0	5	10	0.43	1.202	82.0	118.39	277.38	25.89	1.4333	117.3	1761
JQ02500410	2	50	0.4	10	10	0.70	2.206	79.6	387.27	550.48	41.66	2.0725	149.3	1454
JQ02500610	2	50	0.6	10	10	1.00	2.568	84.4	556.40	559.05	61.79	2.3568	164.4	1378
JQ02500810	2	50	0.8	10	10	1.26	2.827	89.2	712.71	567.62	81.50	2.5773	175.7	1329
JQ02501010	2	50	1.0	10	10	1.49	3.021	94.0	858.16	576.19	100.83	2.7693	185.1	1293
JQ02500415	2	50	0.4	15	10	1.49	3.851	84.4	1251.90	838.57	92.68	2.6558	87.5	1296
JQ02500615	2	50	0.6	15	10	2.06	4.396	91.6	1770.03	857.86	136.82	3.2473	28.4	1187
JQ02500815	2	50	0.8	15	10	2.55	4.759	98.8	2237.25	877.14	179.71	3.4937	27.5	1157
JQ02501015	2	50	1.0	15	10	2.97	5.013	106.0	2663.92	896.43	221.47	3.9018	242.2	1103
JQ02500420	2	50	0.4	20	10	2.51	5.653	89.2	2850.82	1135.24	163.01	2.0431	132.8	1493
JQ02500620	2	50	0.6	20	10	3.40	6.345	98.8	3977.33	1169.52	239.61	2.9458	112.4	1260
JQ02500820	2	50	0.8	20	10	4.13	6.775	108.4	4975.15	1203.81	313.55	3.4368	27.5	1179
JQ02501020	2	50	1.0	20	10	4.75	7.056	118.0	5875.71	1238.10	385.19	3.7935	26.4	1132
JQ02500425	2	50	0.4	25	10	3.72	7.554	94.0	5363.48	1440.48	252.08	1.6496	184.5	1676
JQ02500625	2	50	0.6	25	10	4.95	8.355	106.0	7399.76	1494.05	369.12	2.3612	156.9	1421
JQ02500825	2	50	0.8	25	10	5.93	8.821	118.0	9180.79	1547.62	481.48	2.9873	140.9	1276
JQ02501025	2	50	1.0	25	10	6.73	9.105	130.0	10777.24	1601.19	589.91	3.4870	131.8	1190
JQ02500430	2	50	0.4	30	10	5.10	9.517	98.8	8948.99	1754.29	359.41	1.3879	241.4	1839
JQ02500630	2	50	0.6	30	10	6.68	10.395	113.2	12231.10	1831.43	524.45	1.9624	204.7	1571
JQ02500830	2	50	0.8	30	10	7.90	10.870	127.6	15077.12	1908.57	682.21	2.4781	185.5	1413
JQ02501030	2	50	1.0	30	10	8.87	11.139	142.0	17619.72	1985.71	834.00	2.9072	173.2	1314
JQ02650405	2	65	0.4	5	13	0.15	0.732	95.8	51.29	350.95	10.57	0.6220	121.7	>2000
JQ02650605	2	65	0.6	5	13	0.21	0.877	98.2	75.51	353.10	15.78	0.6837	131.3	>2000
JQ02650805	2	65	0.8	5	13	0.28	0.991	100.6	98.87	355.24	20.94	0.7401	138.3	>2000
JQ02651005	2	65	1.0	5	13	0.34	1.086	103.0	121.44	357.38	26.06	0.7956	145.1	>2000
JQ02650410	2	65	0.4	10	13	0.56	1.983	100.6	395.49	710.48	41.88	1.1505	185.8	>2000
JQ02650610	2	65	0.6	10	13	0.80	2.332	105.4	573.06	719.05	62.27	1.3160	203.8	>2000
JQ02650810	2	65	0.8	10	13	1.02	2.590	110.2	739.50	727.62	82.32	1.4492	217.6	>2000
JQ02651010	2	65	1.0	10	13	1.22	2.792	115.0	896.23	736.19	102.05	1.5672	228.9	>2000
JQ02650415	2	65	0.4	15	13	1.20	3.498	105.4	1289.37	1078.57	93.40	1.7859	243.8	>2000
JQ02650615	2	65	0.6	15	13	1.68	4.045	112.6	1842.76	1097.86	138.33	1.9565	36.7	1976
JQ02650815	2	65	0.8	15	13	2.10	4.429	119.8	2349.92	1117.14	182.21	2.1070	35.8	1923
JQ02651015	2	65	1.0	15	13	2.48	4.711	127.0	2818.70	1136.43	225.14	2.3213	35.0	1848
JQ02650420	2	65	0.4	20	13	2.03	5.181	110.2	2958.02	1455.24	164.63	1.8630	37.0	>2000
JQ02650620	2	65	0.6	20	13	2.80	5.905	119.8	4177.63	1489.52	242.95	1.9528	36.1	1998
JQ02650820	2	65	0.8	20	13	3.46	6.385	129.4	5275.63	1523.81	319.00	2.0995	34.9	1949
JQ02651020	2	65	1.0	20	13	4.03	6.720	139.0	6277.22	1558.10	393.03	2.3221	33.5	1870

J-PLATE   n angle stiffeners
edges pinned
free in-plane

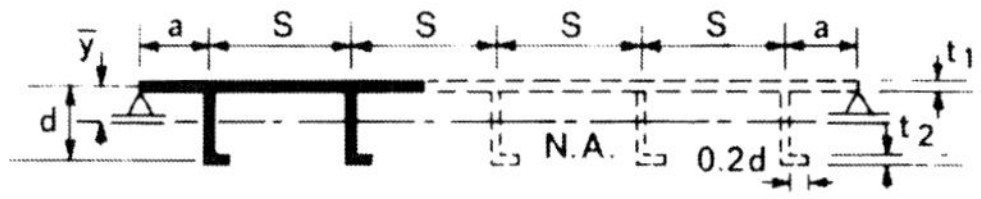

CODE	n	DIMENSIONS				SECTION PROPERTIES						BUCKLING PROPERTIES		
		S	t_2	d	a	$\bar{y}$	rad. gyr.	A	I_{NA}	Z_{plate}	$Z_{stiff.}$	σ_{cr}/E	L_1	L_{euler}
To scale multiply by	–	t_1	t_1	t_1	t_1	t_1	t_1	t_1^2	t_1^4	t_1^3	t_1^3	10^{-3}	t_1	t_1
JQ02650425	2	65	0.4	25	13	3.04	6.979	115.0	5601.45	1840.48	255.12	1.5989	186.2	>2000
JQ02650625	2	65	0.6	25	13	4.13	7.852	127.0	7829.72	1894.05	375.24	1.9245	36.1	>2000
JQ02650825	2	65	0.8	25	13	5.04	8.400	139.0	9808.15	1947.62	491.29	2.0624	35.1	1984
JQ02651025	2	65	1.0	25	13	5.79	8.763	151.0	11596.30	2001.19	603.81	2.2769	33.8	1906
JQ02650430	2	65	0.4	30	13	4.21	8.858	119.8	9399.67	2234.29	364.43	1.3471	242.6	>2000
JQ02650630	2	65	0.6	30	13	5.63	9.850	134.2	13021.16	2311.43	534.39	1.8700	36.1	>2000
JQ02650830	2	65	0.8	30	13	6.78	10.442	148.6	16202.42	2388.57	697.88	2.0120	35.5	>2000
JQ02651030	2	65	1.0	30	13	7.73	10.814	163.0	19060.12	2465.71	855.87	2.2219	34.5	1944
JQ02800405	2	80	0.4	5	16	0.12	0.665	116.8	51.66	430.95	10.58	0.3948	145.2	>2000
JQ02800605	2	80	0.6	5	16	0.18	0.800	119.2	76.30	433.10	15.82	0.4316	155.5	>2000
JQ02800805	2	80	0.8	5	16	0.23	0.908	121.6	100.22	435.24	21.01	0.4654	164.0	>2000
JQ02801005	2	80	1.0	5	16	0.28	0.998	124.0	123.45	437.38	26.17	0.4986	171.7	>2000
JQ02800410	2	80	0.4	10	16	0.46	1.816	121.6	400.88	870.48	42.02	0.7162	219.4	>2000
JQ02800610	2	80	0.6	10	16	0.66	2.150	126.4	584.18	879.05	62.58	0.8210	241.0	>2000
JQ02800810	2	80	0.8	10	16	0.85	2.403	131.2	757.72	887.62	82.84	0.9070	257.1	>2000
JQ02801010	2	80	1.0	10	16	1.03	2.605	136.0	922.55	896.19	102.84	0.9838	270.6	>2000
JQ02800415	2	80	0.4	15	16	1.00	3.225	126.4	1314.40	1318.57	93.86	1.1243	289.9	>2000
JQ02800615	2	80	0.6	15	16	1.41	3.764	133.6	1892.63	1337.86	139.31	1.2908	319.3	>2000
JQ02800815	2	80	0.8	15	16	1.79	4.153	140.8	2428.98	1357.14	183.87	1.4169	341.0	>2000
JQ02801015	2	80	1.0	15	16	2.13	4.449	148.0	2929.56	1376.43	227.60	1.5218	358.5	>2000
JQ02800420	2	80	0.4	20	16	1.71	4.806	131.2	3030.89	1775.24	165.69	1.2452	45.4	>2000
JQ02800620	2	80	0.6	20	16	2.39	5.538	140.8	4318.18	1809.52	245.16	1.3083	44.1	>2000
JQ02800820	2	80	0.8	20	16	2.98	6.043	150.4	5492.20	1843.81	322.67	1.4117	42.7	>2000
JQ02801020	2	80	1.0	20	16	3.50	6.410	160.0	6573.33	1878.10	398.38	1.5654	40.9	>2000
JQ02800425	2	80	0.4	25	16	2.57	6.511	136.0	5765.93	2240.48	257.10	1.2419	45.0	>2000
JQ02800625	2	80	0.6	25	16	3.55	7.415	148.0	8137.67	2294.05	379.33	1.2979	44.0	>2000
JQ02800825	2	80	0.8	25	16	4.38	8.012	160.0	10270.83	2347.62	497.98	1.3945	42.7	>2000
JQ02801025	2	80	1.0	25	16	5.09	8.427	172.0	12215.36	2401.19	613.44	1.5435	40.8	>2000
JQ02800430	2	80	0.4	30	16	3.58	8.307	140.8	9715.91	2714.29	367.74	1.3130	244.4	>2000
JQ02800630	2	80	0.6	30	16	4.87	9.360	155.2	13597.42	2791.43	541.11	1.2810	44.1	>2000
JQ02800830	2	80	0.8	30	16	5.94	10.026	169.6	17049.06	2868.57	708.71	1.3720	43.0	>2000
JQ02801030	2	80	1.0	30	16	6.85	10.470	184.0	20171.74	2945.71	871.27	1.5150	41.3	>2000
JQ03350405	3	35	0.4	5	7	0.23	0.908	91.2	75.16	326.43	15.76	0.9458	143.2	>2000
JQ03350605	3	35	0.6	5	7	0.33	1.075	94.8	109.53	329.64	23.47	1.0540	155.8	>2000
JQ03350805	3	35	0.8	5	7	0.43	1.202	98.4	142.07	332.86	31.07	1.1410	165.8	>2000
JQ03351005	3	35	1.0	5	7	0.51	1.302	102.0	172.98	336.07	38.57	1.2172	174.2	>2000
JQ03350410	3	35	0.4	10	7	0.85	2.403	98.4	568.29	665.71	62.13	1.9468	230.0	1779
JQ03350610	3	35	0.6	10	7	1.19	2.769	105.6	809.66	678.57	91.94	2.1740	253.0	1699
JQ03350810	3	35	0.8	10	7	1.49	3.021	112.8	1029.79	691.43	121.00	2.3275	269.9	1655
JQ03351010	3	35	1.0	10	7	1.75	3.205	120.0	1232.50	704.29	149.39	2.4429	283.6	1627
JQ03350415	3	35	0.4	15	7	1.79	4.153	105.6	1821.73	1017.86	137.90	2.8185	85.5	1490
JQ03350615	3	35	0.6	15	7	2.44	4.680	116.4	2549.52	1046.79	202.92	3.3067	336.8	1391
JQ03350815	3	35	0.8	15	7	2.97	5.013	127.2	3196.70	1075.71	265.76	3.4532	359.0	1373
JQ03351015	3	35	1.0	15	7	3.42	5.235	138.0	3782.20	1104.64	326.73	3.5402	376.8	1366
JQ03350420	3	35	0.4	20	7	2.98	6.043	112.8	4119.15	1382.86	242.00	2.1424	131.2	1724
JQ03350620	3	35	0.6	20	7	3.96	6.684	127.2	5683.02	1434.29	354.35	2.9939	111.8	1475
JQ03350820	3	35	0.8	20	7	4.75	7.056	141.6	7050.85	1485.71	462.22	3.6679	103.6	1343
JQ03351020	3	35	1.0	20	7	5.38	7.284	156.0	8276.92	1537.14	566.32	4.1334	97.9	1273
JQ03350425	3	35	0.4	25	7	4.38	8.012	120.0	7703.13	1760.71	373.48	1.7070	182.5	1946
JQ03350625	3	35	0.6	25	7	5.71	8.725	138.0	10506.11	1841.07	544.54	2.4093	154.1	1658
JQ03350825	3	35	0.8	25	7	6.73	9.105	156.0	12932.69	1921.43	707.89	2.9591	142.0	1509
JQ03351025	3	35	1.0	25	7	7.54	9.316	174.0	15099.68	2001.79	864.97	3.3729	136.0	1422
JQ03350430	3	35	0.4	30	7	5.94	10.026	127.2	12786.79	2151.43	531.53	1.4171	240.3	>2000
JQ03350630	3	35	0.6	30	7	7.62	10.776	148.8	17277.82	2267.14	772.05	2.0290	200.9	1818
JQ03350830	3	35	0.8	30	7	8.87	11.139	170.4	21143.66	2382.96	1000.80	2.4736	184.5	1661
JQ03351030	3	35	1.0	30	7	9.84	11.318	192.0	24595.31	2498.57	1220.23	2.8191	176.8	1565
JQ03500405	3	50	0.4	5	10	0.17	0.776	127.2	76.53	463.57	15.83	0.4247	189.9	>2000
JQ03500605	3	50	0.6	5	10	0.24	0.927	130.8	112.41	466.79	23.62	0.4730	206.7	>2000
JQ03500805	3	50	0.8	5	10	0.31	1.045	134.4	146.88	470.00	31.33	0.5129	220.0	>2000
JQ03501005	3	50	1.0	5	10	0.38	1.142	138.0	180.03	473.21	38.97	0.5485	230.9	>2000
JQ03500410	3	50	0.4	10	10	0.63	2.091	134.4	587.50	940.00	62.67	0.8664	304.6	>2000
JQ03500610	3	50	0.6	10	10	0.89	2.447	141.6	847.88	952.86	93.07	0.9828	334.8	>2000
JQ03500810	3	50	0.8	10	10	1.13	2.707	148.8	1090.32	965.71	122.91	1.0679	357.7	>2000
JQ03501010	3	50	1.0	10	10	1.35	2.906	156.0	1317.31	978.57	152.22	1.1356	376.1	>2000

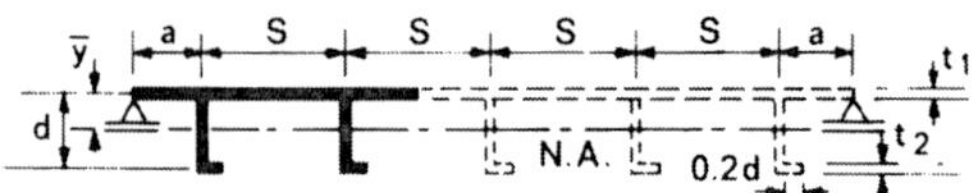

J-PLATE n angle stiffeners / edges pinned / free in-plane

CODE	n	S	t_2	d	a	$\bar{y}$	rad. gyr.	A	I_{NA}	Z_{plate}	$Z_{stiff.}$	σ_{cr}/E	L_1	L_{euler}
To scale multiply by	–	t_1	t_1	t_1	t_1	t_1	t_1	t_1^2	t_1^4	t_1^3	t_1^3	10^{-3}	t_1	t_1
JQ03500415	3	50	0.4	15	10	1.33	3.671	141.6	1907.73	1429.29	139.60	1.3922	407.7	>2000
JQ03500615	3	50	0.6	15	10	1.86	4.219	152.4	2712.62	1458.21	206.44	1.5582	448.5	>2000
JQ03500815	3	50	0.8	15	10	2.32	4.594	163.2	3444.49	1487.14	271.57	1.6660	478.5	>2000
JQ03501015	3	50	1.0	15	10	2.72	4.864	174.0	4116.92	1516.07	335.13	1.7424	502.6	>2000
JQ03500420	3	50	0.4	20	10	2.26	5.414	148.8	4361.29	1931.43	245.82	1.9349	500.0	>2000
JQ03500620	3	50	0.6	20	10	3.09	6.125	163.2	6123.53	1982.86	362.09	2.1252	549.5	>2000
JQ03500820	3	50	0.8	20	10	3.78	6.583	177.6	7697.30	2034.29	474.67	2.2300	586.4	>2000
JQ03501020	3	50	1.0	20	10	4.38	6.894	192.0	9125.00	2085.71	584.00	2.2924	615.5	>2000
JQ03500425	3	50	0.4	25	10	3.37	7.265	156.0	8233.17	2446.43	380.56	1.6545	182.8	>2000
JQ03500625	3	50	0.6	25	10	4.53	8.107	174.0	11435.88	2526.79	558.55	2.1591	36.7	>2000
JQ03500825	3	50	0.8	25	10	5.47	8.617	192.0	14257.81	2607.14	730.00	2.3550	35.3	>2000
JQ03501025	3	50	1.0	25	10	6.25	8.843	210.0	16796.88	2687.50	895.83	2.6562	33.5	>2000
JQ03500430	3	50	0.4	30	10	4.63	9.188	163.2	13777.94	2974.29	543.13	1.3968	240.1	>2000
JQ03500630	3	50	0.6	30	10	6.14	10.129	184.8	18961.36	3090.00	794.57	1.9282	204.7	>2000
JQ03500830	3	50	0.8	30	10	7.33	10.667	206.4	23483.72	3205.71	1035.69	2.2628	36.0	>2000
JQ03501030	3	50	1.0	30	10	8.29	10.989	228.0	27532.89	3321.43	1268.18	2.5576	34.5	>2000
JQ03650405	3	65	0.4	5	13	0.13	0.688	163.2	77.30	600.71	15.87	0.2359	234.6	>2000
JQ03650605	3	65	0.6	5	13	0.19	0.827	166.8	114.05	603.93	23.71	0.2619	255.0	>2000
JQ03650805	3	65	0.8	5	13	0.25	0.937	170.4	149.65	607.14	31.48	0.2838	270.7	>2000
JQ03651005	3	65	1.0	5	13	0.30	1.029	174.0	184.16	610.36	39.20	0.3034	284.2	>2000
JQ03650410	3	65	0.4	10	13	0.49	1.874	170.4	598.59	1214.29	62.96	0.4728	373.8	>2000
JQ03650610	3	65	0.6	10	13	0.71	2.214	177.6	870.61	1227.14	93.71	0.5402	410.8	>2000
JQ03650810	3	65	0.8	10	13	0.91	2.470	184.8	1127.27	1240.00	124.00	0.5915	439.0	>2000
JQ03651010	3	65	1.0	10	13	1.09	2.672	192.0	1370.31	1252.86	153.86	0.6334	461.9	>2000
JQ03650415	3	65	0.4	15	13	1.06	3.321	177.6	1958.87	1840.71	140.56	0.7658	501.2	>2000
JQ03650615	3	65	0.6	15	13	1.50	3.864	188.4	2813.40	1869.64	208.47	0.8696	550.3	>2000
JQ03650815	3	65	0.8	15	13	1.90	4.253	199.2	3602.71	1898.57	274.97	0.9421	588.2	>2000
JQ03651015	3	65	1.0	15	13	2.25	4.544	210.0	4336.88	1927.50	340.15	0.9967	618.3	>2000
JQ03650420	3	65	0.4	20	13	1.82	4.940	184.8	4509.09	2480.00	248.00	1.0845	616.8	>2000
JQ03650620	3	65	0.6	20	13	2.53	5.670	199.2	6404.82	2531.43	366.62	1.2148	677.3	>2000
JQ03650820	3	65	0.8	20	13	3.15	6.168	213.6	8125.84	2582.86	482.13	1.2974	723.1	>2000
JQ03651020	3	65	1.0	20	13	3.68	6.524	228.0	9705.26	2634.29	594.84	1.3541	762.8	>2000
JQ03650425	3	65	0.4	25	13	2.73	6.679	192.0	8564.45	3132.14	384.65	1.2537	48.6	>2000
JQ03650625	3	65	0.6	25	13	3.75	7.574	210.0	12046.88	3212.50	566.91	1.3248	47.0	>2000
JQ03650825	3	65	0.8	25	13	4.61	8.155	228.0	15164.47	3292.86	743.55	1.4510	44.7	>2000
JQ03651025	3	65	1.0	25	13	5.34	8.553	246.0	17997.33	3373.21	915.21	1.6831	889.1	>2000
JQ03650430	3	65	0.4	30	13	3.80	8.506	199.2	14410.84	3797.14	549.93	1.2276	48.1	>2000
JQ03650630	3	65	0.6	30	13	5.14	9.540	220.8	20095.92	3912.86	808.23	1.2998	47.3	>2000
JQ03650830	3	65	0.8	30	13	6.24	10.182	242.4	25128.71	4028.57	1057.50	1.4183	45.3	>2000
JQ03651030	3	65	1.0	30	13	7.16	10.601	264.0	29669.32	4144.29	1298.96	1.6012	42.8	>2000
JQ03800405	3	80	0.4	5	16	0.11	0.625	199.2	77.79	737.86	15.89	0.1484	277.6	>2000
JQ03800605	3	80	0.6	5	16	0.16	0.753	202.8	115.11	741.07	23.76	0.1641	300.6	>2000
JQ03800805	3	80	0.8	5	16	0.20	0.857	206.4	151.45	744.29	31.58	0.1775	319.1	>2000
JQ03801005	3	80	1.0	5	16	0.25	0.943	210.0	186.88	747.50	39.34	0.1896	334.7	>2000
JQ03800410	3	80	0.4	10	16	0.41	1.713	206.4	605.81	1488.57	63.15	0.2917	439.0	>2000
JQ03800610	3	80	0.6	10	16	0.59	2.036	213.6	885.67	1501.43	94.12	0.3346	482.6	>2000
JQ03800810	3	80	0.8	10	16	0.76	2.284	220.8	1152.17	1514.29	124.71	0.3680	515.8	>2000
JQ03801010	3	80	1.0	10	16	0.92	2.484	228.0	1406.58	1527.14	154.93	0.3957	542.9	>2000
JQ03800415	3	80	0.4	15	16	0.88	3.054	213.6	1992.77	2252.14	141.18	0.4733	588.5	>2000
JQ03800615	3	80	0.6	15	16	1.26	3.584	224.4	2881.83	2281.07	209.79	0.5421	647.2	>2000
JQ03800815	3	80	0.8	15	16	1.61	3.973	235.2	3712.50	2310.00	277.20	0.5922	691.6	>2000
JQ03801015	3	80	1.0	15	16	1.92	4.273	246.0	4492.45	2338.93	343.48	0.6312	727.9	>2000
JQ03800420	3	80	0.4	20	16	1.52	4.569	220.8	4608.70	3028.57	249.41	0.6755	725.6	>2000
JQ03800620	3	80	0.6	20	16	2.14	5.297	235.2	6600.00	3080.00	369.60	0.7663	797.4	>2000
JQ03800820	3	80	0.8	20	16	2.69	5.812	249.6	8430.77	3131.43	487.11	0.8279	851.5	>2000
JQ03801020	3	80	1.0	20	16	3.18	6.194	264.0	10127.27	3182.86	602.16	0.8727	894.7	>2000
JQ03800425	3	80	0.4	25	16	2.30	6.209	228.0	8791.12	3817.86	387.32	0.8386	59.6	>2000
JQ03800625	3	80	0.6	25	16	3.20	7.122	246.0	12479.04	3898.21	572.47	0.8906	57.4	>2000
JQ03800825	3	80	0.8	25	16	3.98	7.742	264.0	15823.86	3978.57	752.70	0.9513	54.3	>2000
JQ03801025	3	80	1.0	25	16	4.65	8.185	282.0	18891.29	4058.93	928.51	1.1072	1048.9	>2000
JQ03800430	3	80	0.4	30	16	3.21	7.946	235.2	14850.00	4620.00	554.40	0.8323	59.6	>2000
JQ03800630	3	80	0.6	30	16	4.42	9.024	256.8	20912.38	4735.71	817.40	0.8794	57.6	>2000
JQ03800830	3	80	0.8	30	16	5.43	9.728	278.4	26348.28	4851.43	1072.42	0.9643	54.6	>2000
JQ03801030	3	80	1.0	30	16	6.30	10.213	300.0	31293.00	4967.14	1320.38	1.0926	51.1	>2000

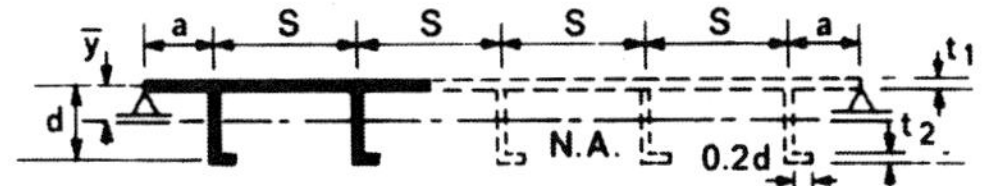

J-PLATE — n angle stiffeners, edges pinned, free in-plane

CODE	n	S	t_2	d	a	$\bar{y}$	rad. gyr.	A	I_{NA}	Z_{plate}	$Z_{stiff.}$	σ_{cr}/E	L_1	L_{euler}
To scale multiply by	–	t_1	t_1	t_1	t_1	t_1	t_1	t_1^2	t_1^4	t_1^3	t_1^3	10^{-3}	t_1	t_1
JQ05650425	5	65	0.4	25	13	2.53	6.463	346.0	14453.88	5715.48	643.22	0.4379	1356.9	>2000
JQ05650625	5	65	0.6	25	13	3.49	7.369	376.0	20418.47	5849.40	949.29	0.4841	1486.7	>2000
JQ05650825	5	65	0.8	25	13	4.31	7.970	406.0	25790.23	5983.33	1246.53	0.5108	1585.1	>2000
JQ05651025	5	65	1.0	25	13	5.02	8.390	436.0	30691.54	6117.26	1535.90	0.5269	1662.3	>2000
JQ05650430	5	65	0.4	30	13	3.52	8.250	358.0	24365.36	6922.86	920.13	0.5454	1548.2	>2000
JQ05650630	5	65	0.6	30	13	4.80	9.308	394.0	34133.76	7115.71	1354.35	0.5945	1694.7	>2000
JQ05650830	5	65	0.8	30	13	5.86	9.980	430.0	42831.63	7308.57	1774.34	0.6196	1803.3	>2000
JQ05651030	5	65	1.0	30	13	6.76	10.431	466.0	50707.08	7501.43	2181.86	0.6322	1890.3	>2000
JQ05800405	5	80	0.4	5	16	0.10	0.598	364.0	129.97	1351.67	26.50	0.0446	514.8	>2000
JQ05800605	5	80	0.6	5	16	0.14	0.721	370.0	192.55	1357.02	39.63	0.0493	558.1	>2000
JQ05800805	5	80	0.8	5	16	0.19	0.821	376.0	253.63	1362.38	52.69	0.0533	592.2	>2000
JQ05801005	5	80	1.0	5	16	0.23	0.906	382.0	313.29	1367.74	65.67	0.0569	624.3	>2000
JQ05800410	5	80	0.4	10	16	0.37	1.643	376.0	1014.54	2724.76	105.38	0.0883	818.5	>2000
JQ05800610	5	80	0.6	10	16	0.54	1.957	388.0	1486.34	2746.19	157.14	0.1012	899.4	>2000
JQ05800810	5	80	0.8	10	16	0.70	2.201	400.0	1937.33	2767.62	208.32	0.1112	961.2	>2000
JQ05801010	5	80	1.0	10	16	0.85	2.398	412.0	2369.34	2789.05	258.93	0.1193	1013.6	>2000
JQ05800415	5	80	0.4	15	16	0.81	2.936	388.0	3344.27	4119.29	235.71	0.1438	1097.5	>2000
JQ05800615	5	80	0.6	15	16	1.16	3.456	406.0	4850.11	4167.50	350.54	0.1646	1206.1	>2000
JQ05800815	5	80	0.8	15	16	1.49	3.844	424.0	6263.92	4215.71	463.51	0.1796	1288.5	>2000
JQ05801015	5	80	1.0	15	16	1.78	4.146	442.0	7596.93	4263.93	574.73	0.1911	1355.1	>2000
JQ05800420	5	80	0.4	20	16	1.40	4.402	400.0	7749.33	5535.24	416.63	0.2061	1353.9	>2000
JQ05800620	5	80	0.6	20	16	1.98	5.125	424.0	11135.85	5620.95	618.01	0.2337	1487.0	>2000
JQ05800820	5	80	0.8	20	16	2.50	5.643	448.0	14266.67	5706.67	815.24	0.2524	1588.4	>2000
JQ05801020	5	80	1.0	20	16	2.97	6.033	472.0	17180.79	5792.38	1008.62	0.2659	1668.7	>2000
JQ05800425	5	80	0.4	25	16	2.12	5.995	412.0	14808.35	6972.62	647.33	0.2727	1594.9	>2000
JQ05800625	5	80	0.6	25	16	2.97	6.910	442.0	21102.59	7106.55	957.88	0.3059	1749.4	>2000
JQ05800825	5	80	0.8	25	16	3.71	7.542	472.0	26844.99	7240.48	1260.78	0.3268	1863.5	>2000
JQ05801025	5	80	1.0	25	16	4.36	8.001	502.0	32134.48	7374.40	1556.72	0.9575	56.4	816
JQ05800430	5	80	0.4	30	16	2.97	7.687	424.0	25055.66	8431.43	927.02	0.3421	1819.4	>2000
JQ05800630	5	80	0.6	30	16	4.11	8.777	460.0	35434.57	8624.29	1368.59	0.7299	65.6	1038
JQ05800830	5	80	0.8	30	16	5.08	9.503	496.0	44796.77	8817.14	1797.67	0.8143	61.3	1071
JQ05801030	5	80	1.0	30	16	5.92	10.014	532.0	53348.68	9010.00	2215.57	0.9383	56.5	1052

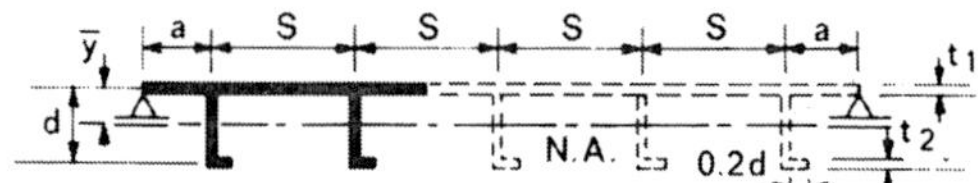

J-PLATE — n angle stiffeners, edges pinned, free in-plane

CODE	n	S	t2	d	a	ȳ	rad. gyr.	A	I_NA	Z_plate	Z_stiff.	σ_{cr}/E	L1	L_euler
To scale multiply by	−	t_1	t_1	t_1	t_1	t_1	t_1	t_1^2	t_1^4	t_1^3	t_1^3	10^{-3}	t_1	t_1
JQ05350405	5	35	0.4	5	7	0.21	0.871	166.0	125.95	597.38	26.30	0.2860	266.5	>2000
JQ05350605	5	35	0.6	5	7	0.31	1.034	172.0	183.98	602.74	39.19	0.3184	290.6	>2000
JQ05350805	5	35	0.8	5	7	0.39	1.159	178.0	239.14	608.10	51.91	0.3440	309.6	>2000
JQ05351005	5	35	1.0	5	7	0.48	1.259	184.0	291.72	613.45	64.48	0.3658	325.0	>2000
JQ05350410	5	35	0.4	10	7	0.79	2.319	178.0	956.55	1216.19	103.82	0.5974	432.1	>2000
JQ05350610	5	35	0.6	10	7	1.11	2.683	190.0	1367.89	1237.62	153.79	0.6669	474.4	>2000
JQ05350810	5	35	0.8	10	7	1.39	2.939	202.0	1745.21	1259.05	202.61	0.7133	505.7	>2000
JQ05351010	5	35	1.0	10	7	1.64	3.128	214.0	2094.24	1280.48	250.37	0.7473	530.7	>2000
JQ05350415	5	35	0.4	15	7	1.66	4.025	190.0	3077.76	1856.43	230.68	0.9559	578.7	>2000
JQ05350615	5	35	0.6	15	7	2.27	4.561	208.0	4326.65	1904.64	339.92	1.0446	633.7	>2000
JQ05350815	5	35	0.8	15	7	2.79	4.908	226.0	5443.81	1952.86	445.76	1.0939	676.3	>2000
JQ05351015	5	35	1.0	15	7	3.23	5.145	244.0	6458.38	2001.07	548.60	1.1234	707.6	>2000
JQ05350420	5	35	0.4	20	7	2.77	5.879	202.0	6980.86	2518.10	405.21	1.3234	712.2	>2000
JQ05350620	5	35	0.6	20	7	3.72	6.544	226.0	9677.88	2603.81	594.35	1.4123	779.0	>2000
JQ05350820	5	35	0.8	20	7	4.48	6.942	250.0	12049.07	2689.52	776.36	1.4485	828.0	>2000
JQ05351020	5	35	1.0	20	7	5.11	7.194	274.0	14180.05	2775.24	952.29	1.4611	867.2	>2000
JQ05350425	5	35	0.4	25	7	4.09	7.821	214.0	13088.98	3201.19	625.93	1.6758	837.1	>2000
JQ05350625	5	35	0.6	25	7	5.38	8.575	244.0	17939.93	3335.12	914.33	1.7486	911.7	>2000
JQ05350825	5	35	0.8	25	7	6.39	8.992	274.0	22156.33	3469.05	1190.36	1.7599	968.4	>2000
JQ05351025	5	35	1.0	25	7	7.20	9.235	304.0	25926.02	3602.98	1456.17	1.7478	1014.1	>2000
JQ05350430	5	35	0.4	30	7	5.58	9.816	226.0	21775.22	3905.71	891.52	1.4180	238.6	>2000
JQ05350630	5	35	0.6	30	7	7.21	10.623	262.0	29566.03	4098.57	1297.54	2.0320	200.7	>2000
JQ05350830	5	35	0.8	30	7	8.46	11.035	298.0	36289.93	4291.43	1684.49	2.0186	1099.5	>2000
JQ05351030	5	35	1.0	30	7	9.43	11.253	334.0	42291.92	4484.29	2056.11	1.9782	1151.2	>2000
JQ05500405	5	50	0.4	5	10	0.15	0.743	232.0	128.05	848.81	26.41	0.1280	353.3	>2000
JQ05500605	5	50	0.6	5	10	0.22	0.890	238.0	188.42	854.17	39.42	0.1425	384.8	>2000
JQ05500805	5	50	0.8	5	10	0.29	1.005	244.0	246.58	859.52	52.32	0.1543	409.4	>2000
JQ05501005	5	50	1.0	5	10	0.35	1.100	250.0	302.71	864.98	65.10	0.1646	429.9	>2000
JQ05500410	5	50	0.4	10	10	0.57	2.011	244.0	986.34	1719.05	104.64	0.2635	569.4	>2000
JQ05500610	5	50	0.6	10	10	0.82	2.362	256.0	1427.73	1740.48	155.53	0.2986	625.3	>2000
JQ05500810	5	50	0.8	10	10	1.04	2.621	268.0	1840.80	1761.90	205.56	0.3239	668.7	>2000
JQ05501010	5	50	1.0	10	10	1.25	2.822	280.0	2229.17	1783.33	254.76	0.3437	702.0	>2000
JQ05500415	5	50	0.4	15	10	1.23	3.542	256.0	3212.40	2610.71	233.30	0.4274	764.8	>2000
JQ05500615	5	50	0.6	15	10	1.72	4.091	274.0	4585.20	2658.93	345.39	0.4785	841.8	>2000
JQ05500815	5	50	0.8	15	10	2.16	4.472	292.0	5840.75	2707.14	454.80	0.5116	894.0	>2000
JQ05501015	5	50	1.0	15	10	2.54	4.752	310.0	6999.50	2755.36	561.77	0.5347	938.6	>2000
JQ05500420	5	50	0.4	20	10	2.09	5.242	268.0	7363.18	3523.81	411.11	0.6042	941.6	>2000
JQ05500620	5	50	0.6	20	10	2.88	5.963	292.0	10383.56	3609.52	606.40	0.6652	1031.5	>2000
JQ05500820	5	50	0.8	20	10	3.54	6.438	316.0	13097.05	3695.24	795.90	0.6999	1098.3	>2000
JQ05501020	5	50	1.0	20	10	4.12	6.767	340.0	15568.63	3780.95	980.25	0.7208	1151.7	>2000
JQ05500425	5	50	0.4	25	10	3.13	7.054	280.0	13932.29	4458.33	636.90	0.7857	1106.3	>2000
JQ05500625	5	50	0.6	25	10	4.23	7.920	310.0	19443.04	4592.26	936.29	0.8499	1210.0	>2000
JQ05500825	5	50	0.8	25	10	5.15	8.459	340.0	24325.98	4726.19	1225.31	0.8805	1287.0	>2000
JQ05501025	5	50	1.0	25	10	5.91	8.812	370.0	28733.81	4860.12	1505.35	0.8947	1348.9	>2000
JQ05500430	5	50	0.4	30	10	4.32	8.945	292.0	23363.01	5414.29	909.60	0.9658	1262.5	>2000
JQ05500630	5	50	0.6	30	10	5.76	9.925	328.0	32309.45	5607.14	1333.02	1.0269	1377.7	>2000
JQ05500830	5	50	0.8	30	10	6.92	10.593	364.0	40153.85	5800.00	1740.00	1.0486	1464.4	>2000
JQ05501030	5	50	1.0	30	10	7.88	10.862	400.0	47193.75	5992.86	2133.05	1.0526	1533.8	>2000
JQ05650405	5	65	0.4	5	13	0.12	0.659	298.0	129.22	1100.24	26.47	0.0710	435.7	>2000
JQ05650605	5	65	0.6	5	13	0.17	0.793	304.0	190.93	1105.60	39.55	0.0788	473.9	>2000
JQ05650805	5	65	0.8	5	13	0.23	0.900	310.0	250.86	1110.95	52.55	0.0853	503.6	>2000
JQ05651005	5	65	1.0	5	13	0.28	0.969	316.0	309.10	1116.31	65.45	0.0911	528.4	>2000
JQ05650410	5	65	0.4	10	13	0.45	1.799	310.0	1003.44	2221.90	105.09	0.1433	697.4	>2000
JQ05650610	5	65	0.6	10	13	0.65	2.132	322.0	1463.04	2243.33	156.51	0.1636	766.8	>2000
JQ05650810	5	65	0.8	10	13	0.84	2.384	334.0	1898.60	2264.76	207.23	0.1789	819.0	>2000
JQ05651010	5	65	1.0	10	13	1.01	2.585	346.0	2312.62	2286.19	257.29	0.1912	861.7	>2000
JQ05650415	5	65	0.4	15	13	0.98	3.197	322.0	3291.85	3365.00	234.77	0.2334	935.9	>2000
JQ05650615	5	65	0.6	15	13	1.39	3.735	340.0	4743.36	3413.21	348.51	0.2648	1027.9	>2000
JQ05650815	5	65	0.8	15	13	1.76	4.125	358.0	6091.34	3461.43	460.06	0.2866	1097.1	>2000
JQ05651015	5	65	1.0	15	13	2.09	4.421	376.0	7350.65	3509.64	569.57	0.3028	1152.8	>2000
JQ05650420	5	65	0.4	20	13	1.68	4.768	334.0	7594.41	4529.52	414.47	0.3329	1153.9	>2000
JQ05650620	5	65	0.6	20	13	2.35	5.500	358.0	10829.05	4615.24	613.42	0.3732	1265.8	>2000
JQ05650820	5	65	0.8	20	13	2.93	6.007	382.0	13782.90	4700.95	807.53	0.3998	1349.4	>2000
JQ05651020	5	65	1.0	20	13	3.45	6.376	406.0	16505.75	4786.67	957.22	0.4162	1418.4	>2000

U-PLATE n trough stiffeners
edges free out-of-plane
free in-plane

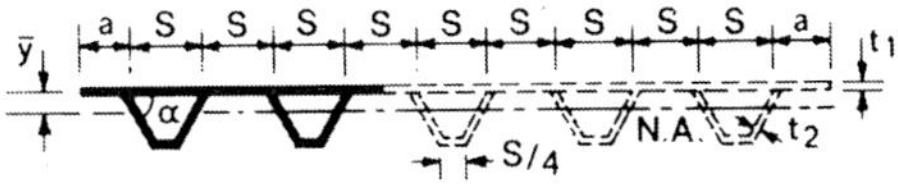

| CODE | n | DIMENSIONS | | | | SECTION PROPERTIES | | | | | | BUCKLING PROPERTIES | | |
| | | S | t_2 | α | a | $\bar{y}$ | rad. gyr. | A | I_{NA} | Z_{plate} | $Z_{stiff.}$ | σ_{cr}/E | L_1 | L_{euler} |
To scale multiply by	–	t_1	t_1	–	t_1	t_1	t_1	t_1^2	t_1^4	t_1^3	t_1^3	10^{-3}	t_1	t_1
UF01350445	1	35	0.4	45	7	2.13	4.133	67.3	1150.33	540.32	104.61	3.7356	32.2	201
UF01350645	1	35	0.6	45	7	2.81	4.542	76.5	1578.90	561.77	153.08	4.1002	29.9	214
UF01350845	1	35	0.8	45	7	3.35	4.772	85.7	1951.58	583.21	199.57	4.7129	27.1	210
UF01351045	1	35	1.0	45	7	3.78	4.907	94.9	2284.60	604.65	244.43	5.5994	24.4	198
UF01500445	1	50	0.4	45	10	3.04	5.904	96.2	3353.75	1102.69	213.50	1.8547	45.7	421
UF01500645	1	50	0.6	45	10	4.02	6.489	109.3	4603.21	1146.46	312.40	2.0293	42.5	446
UF01500845	1	50	0.8	45	10	4.78	6.817	122.4	5689.75	1190.22	407.30	2.3293	38.4	438
UF01501045	1	50	1.0	45	10	5.40	7.010	135.5	6660.64	1233.99	498.84	2.7656	34.7	414
UF01650445	1	65	0.4	45	13	3.95	7.675	125.1	7368.18	1863.55	360.81	1.1028	59.2	719
UF01650645	1	65	0.6	45	13	5.22	8.436	142.1	10113.25	1937.52	527.96	1.2053	55.1	760
UF01650845	1	65	0.8	45	13	6.21	8.862	159.2	12500.38	2011.48	688.33	1.3828	49.8	747
UF01651045	1	65	1.0	45	13	7.02	9.113	176.2	14633.42	2085.44	843.03	1.6413	45.0	705
UF01800445	1	80	0.4	45	16	4.87	9.446	153.9	13736.94	2822.90	546.55	0.7298	72.8	1095
UF01800645	1	80	0.6	45	16	6.42	10.382	174.9	18854.74	2934.94	799.75	0.7971	67.9	1156
UF01800845	1	80	0.8	45	16	7.65	10.908	195.9	23305.22	3046.97	1042.68	0.9143	61.3	1136
UF01801045	1	80	1.0	45	16	8.64	11.216	216.9	27281.97	3159.01	1277.02	1.0851	55.3	1072
UF01350460	1	35	0.4	60	7	4.33	7.421	73.5	4048.24	934.90	219.98	2.1022	15.0	506
UF01350660	1	35	0.6	60	7	5.57	7.995	85.8	5481.74	984.63	319.34	3.6237	29.7	413
UF01350860	1	35	0.8	60	7	6.50	8.280	98.0	6718.36	1034.36	413.74	4.1426	28.7	401
UF01351060	1	35	1.0	60	7	7.22	8.424	110.3	7823.73	1084.09	504.23	4.9127	26.9	374
UF01500460	1	50	0.4	60	10	6.19	10.602	105.0	11802.46	1907.96	448.93	1.0308	21.4	1040
UF01500660	1	50	0.6	60	10	7.95	11.422	122.5	15981.74	2009.45	651.71	1.7831	42.3	851
UF01500860	1	50	0.8	60	10	9.28	11.828	140.0	19587.05	2110.94	844.37	2.0380	40.8	825
UF01501060	1	50	1.0	60	10	10.31	12.034	157.5	22809.71	2212.42	1029.03	2.4167	38.2	771
UF01650460	1	65	0.4	60	13	8.04	13.783	136.5	25929.99	3224.46	758.70	0.6102	28.0	1763
UF01650660	1	65	0.6	60	13	10.34	14.849	159.3	35111.89	3395.97	1101.40	1.0568	54.9	1443
UF01650860	1	65	0.8	60	13	12.06	15.377	182.0	43032.76	3567.48	1426.99	1.2078	53.3	1399
UF01651060	1	65	1.0	60	13	13.40	15.645	204.8	50112.93	3739.00	1739.07	1.4322	49.6	1308
UF01800460	1	80	0.4	60	16	9.90	16.963	168.0	48342.86	4884.38	1149.27	0.4028	34.3	>2000
UF01800660	1	80	0.6	60	16	12.73	18.275	196.0	65461.22	5144.19	1668.39	0.6982	67.7	>2000
UF01800860	1	80	0.8	60	16	14.85	18.925	224.0	90228.57	5404.00	2161.60	0.7979	65.1	>2000
UF01801060	1	80	1.0	60	16	16.50	19.255	252.0	93428.57	5663.81	2634.33	0.9462	61.1	1985
UF01350470	1	35	0.4	70	7	8.17	12.161	83.2	12304.91	1506.11	441.18	1.0517	20.5	1188
UF01350670	1	35	0.6	70	7	10.17	12.796	100.3	16422.56	1615.50	634.20	2.1700	23.7	856
UF01350870	1	35	0.8	70	7	11.58	13.044	117.4	19974.03	1724.88	815.91	2.9267	30.2	721
UF01351070	1	35	1.0	70	7	12.63	13.127	134.5	23175.26	1834.27	989.30	3.5921	31.1	639
UF01500470	1	50	0.4	70	10	11.67	17.373	118.9	35874.37	3073.70	900.38	0.5155	29.4	>2000
UF01500670	1	50	0.6	70	10	14.52	18.280	143.3	47879.18	3296.93	1294.28	1.0642	33.8	1754
UF01500870	1	50	0.8	70	10	16.54	18.634	167.7	58233.31	3520.17	1665.12	1.4365	43.2	1479
UF01501070	1	50	1.0	70	10	18.05	18.752	192.1	67566.35	3743.40	2018.97	1.7675	46.6	1311
UF01650470	1	65	0.4	70	13	15.17	22.585	154.5	78815.98	5194.55	1521.63	0.3051	38.2	>2000
UF01650670	1	65	0.6	70	13	18.88	23.764	186.3	105190.57	5571.81	2187.34	0.6299	44.0	>2000
UF01650870	1	65	0.8	70	13	21.51	24.224	218.0	127938.59	5949.08	2814.05	0.8505	56.1	>2000
UF01651070	1	65	1.0	70	13	23.46	24.378	249.8	148443.26	6326.34	3412.06	1.0444	57.8	>2000
UF01800470	1	80	0.4	70	16	18.67	27.797	190.2	146941.41	7868.67	2304.96	0.2014	46.9	>2000
UF01800670	1	80	0.6	70	16	23.24	29.248	229.3	196113.14	8440.14	3313.36	0.4159	54.1	>2000
UF01800870	1	80	0.8	70	16	26.47	29.814	268.3	238523.65	9011.62	4262.71	0.5617	69.0	>2000
UF01801070	1	80	1.0	70	16	28.88	30.004	307.4	276751.76	9583.10	5168.57	0.6897	71.1	>2000
UF02350445	2	35	0.4	45	7	1.84	3.912	155.7	2383.02	1293.83	211.20	3.3338	32.1	203
UF02350645	2	35	0.6	45	7	2.47	4.357	174.0	3303.68	1336.71	310.10	3.5860	30.3	221
UF02350845	2	35	0.8	45	7	2.98	4.623	192.4	4112.65	1379.60	405.43	4.0299	28.0	222
UF02351045	2	35	1.0	45	7	3.40	4.792	210.7	4839.12	1422.49	497.69	4.6801	25.6	214
UF02500445	2	50	0.4	45	10	2.63	5.589	222.4	6947.57	2640.46	431.02	1.6402	45.8	426
UF02500645	2	50	0.6	45	10	3.53	6.224	248.6	9631.71	2727.99	632.86	1.7621	43.3	460
UF02500845	2	50	0.8	45	10	4.26	6.605	274.9	11990.24	2815.52	827.41	1.9794	40.0	462
UF02501045	2	50	1.0	45	10	4.86	6.846	301.1	14108.22	2903.05	1015.70	2.2990	36.6	445
UF02650445	2	65	0.4	45	13	3.42	7.266	289.2	15263.81	4462.18	728.43	0.9720	59.4	727
UF02650645	2	65	0.6	45	13	4.59	8.091	323.2	21160.86	4610.30	1069.54	1.0438	56.2	785
UF02650845	2	65	0.8	45	13	5.54	8.586	357.3	26342.56	4758.23	1398.31	1.1723	52.0	787
UF02651045	2	65	1.0	45	13	6.32	8.899	391.4	30995.75	4906.15	1716.53	1.3617	47.5	757
UF02800445	2	80	0.4	45	16	4.21	8.942	355.9	28457.24	6759.58	1103.42	0.6423	74.2	1107
UF02800645	2	80	0.6	45	16	5.65	9.958	397.8	39451.47	6983.65	1620.12	0.6894	69.2	1193
UF02800845	2	80	0.8	45	16	6.81	10.568	439.8	49112.02	7207.73	2118.16	0.7743	63.9	1197
UF02801045	2	80	1.0	45	16	7.78	10.953	481.7	57787.26	7431.80	2600.18	0.8994	58.5	1151

U-PLATE n trough stiffeners
edges free out-of-plane
free in-plane

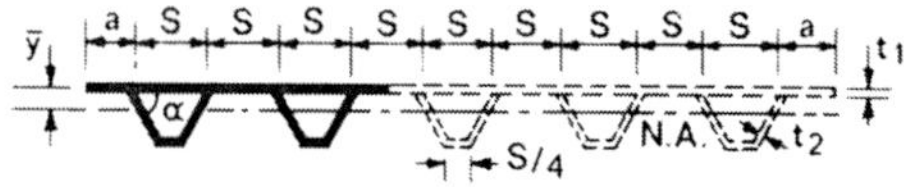

CODE	n	S	t₂	α	a	ȳ	rad gyr.	A	I_{NA}	Z_{plate}	$Z_{stiff.}$	σ_{cr}/E	L_1	L_{euler}
To scale multiply by	−	t_1	t_1	−	t_1	t_1	t_1	t_1^2	t_1^4	t_1^3	t_1^3	10^{-3}	t_1	t_1
UF02350460	2	35	0.4	60	7	3.79	7.088	168.0	8441.02	2227.85	445.57	2.1040	15.0	482
UF02350660	2	35	0.6	60	7	4.96	7.744	192.5	11543.36	2327.31	649.48	3.2886	30.8	420
UF02350860	2	35	0.8	60	7	5.87	8.100	217.0	14236.92	2426.77	844.09	3.6688	29.4	417
UF02351060	2	35	1.0	60	7	6.59	8.302	241.5	16646.10	2526.22	1031.11	4.2484	27.7	397
UF02500460	2	50	0.4	60	10	5.41	10.126	240.0	24609.38	4546.63	909.33	1.0317	21.4	992
UF02500660	2	50	0.6	60	10	7.09	11.062	275.0	33654.12	4749.61	1325.47	1.6137	44.0	866
UF02500860	2	50	0.8	60	10	8.38	11.571	310.0	41507.06	4952.58	1722.64	1.8000	42.1	859
UF02501060	2	50	1.0	60	10	9.41	11.860	345.0	48530.91	5155.56	2104.31	2.0844	39.6	815
UF02650460	2	65	0.4	60	13	7.04	13.164	312.0	54066.80	7683.81	1536.76	0.6107	28.0	1682
UF02650660	2	65	0.6	60	13	9.21	14.381	357.5	73938.10	8026.84	2240.05	0.9554	57.1	1469
UF02650860	2	65	0.8	60	13	10.90	15.043	403.0	91191.00	8369.86	2911.26	1.0656	54.6	1457
UF02651060	2	65	1.0	60	13	12.24	15.419	448.5	106622.41	8712.89	3556.28	1.2340	51.5	1389
UF02800460	2	80	0.4	60	16	8.66	16.202	384.0	100800.00	11639.38	2327.88	0.4032	34.2	>2000
UF02800660	2	80	0.6	60	16	11.34	17.700	440.0	137847.27	12159.00	3393.21	0.6369	70.2	>2000
UF02800860	2	80	0.8	60	16	13.41	18.514	496.0	170012.90	12678.61	4409.95	0.7036	67.4	>2000
UF02801060	2	80	1.0	60	16	15.06	18.977	552.0	198782.61	13198.23	5387.03	0.8148	63.3	>2000
UF02350470	2	35	0.4	70	7	7.25	11.746	187.4	25854.46	3563.94	897.53	1.0537	20.5	1145
UF02350670	2	35	0.6	70	7	9.20	12.533	221.6	34809.61	3782.71	1296.04	2.1892	23.9	845
UF02350870	2	35	0.8	70	7	10.63	12.895	255.8	42532.87	4001.48	1672.46	2.8317	31.0	765
UF02351070	2	35	1.0	70	7	11.72	13.060	290.0	49460.07	4220.25	2031.97	3.3763	31.4	710
UF02500470	2	50	0.4	70	10	10.36	16.780	267.7	75377.43	7273.35	1831.70	0.5165	29.2	>2000
UF02500670	2	50	0.6	70	10	13.15	17.905	316.6	101485.75	7719.82	2644.99	1.0735	34.1	1733
UF02500870	2	50	0.8	70	10	15.18	18.421	365.4	124002.54	8166.29	3413.18	1.3587	44.4	1569
UF02501070	2	50	1.0	70	10	16.74	18.657	414.3	144198.45	8612.75	4146.88	1.6560	44.9	1456
UF02650470	2	65	0.4	70	13	13.47	21.814	348.0	165604.22	12291.96	3095.57	0.3056	38.0	>2000
UF02650670	2	65	0.6	70	13	17.09	23.276	411.5	222964.19	13046.49	4470.03	0.6354	44.4	>2000
UF02650870	2	65	0.8	70	13	19.74	23.947	475.1	272433.57	13801.02	5768.27	0.8220	57.6	>2000
UF02651070	2	65	1.0	70	13	21.77	24.253	538.6	316804.00	14555.55	7008.22	0.9802	58.3	>2000
UF02800470	2	80	0.4	70	16	16.58	26.848	428.3	308745.97	18619.78	4689.14	0.2018	46.8	>2000
UF02800670	2	80	0.6	70	16	21.03	28.648	506.5	415685.63	19762.74	6771.17	0.4195	54.6	>2000
UF02800870	2	80	0.8	70	16	24.30	29.474	584.7	507914.39	20905.69	8737.74	0.5427	70.9	>2000
UF02801070	2	80	1.0	70	16	26.79	29.850	662.9	590636.86	22048.65	10616.01	0.6472	71.7	>2000
UF03350445	3	35	0.4	45	7	1.76	3.845	244.0	3608.61	2047.33	317.59	3.2500	32.1	202
UF03350645	3	35	0.6	45	7	2.38	4.298	271.6	5017.17	2111.66	466.75	3.4815	30.5	221
UF03350845	3	35	0.8	45	7	2.88	4.575	299.1	6259.01	2176.00	610.72	3.8967	28.3	224
UF03351045	3	35	1.0	45	7	3.29	4.752	326.6	7376.28	2240.33	750.19	4.5107	26.0	216
UF03500445	3	50	0.4	45	10	2.52	5.493	348.6	10520.73	4178.23	648.15	1.5959	45.9	424
UF03500645	3	50	0.6	45	10	3.39	6.140	388.0	14627.32	4309.52	952.56	1.7085	43.6	461
UF03500845	3	50	0.8	45	10	4.11	6.535	427.3	18247.85	4440.81	1246.36	1.9119	40.5	465
UF03501045	3	50	1.0	45	10	4.70	6.789	466.6	21505.20	4572.11	1531.01	2.2135	37.1	450
UF03650445	3	65	0.4	45	13	3.27	7.141	453.2	23114.05	7061.20	1095.37	0.9451	59.6	724
UF03650645	3	65	0.6	45	13	4.41	7.982	504.3	32136.21	7283.09	1609.82	1.0115	56.7	786
UF03650845	3	65	0.8	45	13	5.34	8.496	555.5	40090.52	7504.97	2106.35	1.1319	52.6	792
UF03651045	3	65	1.0	45	13	6.11	8.826	606.6	47246.93	7726.86	2587.40	1.3105	48.2	766
UF03800445	3	80	0.4	45	16	4.03	8.789	557.8	43092.92	10696.26	1659.26	0.6243	74.5	1103
UF03800645	3	80	0.6	45	16	5.43	9.824	620.7	59913.49	11032.37	2438.55	0.6679	69.7	1196
UF03800845	3	80	0.8	45	16	6.57	10.456	683.6	74743.18	11368.48	3190.69	0.7474	64.7	1205
UF03801045	3	80	1.0	45	16	7.53	10.862	746.6	88085.30	11704.60	3919.38	0.8654	59.4	1164
UF03350460	3	35	0.4	60	7	3.64	6.985	262.5	12806.23	3520.80	670.63	2.1045	15.0	475
UF03350660	3	35	0.6	60	7	4.79	7.661	299.3	17564.29	3669.99	978.66	3.2127	31.0	420
UF03350860	3	35	0.8	60	7	5.68	8.037	336.0	21705.47	3819.17	1273.06	3.5669	29.7	420
UF03351060	3	35	1.0	60	7	6.40	8.257	372.8	25412.21	3968.36	1556.22	4.1129	28.0	402
UF03500460	3	50	0.4	60	10	5.20	9.978	375.0	37335.94	7185.30	1368.63	1.0319	21.4	977
UF03500660	3	50	0.6	60	10	6.84	10.945	427.5	51207.85	7489.77	1997.27	1.5755	44.3	867
UF03500860	3	50	0.8	60	10	8.12	11.482	480.0	63281.25	7794.23	2593.08	1.7491	42.5	865
UF03501060	3	50	1.0	60	10	9.15	11.795	532.5	74088.08	8098.69	3175.96	2.0168	40.0	828
UF03650460	3	65	0.4	60	13	6.75	12.972	487.5	82027.05	12143.16	2312.98	0.6108	28.0	1657
UF03650660	3	65	0.6	60	13	8.89	14.228	555.8	112503.65	12657.71	3375.39	0.9335	57.5	1471
UF03650860	3	65	0.8	60	13	10.55	14.927	624.0	139028.91	13172.25	4390.75	1.0352	55.1	1467
UF03651060	3	65	1.0	60	13	11.89	15.334	692.3	162771.52	13686.79	5367.37	1.1937	52.0	1404
UF03800460	3	80	0.4	60	16	8.31	15.965	600.0	152928.00	18394.38	3503.69	0.4033	34.2	>2000
UF03800660	3	80	0.6	60	16	10.94	17.511	684.0	209764.37	19173.80	5113.01	0.6157	70.8	>2000
UF03800860	3	80	0.8	60	16	12.99	18.371	768.0	259200.00	19953.23	6651.08	0.6835	68.0	>2000
UF03801060	3	80	1.0	60	16	14.64	18.873	852.0	303464.79	20732.65	8130.45	0.7882	64.0	>2000

U-PLATE n trough stiffeners / edges free out-of-plane / free in-plane

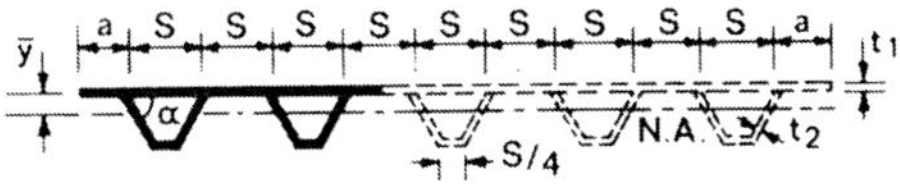

CODE	n	S	t_2	α	a	$\bar{y}$	rad. gyr.	A	I_{NA}	Z_{plate}	$Z_{stiff.}$	σ_{cr}/E	L_1	L_{euler}
To scale multiply by	–	t_1	t_1	–	t_1	t_1	t_1	t_1^2	t_1^4	t_1^3	t_1^3	10^{-3}	t_1	t_1
UF03350470	3	35	0.4	70	7	6.99	11.611	291.6	39314.38	5621.77	1352.52	1.0540	20.5	1131
UF03350670	3	35	0.6	70	7	8.92	12.441	342.9	53076.35	5949.93	1955.64	2.2011	25.7	837
UF03350870	3	35	0.8	70	7	10.35	12.836	394.2	64954.01	6278.08	2525.97	2.8045	31.2	765
UF03351070	3	35	1.0	70	7	11.44	13.027	445.5	75598.30	6606.23	3070.96	3.3194	31.5	714
UF03500470	3	50	0.4	70	10	9.99	16.588	416.6	114619.18	11473.00	2760.25	0.5166	29.2	>2000
UF03500670	3	50	0.6	70	10	12.74	17.773	489.9	154741.55	12142.71	3991.10	1.0756	34.0	1718
UF03500870	3	50	0.8	70	10	14.78	18.338	563.1	189370.30	12812.41	5155.04	1.3776	46.5	1567
UF03501070	3	50	1.0	70	10	16.35	18.609	636.4	220403.21	13482.11	6267.26	1.6276	45.0	1464
UF03650470	3	65	0.4	70	13	12.99	21.564	541.5	251818.33	19389.38	4664.83	0.3057	38.0	>2000
UF03650670	3	65	0.6	70	13	16.57	23.105	636.8	339967.20	20521.17	6744.97	0.6367	44.3	>2000
UF03650870	3	65	0.8	70	13	19.21	23.839	732.1	416046.55	21652.97	8712.02	0.8139	58.0	>2000
UF03651070	3	65	1.0	70	13	21.25	24.192	827.4	484225.86	22784.76	10591.68	0.9633	58.5	>2000
UF03800470	3	80	0.4	70	16	15.98	26.540	666.5	469480.14	29370.89	7066.25	0.2018	46.8	>2000
UF03800670	3	80	0.6	70	16	20.39	28.437	783.8	633821.41	31085.33	10217.23	0.4203	54.5	>2000
UF03800870	3	80	0.8	70	16	23.65	29.340	901.0	775660.75	32799.76	13196.91	0.5374	71.4	>2000
UF03801070	3	80	1.0	70	16	26.16	29.775	1018.3	902771.56	34514.20	16044.20	0.6360	72.0	>2000
UF05350445	5	35	0.4	45	7	1.70	3.794	420.7	6056.40	3554.34	530.28	3.1980	32.2	201
UF05350645	5	35	0.6	45	7	2.30	4.253	466.6	8438.61	3661.56	779.88	3.4185	30.7	221
UF05350845	5	35	0.8	45	7	2.80	4.536	512.5	10544.35	3768.79	1021.43	3.8196	28.6	224
UF05351045	5	35	1.0	45	7	3.21	4.720	558.4	12441.77	3876.01	1254.84	4.4171	26.2	217
UF05500445	5	50	0.4	45	10	2.43	5.420	601.1	17657.14	7253.76	1082.21	1.5686	46.0	422
UF05500645	5	50	0.6	45	10	3.29	6.075	666.6	24602.36	7472.58	1591.60	1.6763	43.9	461
UF05500845	5	50	0.8	45	10	4.00	6.480	732.1	30741.55	7691.40	2083.73	1.8729	40.8	466
UF05501045	5	50	1.0	45	10	4.59	6.743	797.7	36273.38	7910.23	2560.89	2.1663	37.5	452
UF05650445	5	65	0.4	45	13	3.16	7.046	781.4	38792.74	12258.85	1828.94	0.9285	59.8	721
UF05650645	5	65	0.6	45	13	4.28	7.898	866.6	54051.38	12628.66	2689.80	0.9921	57.0	785
UF05650845	5	65	0.8	45	13	5.20	8.424	951.8	67539.19	12998.47	3521.51	1.1086	53.4	794
UF05651045	5	65	1.0	45	13	5.96	8.767	1037.0	79692.61	13368.28	4327.90	1.2823	48.7	769
UF05800445	5	80	0.4	45	16	3.89	8.672	961.7	72323.65	18669.01	2770.46	0.6131	73.5	1098
UF05800645	5	80	0.6	45	16	5.27	9.720	1066.6	100771.26	19129.80	4074.49	0.6550	70.1	1195
UF05800845	5	80	0.8	45	16	6.39	10.368	1171.4	125917.39	19689.99	5334.35	0.7319	65.2	1207
UF05801045	5	80	1.0	45	16	7.34	10.790	1276.3	148575.75	20250.18	6555.88	0.8466	60.0	1169
UF05350460	5	35	0.4	60	7	3.52	6.904	451.5	21523.19	6106.70	1120.49	2.1046	15.0	469
UF05350660	5	35	0.6	60	7	4.66	7.596	512.8	29585.74	6355.34	1636.57	3.1646	31.2	420
UF05350860	5	35	0.8	60	7	5.54	7.987	574.0	36616.95	6603.98	2130.32	3.5052	29.9	421
UF05351060	5	35	1.0	60	7	6.26	8.219	635.3	42915.14	6852.63	2605.56	4.0347	28.2	404
UF05500460	5	50	0.4	60	10	5.04	9.863	645.0	62749.82	12462.65	2286.72	1.0320	21.4	965
UF05500660	5	50	0.6	60	10	6.65	10.852	732.5	86255.80	12970.08	3339.94	1.5513	44.6	866
UF05500860	5	50	0.8	60	10	7.92	11.410	820.0	106754.95	13477.52	4347.59	1.7182	42.7	867
UF05501060	5	50	1.0	60	10	8.95	11.742	907.5	125117.03	13984.96	5317.47	1.9779	40.3	832
UF05650460	5	65	0.4	60	13	6.55	12.822	838.5	137861.35	21061.87	3864.56	0.6109	28.0	1637
UF05650660	5	65	0.6	60	13	8.65	14.107	952.3	189503.59	21919.44	5644.49	0.9181	57.9	1470
UF05650860	5	65	0.8	60	13	10.30	14.833	1066.0	234540.63	22777.01	7347.42	1.0169	55.5	1471
UF05651060	5	65	1.0	60	13	11.63	15.264	1179.8	274882.11	23634.58	8986.53	1.1705	52.4	1412
UF05800460	5	80	0.4	60	16	8.06	15.781	1032.0	257023.26	31904.38	5854.01	0.4033	34.2	>2000
UF05800660	5	80	0.6	60	16	10.64	17.362	1172.0	353303.75	33203.41	8550.24	0.6061	71.2	>2000
UF05800860	5	80	0.8	60	16	12.67	18.256	1312.0	437268.29	34502.45	11129.82	0.6714	68.5	>2000
UF05801060	5	80	1.0	60	16	14.31	18.787	1452.0	512479.34	35801.49	13612.73	0.7728	64.5	>2000
UF05350470	5	35	0.4	70	7	6.80	11.506	500.0	66189.47	9737.43	2261.87	1.0541	20.5	1120
UF05350670	5	35	0.6	70	7	8.71	12.367	585.5	89548.01	10284.36	3273.74	2.2044	25.8	831
UF05350870	5	35	0.8	70	7	10.13	12.788	671.0	109724.02	10831.28	4231.49	2.7853	31.4	765
UF05351070	5	35	1.0	70	7	11.23	12.997	756.5	127796.59	11378.20	5147.08	3.2920	31.6	716
UF05500470	5	50	0.4	70	10	9.71	16.437	714.3	192972.20	19872.31	4616.05	0.5167	29.2	>2000
UF05500670	5	50	0.6	70	10	12.44	17.667	836.4	261072.91	20988.48	6681.10	1.0767	33.8	1706
UF05500870	5	50	0.8	70	10	14.47	18.268	958.6	319895.11	22104.65	8635.70	1.3655	44.8	1568
UF05501070	5	50	1.0	70	10	16.05	18.568	1080.7	372584.80	23220.82	10504.25	1.6090	45.2	1469
UF05650470	5	65	0.4	70	13	12.62	21.368	928.6	423959.93	33584.21	7801.13	0.3058	38.0	>2000
UF05650670	5	65	0.6	70	13	16.17	22.967	1087.4	573577.17	35470.53	11291.06	0.6373	44.0	>2000
UF05650870	5	65	0.8	70	13	18.81	23.748	1246.1	702809.56	37356.86	14594.32	0.8081	58.2	>2000
UF05651070	5	65	1.0	70	13	20.86	24.138	1404.9	818568.81	39243.18	17752.18	0.9522	58.7	>2000
UF05800470	5	80	0.4	70	16	15.54	26.299	1142.9	790414.15	50873.12	11817.09	0.2019	46.8	>2000
UF05800670	5	80	0.6	70	16	19.90	28.267	1338.3	1069354.62	53730.51	17103.62	0.4248	54.1	>2000
UF05800870	5	80	0.8	70	16	23.15	29.229	1533.7	1310290.38	56587.90	22107.38	0.5335	71.7	>2000
UF05801070	5	80	1.0	70	16	25.67	29.708	1729.1	1526107.35	59445.30	26890.88	0.6287	72.2	>2000

U-PLATE   n trough stiffeners
edges pinned
free in-plane

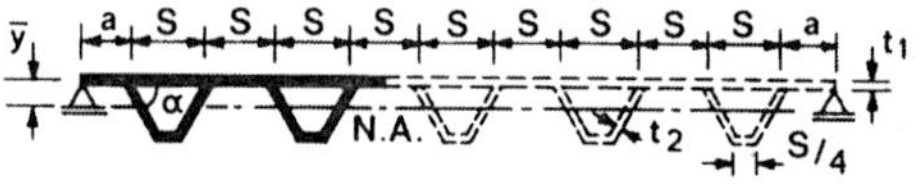

CODE	n	S	t_2	α	a	$\bar{y}$	rad. gyr.	A	I_{NA}	Z_{plate}	$Z_{stiff.}$	σ_{cr}/E	L_1	L_{euler}
To scale multiply by	$-$	t_1	t_1	$-$	t_1	t_1	t_1	t_1^2	t_1^4	t_1^3	t_1^3	10^{-3}	t_1	t_1
UP01350445	1	35	0.4	45	35	1.16	3.233	123.3	1288.92	1108.81	107.75	2.0571	379.4	1962
UP01350645	1	35	0.6	45	35	1.62	3.720	132.5	1834.34	1130.26	159.48	2.4044	397.1	1758
UP01350845	1	35	0.8	45	35	2.02	4.056	141.7	2330.83	1151.70	209.96	2.7065	402.8	1610
UP01351045	1	35	1.0	45	35	2.38	4.298	150.9	2787.32	1173.15	259.31	2.9233	406.5	1508
UP01500445	1	50	0.4	45	50	1.66	4.618	176.2	3757.80	2262.88	219.89	1.4178	651.1	>2000
UP01500645	1	50	0.6	45	50	2.32	5.315	189.3	5347.94	2306.65	325.47	1.6628	681.5	>2000
UP01500845	1	50	0.8	45	50	2.89	5.794	202.4	6795.42	2350.41	428.49	1.8755	691.1	>2000
UP01501045	1	50	1.0	45	50	3.39	6.140	215.5	8126.29	2394.18	529.20	1.8873	41.0	>2000
UP01650445	1	65	0.4	45	65	2.16	6.003	229.1	8255.88	3824.27	371.62	0.9135	60.7	>2000
UP01650645	1	65	0.6	45	65	3.01	6.909	246.1	11749.41	3898.23	550.04	0.9557	60.3	>2000
UP01650845	1	65	0.8	45	65	3.76	7.532	263.2	14929.55	3972.20	724.16	1.0260	56.1	>2000
UP01651045	1	65	1.0	45	65	4.41	7.982	280.2	17853.45	4046.16	894.35	1.1171	53.5	>2000
UP01800445	1	80	0.4	45	80	2.66	7.389	281.9	15391.94	5792.98	562.92	0.6030	74.6	>2000
UP01800645	1	80	0.6	45	80	3.71	8.504	302.9	21905.14	5905.02	833.20	0.6304	72.4	>2000
UP01800845	1	80	0.8	45	80	4.63	9.270	323.9	27834.06	6017.06	1096.95	0.6774	69.0	>2000
UP01801045	1	80	1.0	45	80	5.43	9.824	344.9	33285.27	6129.09	1354.75	0.7375	65.6	>2000
UP01350460	1	35	0.4	60	35	2.46	5.989	129.5	4644.19	1889.69	229.05	2.1080	15.0	1898
UP01350660	1	35	0.6	60	35	3.37	6.788	141.8	6531.74	1939.42	337.29	3.1311	31.9	1496
UP01350860	1	35	0.8	60	35	4.13	7.307	154.0	8221.77	1989.15	442.03	3.3583	31.1	1395
UP01351060	1	35	1.0	60	35	4.79	7.661	166.3	9757.94	2038.88	543.70	3.6723	29.8	1292
UP01500460	1	50	0.4	60	50	3.51	8.555	185.0	13539.91	3856.52	467.46	1.0337	21.3	>2000
UP01500660	1	50	0.6	60	50	4.81	9.697	202.5	19042.97	3958.01	688.35	1.5344	45.7	>2000
UP01500860	1	50	0.8	60	50	5.90	10.438	220.0	23970.17	4059.49	902.11	1.6457	44.4	>2000
UP01501060	1	50	1.0	60	50	6.84	10.945	237.5	28448.81	4160.98	1109.60	1.7997	42.7	>2000
UP01650460	1	65	0.4	60	65	4.56	11.122	240.5	29747.18	6517.52	790.00	0.6118	27.9	>2000
UP01650660	1	65	0.6	60	65	6.25	12.607	263.3	41837.40	6689.03	1163.31	0.9080	59.3	>2000
UP01650860	1	65	0.8	60	65	7.68	13.570	286.0	52662.46	6860.54	1524.57	0.9738	57.7	>2000
UP01651060	1	65	1.0	60	65	8.89	14.228	308.8	62502.03	7032.06	1875.22	1.0650	55.4	>2000
UP01800460	1	80	0.4	60	80	5.62	13.688	296.0	55459.46	9872.69	1196.69	0.4040	34.1	>2000
UP01800660	1	80	0.6	60	80	7.70	15.516	324.0	78000.00	10132.50	1762.17	0.5996	74.1	>2000
UP01800860	1	80	0.8	60	80	9.45	16.701	352.0	98181.82	10392.30	2309.40	0.6429	71.0	>2000
UP01801060	1	80	1.0	60	80	10.94	17.511	380.0	116526.32	10652.11	2840.56	0.7031	68.4	>2000
UP01350470	1	35	0.4	70	35	4.88	10.220	139.2	14539.07	2977.36	466.33	1.0570	20.4	>2000
UP01350670	1	35	0.6	70	35	6.52	11.350	156.3	20136.20	3086.75	681.72	2.2268	22.9	1707
UP01350870	1	35	0.8	70	35	7.84	12.021	173.4	25058.19	3196.13	887.94	2.8363	31.6	1448
UP01351070	1	35	1.0	70	35	8.92	12.441	190.5	29486.86	3305.51	1086.47	3.2249	32.1	1306
UP01500470	1	50	0.4	70	50	6.98	14.600	198.9	42387.95	6076.25	951.70	0.5181	29.1	>2000
UP01500670	1	50	0.6	70	50	9.32	16.215	223.3	58706.12	6299.48	1391.27	1.0920	32.7	>2000
UP01500870	1	50	0.8	70	50	11.20	17.173	247.7	73055.96	6522.71	1812.13	1.3904	45.1	>2000
UP01501070	1	50	1.0	70	50	12.74	17.773	272.1	85967.53	6745.95	2217.28	1.5806	45.9	>2000
UP01650470	1	65	0.4	70	65	9.07	18.980	258.5	93126.33	10268.86	1608.37	0.3066	37.9	>2000
UP01650670	1	65	0.6	70	65	12.11	21.079	290.3	128977.34	10646.12	2351.25	0.6463	42.6	>2000
UP01650870	1	65	0.8	70	65	14.56	22.325	322.0	160503.94	11023.39	3062.50	0.8228	58.6	>2000
UP01651070	1	65	1.0	70	65	16.57	23.105	353.8	188870.66	11400.65	3747.20	0.9354	59.6	>2000
UP01800470	1	80	0.4	70	80	11.16	23.360	318.2	173621.06	15555.19	2436.35	0.2024	46.6	>2000
UP01800670	1	80	0.6	70	80	14.91	25.944	357.3	240460.26	16126.67	3561.66	0.4267	52.3	>2000
UP01800870	1	80	0.8	70	80	17.92	27.477	396.3	299237.21	16698.15	4639.05	0.5433	72.1	>2000
UP01801070	1	80	1.0	70	80	20.39	28.437	435.4	352123.01	17269.63	5676.24	0.6175	73.3	>2000
UP02350460	2	35	0.4	60	35	2.84	6.354	224.0	9043.95	3182.64	454.66	2.4362	660.6	>2000
UP02350660	2	35	0.6	60	35	3.84	7.124	248.5	12610.57	3282.10	667.55	2.6975	692.2	>2000
UP02350860	2	35	0.8	60	35	4.66	7.600	273.0	15768.93	3381.56	872.66	2.8734	707.3	>2000
UP02351060	2	35	1.0	60	35	5.35	7.911	297.5	18619.89	3481.02	1071.08	2.9876	716.7	>2000
UP02500460	2	50	0.4	60	50	4.06	9.077	320.0	26367.19	6495.19	927.88	1.0332	21.3	>2000
UP02500660	2	50	0.6	60	50	5.49	10.177	355.0	36765.51	6698.17	1362.34	1.5338	45.3	>2000
UP02500860	2	50	0.8	60	50	6.66	10.857	390.0	45973.56	6901.14	1780.94	1.6643	43.9	>2000
UP02501060	2	50	1.0	60	50	7.64	11.302	425.0	54285.39	7104.11	2185.88	1.8421	42.1	>2000
UP02650460	2	65	0.4	60	65	5.28	11.800	416.0	57928.71	10976.87	1568.12	0.6115	27.9	>2000
UP02650660	2	65	0.6	60	65	7.14	13.230	461.5	80773.84	11319.90	2302.35	0.9076	58.9	>2000
UP02650860	2	65	0.8	60	65	8.66	14.114	507.0	101003.91	11662.93	3009.79	0.9849	57.1	>2000
UP02651060	2	65	1.0	60	65	9.93	14.692	552.5	119264.99	12005.95	3694.14	1.0901	54.6	>2000
UP02800460	2	80	0.4	60	80	6.50	14.524	512.0	108000.00	16627.69	2375.38	0.4038	34.2	>2000
UP02800660	2	80	0.6	60	80	8.78	16.283	568.0	150591.55	17147.30	3487.59	0.5992	72.4	>2000
UP02800860	2	80	0.8	60	80	10.66	17.372	624.0	188307.69	17666.92	4559.20	0.6502	70.2	>2000
UP02801060	2	80	1.0	60	80	12.23	18.083	680.0	222352.94	18186.53	5595.86	0.7197	67.4	>2000

U-PLATE n trough stiffeners / edges pinned / free in-plane

CODE	n	S	t_2	α	a	$\bar{y}$	rad. gyr.	A	I_{NA}	Z_{plate}	$Z_{stiff.}$	σ_{cr}/E	L_1	L_{euler}
To scale multiply by	$-$	t_1	t_1	$-$	t_1	t_1	t_1	t_1^2	t_1^4	t_1^3	t_1^3	10^{-3}	t_1	t_1
UP02350470	2	35	0.4	70	35	5.59	10.749	243.4	28123.52	5035.19	922.83	1.0564	20.4	>2000
UP02350670	2	35	0.6	70	35	7.35	11.791	277.6	38595.17	5253.96	1344.09	2.2182	23.5	>2000
UP02350870	2	35	0.8	70	35	8.72	12.372	311.8	47723.50	5472.73	1745.53	2.8032	31.5	>2000
UP02351070	2	35	1.0	70	35	9.82	12.711	346.0	55906.85	5691.50	2130.78	3.2366	31.9	>2000
UP02500470	2	50	0.4	70	50	7.98	15.356	347.7	81992.78	10275.90	1883.33	0.5178	29.1	>2000
UP02500670	2	50	0.6	70	50	10.49	16.845	396.6	112522.37	10722.37	2743.04	1.0876	33.6	>2000
UP02500870	2	50	0.8	70	50	12.46	17.674	445.4	139135.58	11168.84	3562.31	1.3741	45.0	>2000
UP02501070	2	50	1.0	70	50	14.03	18.159	494.3	162993.74	11615.30	4348.53	1.5864	45.7	>2000
UP02650470	2	65	0.4	70	65	10.37	19.963	452.0	180138.14	17366.27	3182.83	0.3064	37.9	>2000
UP02650670	2	65	0.6	70	65	13.64	21.898	515.5	247211.65	18120.80	4635.74	0.6438	47.8	>2000
UP02650870	2	65	0.8	70	65	16.19	22.976	579.1	305680.87	18875.33	6020.30	0.8132	58.5	>2000
UP02651070	2	65	1.0	70	65	18.24	23.607	642.6	358097.24	19629.86	7349.01	0.9388	59.3	>2000
UP02800470	2	80	0.4	70	80	12.77	24.570	556.3	335842.43	26306.30	4821.32	0.2023	46.6	>2000
UP02800670	2	80	0.6	70	80	16.79	26.951	634.5	460891.63	27449.26	7022.19	0.4250	53.8	>2000
UP02800870	2	80	0.8	70	80	19.93	28.278	712.7	569899.33	28592.22	9119.51	0.5369	72.0	>2000
UP02801070	2	80	1.0	70	80	22.45	29.055	790.9	667622.34	29735.18	11132.23	0.6290	74.2	>2000
UP03350445	3	35	0.4	45	35	1.43	3.535	300.0	3750.12	2615.82	320.76	1.0017	621.5	>2000
UP03350645	3	35	0.6	45	35	1.97	4.015	327.6	5279.25	2680.16	473.25	1.1530	640.9	>2000
UP03350845	3	35	0.8	45	35	2.42	4.327	355.1	6649.27	2744.49	621.30	1.2554	651.9	>2000
UP03351045	3	35	1.0	45	35	2.81	4.542	382.6	7894.50	2808.83	765.39	1.3214	661.9	>2000
UP03500445	3	50	0.4	45	50	2.05	5.050	428.6	10933.29	5338.41	654.61	0.7511	971.3	>2000
UP03500645	3	50	0.6	45	50	2.81	5.735	468.0	15391.40	5469.71	965.82	0.8586	1016.2	>2000
UP03500845	3	50	0.8	45	50	3.46	6.182	507.3	19385.61	5601.00	1267.95	0.9316	1036.8	>2000
UP03501045	3	50	1.0	45	50	4.02	6.489	546.6	23016.04	5732.30	1562.01	0.9792	1054.9	>2000
UP03650445	3	65	0.4	45	65	2.66	6.566	557.2	24020.43	9021.92	1106.29	0.5995	1374.7	>2000
UP03650645	3	65	0.6	45	65	3.66	7.456	608.3	33814.90	9243.81	1632.24	0.6827	1441.6	>2000
UP03650845	3	65	0.8	45	65	4.50	8.036	659.5	42590.19	9465.69	2142.84	0.7396	1480.2	>2000
UP03651045	3	65	1.0	45	65	5.22	8.436	710.6	50566.25	9687.58	2639.80	0.7769	1510.4	>2000
UP03800445	3	80	0.4	45	80	3.28	8.081	685.8	44782.75	13666.34	1675.80	0.4983	1828.2	>2000
UP03800645	3	80	0.6	45	80	4.50	9.176	748.7	63043.17	14002.45	2472.50	0.6385	71.5	1450
UP03800845	3	80	0.8	45	80	5.54	9.891	811.6	79403.47	14338.57	3245.96	0.6962	67.8	1481
UP03801045	3	80	1.0	45	80	6.42	10.382	874.6	94273.71	14674.68	3998.76	0.7635	64.1	1443
UP03350460	3	35	0.4	60	35	3.00	6.490	318.5	13416.84	4475.59	679.84	1.8248	805.9	>2000
UP03350660	3	35	0.6	60	35	4.03	7.245	355.3	18644.78	4624.78	996.96	2.0303	838.3	>2000
UP03350860	3	35	0.8	60	35	4.87	7.702	392.0	23255.86	4773.97	1301.99	2.1536	849.1	>2000
UP03351060	3	35	1.0	60	35	5.57	7.995	428.8	27408.69	4923.15	1596.70	2.2275	859.7	>2000
UP03500460	3	50	0.4	60	50	4.28	9.272	455.0	39116.16	9133.86	1387.42	1.0330	21.3	>2000
UP03500660	3	50	0.6	60	50	5.76	10.349	507.5	54357.97	9438.32	2034.61	1.4938	1345.7	>2000
UP03500860	3	50	0.8	60	50	6.96	11.003	560.0	67801.34	9742.79	2657.12	1.5784	1375.4	>2000
UP03501060	3	50	1.0	60	50	7.95	11.422	612.5	79908.72	10047.25	3258.57	1.6291	1396.8	>2000
UP03650460	3	65	0.4	60	65	5.57	12.054	591.5	85938.20	15436.23	2344.74	0.6114	27.9	>2000
UP03650660	3	65	0.6	60	65	7.49	13.454	659.8	119424.46	15950.77	3438.49	0.9074	58.7	>2000
UP03650860	3	65	0.8	60	65	9.05	14.304	728.0	148959.54	16465.31	4490.54	0.9893	56.8	>2000
UP03651060	3	65	1.0	60	65	10.34	14.849	796.3	175559.46	16979.85	5506.98	1.0985	54.3	>2000
UP03800460	3	80	0.4	60	80	6.85	14.835	728.0	160219.78	23382.69	3551.80	0.4037	34.2	>2000
UP03800660	3	80	0.6	60	80	9.21	16.559	812.0	222650.25	24162.11	5208.60	0.5991	72.2	>2000
UP03800860	3	80	0.8	60	80	11.13	17.605	896.0	277714.29	24941.53	6802.24	0.6531	69.9	>2000
UP03801060	3	80	1.0	60	80	12.73	18.275	980.0	327306.12	25720.95	8341.93	0.7253	67.1	>2000
UP03350470	3	35	0.4	70	35	5.87	10.941	347.6	41611.86	7093.02	1378.15	1.0562	20.4	>2000
UP03350670	3	35	0.6	70	35	7.67	11.944	398.9	56906.99	7421.17	2004.30	2.2142	23.8	>2000
UP03350870	3	35	0.8	70	35	9.06	12.487	450.2	70202.78	7749.33	2599.97	2.7905	31.5	>2000
UP03351070	3	35	1.0	70	35	10.17	12.796	501.5	82112.80	8077.48	3170.99	3.2412	31.9	>2000
UP03500470	3	50	0.4	70	50	8.38	15.630	496.6	121317.37	14475.55	2812.55	0.5177	29.1	>2000
UP03500670	3	50	0.6	70	50	10.95	17.063	569.9	165909.58	15145.25	4090.41	1.0856	34.0	>2000
UP03500870	3	50	0.8	70	50	12.94	17.839	643.1	204672.83	15814.96	5306.05	1.3679	45.0	>2000
UP03501070	3	50	1.0	70	50	14.52	18.280	716.4	239395.92	16484.66	6471.41	1.5887	45.6	>2000
UP03650470	3	65	0.4	70	65	10.90	20.320	645.5	266534.27	24463.68	4753.20	0.3064	37.9	>2000
UP03650670	3	65	0.6	70	65	14.24	22.182	740.8	364503.35	25595.48	6912.79	0.6426	44.3	>2000
UP03650870	3	65	0.8	70	65	16.82	23.191	836.1	449666.22	26727.28	8967.23	0.8095	58.5	>2000
UP03651070	3	65	1.0	70	65	18.88	23.764	931.4	525952.84	27859.07	10936.68	0.9401	59.2	>2000
UP03800470	3	80	0.4	70	80	13.41	25.009	794.5	496915.96	37057.42	7200.12	0.2023	46.6	>2000
UP03800670	3	80	0.6	70	80	17.53	27.301	911.8	679565.64	38771.85	10471.44	0.4242	54.5	>2000
UP03800870	3	80	0.8	70	80	20.71	28.543	1029.0	838339.93	40486.29	13583.49	0.5344	71.9	>2000
UP03801070	3	80	1.0	70	80	23.24	29.248	1146.3	980565.69	42200.72	16566.81	0.6207	72.8	>2000

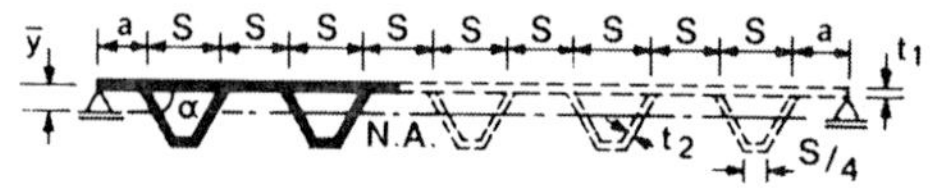

CODE	n	S	t_2	α	a	$\bar{y}$	rad. gyr.	A	I_{NA}	Z_{plate}	$Z_{stiff.}$	σ_{cr}/E	L_1	L_{euler}
To scale multiply by	−	t_1	t_1	−	t_1	t_1	t_1	t_1^2	t_1^4	t_1^3	t_1^3	10^{-3}	t_1	t_1
UP05350445	5	35	0.4	45	35	1.50	3.606	476.7	6199.89	4122.83	533.50	0.5758	1033.8	>2000
UP05350645	5	35	0.6	45	35	2.06	4.081	522.6	8704.18	4230.06	786.48	0.6760	1031.8	>2000
UP05350845	5	35	0.8	45	35	2.52	4.387	568.5	10939.52	4337.28	1031.76	0.7386	1034.6	>2000
UP05351045	5	35	1.0	45	35	2.92	4.594	614.4	12966.18	4444.50	1270.24	0.7767	1044.3	>2000
UP05500445	5	50	0.4	45	50	2.15	5.152	681.1	18075.49	8413.94	1088.77	0.4730	1419.2	>2000
UP05500645	5	50	0.6	45	50	2.94	5.830	746.6	25376.61	8632.77	1605.05	0.5473	1435.5	>2000
UP05500845	5	50	0.8	45	50	3.60	6.267	812.1	31893.66	8851.59	2105.63	0.5936	1452.4	>2000
UP05501045	5	50	1.0	45	50	4.17	6.563	877.7	37802.28	9070.42	2592.33	0.6224	1474.0	>2000
UP05650445	5	65	0.4	45	65	2.79	6.697	885.4	39711.85	14219.57	1840.02	0.3963	1856.7	>2000
UP05650645	5	65	0.6	45	65	3.82	7.579	970.6	55752.40	14589.38	2712.54	0.4543	1912.0	>2000
UP05650845	5	65	0.8	45	65	4.68	8.147	1055.8	70070.37	14959.19	3558.52	1.0597	54.8	842
UP05651045	5	65	1.0	45	65	5.42	8.532	1141.0	83051.61	15329.00	4381.04	1.1589	52.1	838
UP05800445	5	80	0.4	45	80	3.44	8.243	1089.7	74037.20	21539.70	2787.26	0.6041	74.2	1162
UP05800645	5	80	0.6	45	80	4.70	9.328	1194.6	103942.58	22099.89	4108.94	0.6406	71.2	1287
UP05800845	5	80	0.8	45	80	5.77	10.027	1299.4	130636.43	22660.07	5390.41	0.6997	67.6	1317
UP05801045	5	80	1.0	45	80	6.67	10.501	1404.3	154838.14	23220.26	6636.37	0.7653	64.1	1306
UP05350460	5	35	0.4	60	35	3.14	6.605	507.5	22142.07	7061.49	1129.84	1.1223	1225.8	>2000
UP05350660	5	35	0.6	60	35	4.20	7.345	568.8	30679.85	7310.13	1655.12	1.2775	1211.1	>2000
UP05350860	5	35	0.8	60	35	5.05	7.785	630.0	38185.55	7558.78	2159.65	1.3630	1208.8	>2000
UP05351060	5	35	1.0	60	35	5.76	8.062	691.3	44933.53	7807.42	2646.58	1.4106	1212.8	>2000
UP05500460	5	50	0.4	60	50	4.48	9.436	725.0	64554.15	14411.20	2305.79	0.8933	1745.6	>2000
UP05500660	5	50	0.6	60	50	6.00	10.492	812.5	89445.61	14918.64	3377.81	0.9992	1773.7	>2000
UP05500860	5	50	0.8	60	50	7.22	11.122	900.0	111328.13	15426.08	4407.45	1.0565	1799.3	>2000
UP05501060	5	50	1.0	60	50	8.22	11.518	987.5	131001.53	15933.51	5401.19	1.0884	1822.6	>2000
UP05650460	5	65	0.4	60	65	5.82	12.267	942.5	141825.46	24354.93	3896.79	0.6113	27.9	>2000
UP05650660	5	65	0.6	60	65	7.79	13.640	1056.3	196512.01	25212.50	5708.49	0.9072	58.5	1821
UP05650860	5	65	0.8	60	65	9.38	14.459	1170.0	244587.89	26070.07	7448.59	0.9930	56.6	1784
UP05651060	5	65	1.0	60	65	10.69	14.973	1283.8	287810.37	26927.64	9128.01	1.1029	54.2	1680
UP05800460	5	80	0.4	60	80	7.17	15.098	1160.0	264413.79	36892.68	5902.83	0.4036	34.2	>2000
UP05800660	5	80	0.6	60	80	9.59	16.788	1300.0	366369.23	38191.72	8647.18	0.5989	72.0	>2000
UP05800860	5	80	0.8	60	80	11.55	17.795	1440.0	456000.00	39490.76	11283.07	0.6556	69.6	>2000
UP05801060	5	80	1.0	60	80	13.15	18.428	1580.0	536582.28	40789.80	13827.05	0.7282	66.9	>2000
UP05350470	5	35	0.4	70	35	6.11	11.101	556.0	68516.34	11208.68	2287.85	1.0560	20.4	>2000
UP05350670	5	35	0.6	70	35	7.95	12.068	641.5	93423.04	11755.60	3323.06	2.0142	1525.3	>2000
UP05350870	5	35	0.8	70	35	9.35	12.579	727.0	115028.23	12302.53	4306.45	2.0892	1517.7	>2000
UP05351070	5	35	1.0	70	35	10.46	12.860	812.5	134374.15	12849.45	5248.36	2.1179	1516.0	>2000
UP05500470	5	50	0.4	70	50	8.73	15.858	794.3	199756.08	22874.86	4669.09	0.5176	29.1	>2000
UP05500670	5	50	0.6	70	50	11.35	17.240	916.4	272370.39	23991.03	6781.76	1.0840	34.4	>2000
UP05500870	5	50	0.8	70	50	13.36	17.970	1038.6	335359.26	25107.20	8788.68	1.3627	45.0	>2000
UP05500670	5	50	0.6	70	50	11.35	17.240	916.4	272370.39	23991.03	6781.76	1.0840	34.4	>2000
UP05650470	5	65	0.4	70	65	11.35	20.616	1032.6	438864.11	38658.51	7890.77	0.3063	37.9	>2000
UP05650670	5	65	0.6	70	65	14.76	22.412	1191.4	598397.75	40544.84	11461.18	0.6416	44.8	>2000
UP05650870	5	65	0.8	70	65	17.36	23.360	1350.1	736784.30	42431.17	14852.86	0.8064	58.5	>2000
UP05651070	5	65	1.0	70	65	19.42	23.883	1508.9	860699.72	44317.49	18101.50	0.9413	59.1	>2000
UP05800470	5	80	0.4	70	80	13.97	25.374	1270.9	818200.91	58559.64	11952.87	0.2022	46.6	>2000
UP05800670	5	80	0.6	70	80	18.16	27.584	1466.3	1115629.12	61417.03	17361.31	0.4236	55.0	>2000
UP05800870	5	80	0.8	70	80	21.37	28.751	1661.7	1373631.53	64274.43	22499.01	0.5324	71.9	>2000
UP05801070	5	80	1.0	70	80	23.90	29.395	1857.1	1604654.55	67131.82	27420.02	0.6214	72.7	>2000

U-PLATE — n trough stiffeners, edges pinned, free in-plane

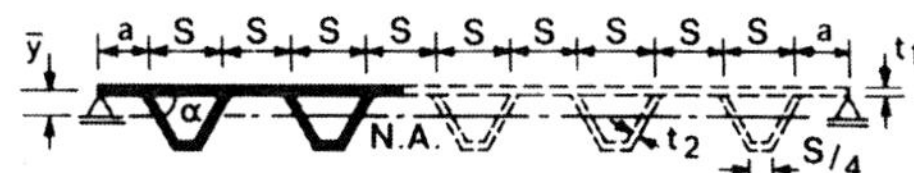

CODE	DIMENSIONS					SECTION PROPERTIES						BUCKLING PROPERTIES		
	n	S	t_2	α	a	$\bar{y}$	rad. gyr.	A	I_{NA}	Z_{plate}	$Z_{stiff.}$	σ_{cr}/E	L_1	L_{euler}
To scale multiply by	$-$	t_1	t_1	$-$	t_1	t_1	t_1	t_1^2	t_1^4	t_1^3	t_1^3	10^{-3}	t_1	t_1
UQ02350445	2	35	0.4	45	7	1.84	3.912	155.7	2383.02	1293.83	211.20	3.6361	29.2	1810
UQ02350645	2	35	0.6	45	7	2.47	4.357	174.0	3303.68	1336.71	310.10	3.8756	28.1	1760
UQ02350845	2	35	0.8	45	7	2.98	4.623	192.4	4112.65	1379.60	405.43	4.2789	26.4	1679
UQ02351045	2	35	1.0	45	7	3.40	4.792	210.7	4839.12	1422.49	497.69	4.8854	24.6	1573
UQ02500445	2	50	0.4	45	10	2.63	5.589	222.4	6947.57	2640.46	431.02	1.8063	41.2	>2000
UQ02500645	2	50	0.6	45	10	3.53	6.224	248.6	9631.71	2727.99	632.86	1.9161	39.8	>2000
UQ02500845	2	50	0.8	45	10	4.26	6.605	274.9	11990.24	2815.52	827.41	2.1112	37.5	>2000
UQ02501045	2	50	1.0	45	10	4.86	6.846	301.1	14108.22	2903.05	1015.70	2.4086	35.0	>2000
UQ02650445	2	65	0.4	45	13	3.42	7.266	289.2	15263.81	4462.38	728.43	1.0745	53.5	>2000
UQ02650645	2	65	0.6	45	13	4.59	8.091	323.2	21150.86	4610.30	1069.54	1.1377	51.6	>2000
UQ02650845	2	65	0.8	45	13	5.54	8.586	357.3	26342.56	4758.23	1398.31	1.2526	48.7	>2000
UQ02651045	2	65	1.0	45	13	6.32	8.899	391.4	30995.75	4906.15	1716.53	1.4287	45.4	>2000
UQ02800445	2	80	0.4	45	16	4.21	8.942	355.9	28457.24	6759.58	1103.42	0.7111	65.4	>2000
UQ02800645	2	80	0.6	45	16	5.65	9.958	397.8	39451.47	6983.65	1620.12	0.7524	63.3	>2000
UQ02800845	2	80	0.8	45	16	6.81	10.568	439.8	49112.02	7207.73	2118.16	0.8280	59.9	>2000
UQ02801045	2	80	1.0	45	16	7.78	10.953	481.7	57787.26	7431.80	2600.18	0.9443	55.9	>2000
UQ02350460	2	35	0.4	60	7	3.79	7.088	168.0	8441.02	2227.85	445.57	2.1056	15.0	>2000
UQ02350660	2	35	0.6	60	7	4.96	7.744	192.5	11543.36	2327.31	649.48	3.5877	27.9	1836
UQ02350860	2	35	0.8	60	7	5.87	8.100	217.0	14236.92	2426.77	844.09	3.9436	27.3	1756
UQ02351060	2	35	1.0	60	7	6.59	8.302	241.5	16646.10	2526.22	1031.11	4.4859	26.3	1648
UQ02500460	2	50	0.4	60	10	5.41	10.126	240.0	24609.38	4546.63	909.33	1.0326	21.3	>2000
UQ02500660	2	50	0.6	60	10	7.09	11.062	275.0	33654.12	4749.61	1325.47	1.7657	39.7	>2000
UQ02500860	2	50	0.8	60	10	8.38	11.571	310.0	41507.06	4952.58	1722.64	1.9401	39.5	>2000
UQ02501060	2	50	1.0	60	10	9.41	11.860	345.0	48530.91	5155.56	2104.31	2.2055	37.4	>2000
UQ02650460	2	65	0.4	60	13	7.04	13.164	312.0	54066.80	7683.81	1536.76	0.6112	27.8	>2000
UQ02650660	2	65	0.6	60	13	9.21	14.381	357.5	73938.10	8026.84	2240.05	1.0466	51.6	>2000
UQ02650860	2	65	0.8	60	13	10.90	15.043	403.0	91191.00	8369.86	2911.26	1.1493	50.6	>2000
UQ02651060	2	65	1.0	60	13	12.24	15.419	448.5	106622.41	8712.89	3556.28	1.3068	48.6	>2000
UQ02800460	2	80	0.4	60	16	8.66	16.202	384.0	100800.00	11639.38	2327.88	0.4035	34.1	>2000
UQ02800660	2	80	0.6	60	16	11.34	17.700	440.0	137047.27	12159.00	3545.21	0.6915	63.4	>2000
UQ02800860	2	80	0.8	60	16	13.41	18.514	496.0	170012.90	12678.61	4409.95	0.7593	62.3	>2000
UQ02801060	2	80	1.0	60	16	15.06	18.977	552.0	198782.61	13198.23	5387.03	0.8633	59.8	>2000
UQ02350470	2	35	0.4	70	7	7.25	11.746	187.4	25854.46	3563.94	897.53	1.0567	20.3	>2000
UQ02350670	2	35	0.6	70	7	9.20	12.533	221.6	34809.61	3782.71	1296.04	2.2468	22.2	>2000
UQ02350870	2	35	0.8	70	7	10.63	12.895	255.8	42532.87	4001.48	1672.46	3.0598	28.4	>2000
UQ02351070	2	35	1.0	70	7	11.72	13.060	290.0	49460.07	4220.25	2031.97	3.6113	29.3	1846
UQ02500470	2	50	0.4	70	10	10.36	16.780	267.7	75377.43	7273.35	1831.70	0.5180	29.1	>2000
UQ02500670	2	50	0.6	70	10	13.15	17.905	316.6	101485.75	7719.82	2644.99	1.1022	31.7	>2000
UQ02500870	2	50	0.8	70	10	15.18	18.421	365.4	124002.54	8166.29	3413.18	1.5035	40.5	>2000
UQ02501070	2	50	1.0	70	10	16.74	18.657	414.3	144198.45	8612.75	4146.88	1.7749	41.8	>2000
UQ02650470	2	65	0.4	70	13	13.47	21.814	348.0	165604.22	12291.96	3095.57	0.3065	37.8	>2000
UQ02650670	2	65	0.6	70	13	17.09	23.276	411.5	222964.19	13046.49	4470.03	0.6525	41.3	>2000
UQ02650870	2	65	0.8	70	13	19.74	23.947	475.1	272433.57	13801.02	5768.27	0.8907	52.9	>2000
UQ02651070	2	65	1.0	70	13	21.77	24.253	538.6	316804.00	14555.55	7008.22	1.0515	54.3	>2000
UQ02800470	2	80	0.4	70	16	16.58	26.848	428.3	308745.97	18619.78	4689.14	0.2024	46.5	>2000
UQ02800670	2	80	0.6	70	16	21.03	28.648	506.5	415685.63	19762.74	6771.17	0.4308	50.7	>2000
UQ02800870	2	80	0.8	70	16	24.30	29.474	584.7	507914.39	20905.69	8737.74	0.5883	64.7	>2000
UQ02801070	2	80	1.0	70	16	26.79	29.850	662.9	590636.86	22048.65	10616.01	0.6946	67.0	>2000
UQ03350445	3	35	0.4	45	7	1.76	3.845	244.0	3608.61	2047.33	317.59	2.7568	381.4	>2000
UQ03350645	3	35	0.6	45	7	2.38	4.298	271.6	5017.17	2111.66	466.75	3.3067	394.1	>2000
UQ03350845	3	35	0.8	45	7	2.88	4.575	299.1	6259.01	2176.00	610.72	3.7363	401.0	>2000
UQ03351045	3	35	1.0	45	7	3.29	4.752	326.6	7376.28	2240.33	750.19	4.0591	408.6	>2000
UQ03500445	3	50	0.4	45	10	2.52	5.493	348.6	10520.73	4178.23	648.15	1.6534	44.1	>2000
UQ03500645	3	50	0.6	45	10	3.39	6.140	388.0	14627.32	4309.52	952.56	1.7600	42.3	>2000
UQ03500845	3	50	0.8	45	10	4.11	6.535	427.3	18247.85	4440.81	1246.36	1.9542	39.6	>2000
UQ03501045	3	50	1.0	45	10	4.70	6.789	466.6	21505.20	4572.11	1531.01	2.2474	36.6	>2000
UQ03650445	3	65	0.4	45	13	3.27	7.141	453.2	23114.05	7061.20	1095.37	0.9804	57.1	>2000
UQ03650545	3	65	0.6	45	13	4.41	7.982	504.3	32136.21	7283.09	1609.82	1.0429	54.8	>2000
UQ03650845	3	65	0.8	45	13	5.34	8.496	555.5	40090.52	7504.97	2106.35	1.1576	51.3	>2000
UQ03651045	3	65	1.0	45	13	6.11	8.826	606.6	47246.93	7726.86	2587.40	1.3312	47.5	>2000
UQ03800445	3	80	0.4	45	16	4.03	8.789	557.8	43092.92	10696.26	1659.26	0.6479	70.2	>2000
UQ03800645	3	80	0.6	45	16	5.43	9.824	620.7	59913.49	11032.37	2438.55	0.6889	67.6	>2000
UQ03800845	3	80	0.8	45	16	6.57	10.456	683.6	74743.18	11368.48	3190.69	0.7646	63.2	>2000
UQ03801045	3	80	1.0	45	16	7.53	10.862	746.6	88085.30	11704.60	3919.38	0.8792	58.5	>2000

U-PLATE — n trough stiffeners, edges pinned, free in-plane

CODE	n	S	t₂	α	a	ȳ	rad. gyr.	A	I_NA	Z_plate	Z_stiff.	σ_cr/E	L₁	L_euler
To scale multiply by	−	t_1	t_1	−	t_1	t_1	t_1	t_1^2	t_1^4	t_1^3	t_1^3	10^{-3}	t_1	t_1
UG03350460	3	35	0.4	60	7	3.64	6.985	262.5	12806.23	3520.80	670.63	2.1058	15.0	>2000
UG03350660	3	35	0.6	60	7	4.79	7.661	299.3	17564.29	3669.99	978.66	3.3169	29.8	>2000
UG03350860	3	35	0.8	60	7	5.68	8.037	336.0	21705.47	3819.17	1273.06	3.6580	28.9	>2000
UG03351060	3	35	1.0	60	7	6.40	8.257	372.8	25412.21	3968.36	1556.22	4.1882	27.5	>2000
UG03500460	3	50	0.4	60	10	5.20	9.978	375.0	37335.94	7185.30	1368.63	1.0326	21.3	>2000
UG03500660	3	50	0.6	60	10	6.84	10.945	427.5	51207.85	7489.77	1997.27	1.6282	42.6	>2000
UG03500860	3	50	0.8	60	10	8.12	11.482	480.0	63281.25	7794.23	2598.08	1.7951	41.3	>2000
UG03501060	3	50	1.0	60	10	9.15	11.795	532.5	74088.08	8098.69	3175.96	2.0551	39.3	>2000
UG03650460	3	65	0.4	60	13	6.75	12.972	487.5	82027.05	12143.16	2312.98	0.6112	27.9	>2000
UG03650660	3	65	0.6	60	13	8.89	14.228	555.8	112503.65	12657.71	3375.39	0.9641	55.3	>2000
UG03650860	3	65	0.8	60	13	10.55	14.927	624.0	139028.91	13172.25	4390.75	1.0628	53.6	>2000
UG03651060	3	65	1.0	60	13	11.89	15.334	692.3	162771.52	13686.79	5367.37	1.2167	50.9	>2000
UG03800460	3	80	0.4	60	16	8.31	15.965	600.0	152928.00	18394.38	3503.69	0.4035	34.2	>2000
UG03800660	3	80	0.6	60	16	10.94	17.511	684.0	209747.37	19173.80	5113.01	0.6367	68.2	>2000
UG03800860	3	80	0.8	60	16	12.99	18.371	768.0	259200.00	19953.23	6651.08	0.7018	66.2	>2000
UG03801060	3	80	1.0	60	16	14.64	18.873	852.0	303464.79	20732.65	8130.45	0.8034	62.7	>2000
UG03350470	3	35	0.4	70	7	6.99	11.611	291.6	39314.38	5621.77	1352.52	1.0563	20.4	>2000
UG03350670	3	35	0.6	70	7	8.92	12.441	342.9	53076.35	5949.93	1955.64	2.2235	23.2	>2000
UG03350870	3	35	0.8	70	7	10.35	12.836	394.2	64954.01	6278.08	2525.97	2.8882	30.1	>2000
UG03351070	3	35	1.0	70	7	11.44	13.027	445.5	75598.30	6606.23	3070.96	3.3989	30.7	>2000
UG03500470	3	50	0.4	70	10	9.99	16.588	416.6	114619.18	11473.00	2760.25	0.5178	29.1	>2000
UG03500670	3	50	0.6	70	10	12.74	17.773	469.9	154741.55	12142.71	3991.10	1.0904	33.1	>2000
UG03500870	3	50	0.8	70	10	14.78	18.338	563.1	189370.30	12812.41	5155.04	1.4170	43.1	>2000
UG03501070	3	50	1.0	70	10	16.35	18.609	636.4	220403.21	13482.11	6267.26	1.6676	43.9	>2000
UG03650470	3	65	0.4	70	13	12.99	21.564	541.5	251818.33	19389.38	4664.83	0.3064	37.8	>2000
UG03650670	3	65	0.6	70	13	16.57	23.105	636.8	339967.20	20521.17	6744.97	0.6454	43.1	>2000
UG03650870	3	65	0.8	70	13	19.21	23.839	732.1	416046.55	21652.97	8712.02	0.8388	55.9	>2000
UG03651070	3	65	1.0	70	13	21.25	24.192	827.4	484225.86	22784.76	10591.68	0.9872	57.0	>2000
UG03800470	3	80	0.4	70	16	15.98	26.540	666.5	469480.14	29370.89	7066.25	0.2023	46.5	>2000
UG03800670	3	80	0.6	70	16	20.39	28.437	783.8	633821.41	31085.33	10217.23	0.4261	52.9	>2000
UG03800870	3	80	0.8	70	16	23.65	29.340	901.0	775660.75	32799.76	13196.91	0.5539	68.8	>2000
UG03801070	3	80	1.0	70	16	26.16	29.775	1018.3	902771.56	34514.20	16044.20	0.6519	70.1	>2000
UG05350445	5	35	0.4	45	7	1.70	3.794	420.7	6056.40	3554.34	530.28	1.0846	894.6	>2000
UG05350645	5	35	0.6	45	7	2.30	4.253	466.6	8438.61	3661.56	779.88	1.3000	895.1	>2000
UG05350845	5	35	0.8	45	7	2.80	4.536	512.5	10544.35	3768.79	1021.03	1.4384	906.8	>2000
UG05351045	5	35	1.0	45	7	3.21	4.720	558.4	12441.77	3876.01	1254.84	1.5260	926.5	>2000
UG05500445	5	50	0.4	45	10	2.43	5.420	601.1	17657.14	7253.76	1082.21	0.9366	1113.3	>2000
UG05500645	5	50	0.6	45	10	3.29	6.075	666.6	24602.36	7472.58	1591.60	1.1104	1112.8	>2000
UG05500845	5	50	0.8	45	10	4.00	6.480	732.1	30741.55	7691.40	2083.73	1.2243	1127.0	>2000
UG05501045	5	50	1.0	45	10	4.59	6.743	797.7	36273.38	7910.23	2560.89	1.3008	1150.1	>2000
UG05650445	5	65	0.4	45	13	3.16	7.046	781.4	38792.74	12258.85	1828.94	0.8062	1376.9	>2000
UG05650645	5	65	0.6	45	13	4.28	7.898	866.6	54051.38	12628.66	2689.80	0.9481	1410.4	>2000
UG05650845	5	65	0.8	45	13	5.20	8.424	951.8	67539.19	12998.47	3521.51	1.0436	1438.3	>2000
UG05651045	5	65	1.0	45	13	5.96	8.767	1037.0	79692.61	13368.28	4327.90	1.1104	1462.7	>2000
UG05800445	5	80	0.4	45	16	3.89	8.672	961.7	72323.65	18569.61	2770.46	0.6191	72.6	>2000
UG05800645	5	80	0.6	45	16	5.27	9.720	1066.6	100771.26	19129.80	4074.49	0.6602	69.5	>2000
UG05800845	5	80	0.8	45	16	6.39	10.368	1171.4	125917.39	19689.99	5334.35	0.7359	64.9	>2000
UG05801045	5	80	1.0	45	16	7.34	10.790	1276.3	148575.75	20250.18	6555.88	0.8498	59.8	>2000
UG05350460	5	35	0.4	60	7	3.52	6.904	451.5	21523.19	6106.70	1120.49	2.1852	1006.8	>2000
UG05350660	5	35	0.6	60	7	4.66	7.596	512.8	29585.74	6355.34	1636.57	2.5574	971.0	>2000
UG05350860	5	35	0.8	60	7	5.54	7.987	574.0	36616.95	6603.98	2130.32	2.7738	961.7	>2000
UG05351060	5	35	1.0	60	7	6.26	8.219	635.3	42915.14	6852.63	2605.56	2.9182	965.5	>2000
UG05500460	5	50	0.4	60	10	5.04	9.863	645.0	62749.82	12462.65	2286.72	1.0326	21.3	>2000
UG05500660	5	50	0.6	60	10	6.65	10.852	732.5	86255.80	12970.08	3339.94	1.5647	44.1	>2000
UG05500860	5	50	0.8	60	10	7.92	11.410	820.0	106754.95	13477.52	4347.59	1.7294	42.4	>2000
UG05501060	5	50	1.0	60	10	8.95	11.742	907.5	125117.03	13984.96	5317.47	1.9867	40.1	>2000
UG05650460	5	65	0.4	60	13	6.55	12.822	838.5	137861.35	21061.87	3864.56	0.6112	27.9	>2000
UG05650660	5	65	0.6	60	13	8.65	14.107	952.3	199503.99	21919.44	5644.49	0.9261	57.2	>2000
UG05650860	5	65	0.8	60	13	10.30	14.833	1066.0	234540.63	22777.01	7347.42	1.0235	55.1	>2000
UG05651060	5	65	1.0	60	13	11.63	15.264	1179.8	274882.11	23634.58	8986.53	1.1758	52.2	>2000
UG05800460	5	80	0.4	60	16	8.06	15.781	1032.0	257023.26	31904.38	5854.01	0.4035	34.2	>2000
UG05800660	5	80	0.6	60	16	10.64	17.362	1172.0	353303.75	33203.41	8550.24	0.6115	70.4	>2000
UG05800860	5	80	0.8	60	16	12.67	18.256	1312.0	437268.29	34502.45	11129.82	0.6758	68.0	>2000
UG05801060	5	80	1.0	60	16	14.31	18.787	1452.0	512479.34	35801.49	13612.73	0.7763	64.2	>2000

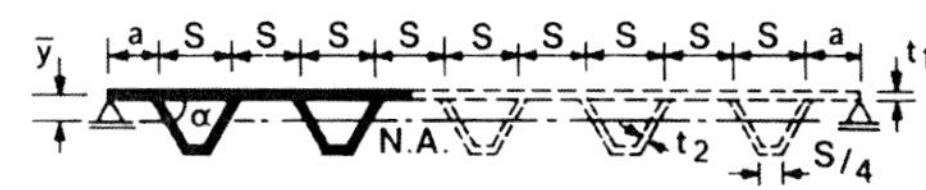

U-PLATE — n trough stiffeners / edges pinned / free in-plane

CODE	n	S	t_2	α	a	$\bar{y}$	rad. gyr.	A	I_{NA}	Z_{plate}	$Z_{stiff.}$	σ_{cr}/E	L_1	L_{euler}
To scale multiply by	–	t_1	t_1	–	t_1	t_1	t_1	t_1^2	t_1^4	t_1^3	t_1^3	10^{-3}	t_1	t_1
UQ05350470	5	35	0.4	70	7	6.80	11.506	500.0	66189.47	9737.43	2261.87	1.0561	20.4	>2000
UQ05350670	5	35	0.6	70	7	8.71	12.367	585.5	89548.01	10284.36	3273.74	2.2134	23.8	>2000
UQ05350870	5	35	0.8	70	7	10.13	12.788	671.0	109724.02	10831.28	4231.49	2.8083	31.0	>2000
UQ05351070	5	35	1.0	70	7	11.23	12.997	756.5	127796.59	11378.20	5147.08	3.3019	31.4	>2000
UQ05500470	5	50	0.4	70	10	9.71	16.437	714.3	192972.20	19872.31	4616.05	0.5176	29.1	>2000
UQ05500670	5	50	0.6	70	10	12.44	17.667	836.4	261072.91	20988.48	6681.10	1.0853	34.1	>2000
UQ05500870	5	50	0.8	70	10	14.47	18.268	958.6	319895.11	22104.65	8635.70	1.3769	44.4	>2000
UQ05501070	5	50	1.0	70	10	16.05	18.568	1080.7	372584.80	23220.82	10504.25	1.6189	44.8	>2000
UQ05650470	5	65	0.4	70	13	12.62	21.368	928.6	423959.93	33584.21	7801.13	0.3063	37.9	>2000
UQ05650670	5	65	0.6	70	13	16.17	22.967	1087.4	573577.17	35470.53	11291.06	0.6424	44.4	>2000
UQ05650870	5	65	0.8	70	13	18.81	23.748	1246.1	702809.56	37356.86	14594.32	0.8150	57.6	>2000
UQ05651070	5	65	1.0	70	13	20.86	24.138	1404.9	818568.81	39243.18	17752.18	0.9582	58.2	>2000
UQ05800470	5	80	0.4	70	16	15.54	26.299	1142.9	790414.15	50873.12	11817.09	0.2022	46.6	>2000
UQ05800670	5	80	0.6	70	16	19.90	28.267	1338.3	1069354.62	53730.51	17103.62	0.4241	54.5	>2000
UQ05800870	5	80	0.8	70	16	23.15	29.229	1533.7	1310290.38	56587.90	22107.38	0.5381	70.9	>2000
UQ05801070	5	80	1.0	70	16	25.67	29.708	1729.1	1526107.35	59445.30	26890.88	0.6326	71.6	>2000

n V-stiffeners
edges free out-of-plane
free in-plane

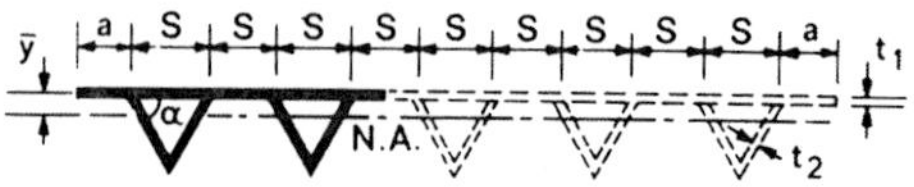

CODE	n	S	t_2	α	a	$\bar{y}$	rad. gyr.	A	I_{NA}	Z_{plate}	$Z_{stiff.}$	σ_{cr}/E	L_1	L_{euler}
To scale multiply by	$-$	t_1	t_1	$-$	t_1	t_1	t_1	t_1^2	t_1^4	t_1^3	t_1^3	10^{-3}	t_1	t_1
VF01350445	1	35	0.4	45	7	2.52	4.800	68.8	1584.91	629.41	105.79	1.6217	17.5	373
VF01350645	1	35	0.6	45	7	3.30	5.255	78.7	2173.66	658.29	153.10	3.4818	18.3	275
VF01350845	1	35	0.8	45	7	3.91	5.507	88.6	2687.30	687.16	197.75	4.4499	27.9	254
VF01351045	1	35	1.0	45	7	4.40	5.654	98.5	3148.47	716.03	240.29	5.2469	25.5	241
VF01500445	1	50	0.4	45	10	3.60	6.857	98.3	4620.74	1284.52	215.89	0.7952	25.0	769
VF01500645	1	50	0.6	45	10	4.72	7.508	112.4	6337.20	1343.44	312.44	1.7089	26.1	571
VF01500845	1	50	0.8	45	10	5.59	7.868	126.6	7834.68	1402.37	403.57	2.1979	39.6	527
VF01501045	1	50	1.0	45	10	6.28	8.077	140.7	9179.22	1461.29	490.38	2.5905	36.2	499
VF01650445	1	65	0.4	45	13	4.68	8.914	127.8	10151.76	2170.84	364.86	0.4706	32.4	1304
VF01650645	1	65	0.6	45	13	6.13	9.760	146.2	13922.82	2270.42	528.03	1.0119	33.9	971
VF01650845	1	65	0.8	45	13	7.26	10.228	164.5	17212.80	2370.00	682.04	1.3045	51.5	896
VF01651045	1	65	1.0	45	13	8.17	10.500	182.9	20166.74	2469.59	828.75	1.5372	47.0	849
VF01800445	1	80	0.4	45	16	5.76	10.971	157.3	18926.53	3288.37	552.69	0.3107	40.0	1979
VF01800645	1	80	0.6	45	16	7.55	12.013	179.9	25957.16	3439.22	799.85	0.6682	41.8	1475
VF01800845	1	80	0.8	45	16	8.94	12.588	202.5	32090.87	3590.06	1033.15	0.8624	63.2	1362
VF01801045	1	80	1.0	45	16	10.05	12.923	225.1	37598.08	3740.91	1255.38	1.0162	57.9	1289
VF01350460	1	35	0.4	60	7	5.51	9.000	77.0	6236.36	1131.61	251.47	0.8095	24.7	1006
VF01350660	1	35	0.6	60	7	6.99	9.613	91.0	8410.10	1202.33	360.70	1.7423	25.5	731
VF01350860	1	35	0.8	60	7	8.08	9.899	105.0	10290.00	1273.06	462.93	2.8678	26.9	586
VF01351060	1	35	1.0	60	7	8.91	10.033	119.0	11979.78	1343.78	559.91	4.0778	28.4	497
VF01500460	1	50	0.4	60	10	7.87	12.856	110.0	18181.82	2309.40	513.20	0.3969	35.2	>2000
VF01500660	1	50	0.6	60	10	9.99	13.734	130.0	24519.23	2453.74	736.12	0.8548	36.4	1497
VF01500860	1	50	0.8	60	10	11.55	14.142	150.0	30000.00	2598.08	944.75	1.4085	38.3	1201
VF01501060	1	50	1.0	60	10	12.74	14.334	170.0	34926.47	2742.41	1142.67	2.0052	40.5	1020
VF01650460	1	65	0.4	60	13	10.23	16.713	143.0	39945.45	3902.89	867.31	0.2349	45.8	>2000
VF01650660	1	65	0.6	60	13	12.99	17.854	169.0	53868.75	4146.82	1244.05	0.5061	47.3	>2000
VF01650860	1	65	0.8	60	13	15.01	18.385	195.0	65910.00	4390.75	1596.64	0.8342	49.8	>2000
VF01651060	1	65	1.0	60	13	16.56	18.634	221.0	76733.46	4634.68	1931.12	1.1882	53.0	1728
VF01800460	1	80	0.4	60	16	12.60	20.570	176.0	74472.73	5912.07	1313.79	0.1551	56.4	>2000
VF01800660	1	80	0.6	60	16	15.99	21.974	208.0	100430.77	6281.57	1884.47	0.3342	58.3	>2000
VF01800860	1	80	0.8	60	16	18.48	22.627	240.0	122880.00	6651.08	2418.57	0.5509	61.3	>2000
VF01801060	1	80	1.0	60	16	20.38	22.934	272.0	143058.82	7020.58	2925.24	0.7849	64.8	>2000
VF01350470	1	35	0.4	70	7	10.94	15.199	89.9	20775.20	1898.66	559.39	0.3788	36.0	>2000
VF01350670	1	35	0.6	70	7	13.37	15.805	110.4	27578.49	2062.67	794.53	0.8177	36.9	1342
VF01350870	1	35	0.8	70	7	15.04	15.996	130.9	33487.07	2226.68	1013.48	1.3601	38.3	1012
VF01351070	1	35	1.0	70	7	16.26	16.025	151.3	38863.95	2390.69	1221.20	1.9635	40.2	824
VF01500470	1	50	0.4	70	10	15.63	21.713	128.5	60569.09	3874.81	1141.62	0.1857	51.3	>2000
VF01500670	1	50	0.6	70	10	19.10	22.579	157.7	80403.75	4209.53	1621.48	0.4013	53.0	>2000
VF01500870	1	50	0.8	70	10	21.48	22.852	187.0	97629.94	4544.24	2069.32	0.6681	54.7	>2000
VF01501070	1	50	1.0	70	10	23.22	22.893	216.2	113305.97	4878.95	2492.24	0.9657	57.2	1687
VF01650470	1	65	0.4	70	13	20.32	28.227	167.0	133070.29	6548.44	1929.34	0.1099	67.0	>2000
VF01650670	1	65	0.6	70	13	24.83	29.353	205.0	176647.04	7114.10	2740.31	0.2375	68.4	>2000
VF01650870	1	65	0.8	70	13	27.93	29.708	243.0	214492.97	7679.76	3495.46	0.3957	71.0	>2000
VF01651070	1	65	1.0	70	13	30.19	29.761	281.0	248933.21	8245.43	4211.89	0.5723	74.3	>2000
VF01800470	1	80	0.4	70	16	25.01	34.740	205.6	248090.98	9919.52	2922.54	0.0726	82.3	>2000
VF01800670	1	80	0.6	70	16	30.56	36.126	252.3	329333.77	10776.39	4151.00	0.1569	84.1	>2000
VF01800870	1	80	0.8	70	16	34.37	36.563	299.1	399892.22	11633.25	5294.89	0.2614	88.4	>2000
VF01801070	1	80	1.0	70	16	37.16	36.629	345.9	464101.25	12490.11	6380.13	0.3781	91.6	>2000
VF02350445	2	35	0.4	45	7	2.18	4.551	158.6	3285.35	1503.83	214.51	1.6246	17.4	353
VF02350645	2	35	0.6	45	7	2.91	5.050	178.4	4549.33	1561.57	311.88	3.4563	30.5	265
VF02350845	2	35	0.8	45	7	3.50	5.345	198.2	5661.73	1619.32	404.30	3.8421	28.7	267
VF02351045	2	35	1.0	45	7	3.97	5.529	218.0	6663.85	1677.07	492.65	4.4343	26.7	257
VF02500445	2	50	0.4	45	10	3.12	6.502	226.6	9578.28	3069.04	437.78	0.7967	24.9	727
VF02500645	2	50	0.6	45	10	4.16	7.214	254.9	13263.36	3186.89	636.49	1.6972	43.5	550
VF02500845	2	50	0.8	45	10	4.99	7.635	283.1	16506.51	3304.74	825.11	1.8863	41.0	553
VF02501045	2	50	1.0	45	10	5.68	7.898	311.4	19428.15	3422.59	1005.41	2.1767	38.1	533
VF02650445	2	65	0.4	45	13	4.06	8.453	294.5	21043.47	5186.67	739.85	0.4715	32.4	1234
VF02650645	2	65	0.6	45	13	5.41	9.378	331.3	29139.60	5385.84	1075.68	1.0051	56.6	935
VF02650845	2	65	0.8	45	13	6.49	9.926	368.1	36264.80	5585.01	1394.44	1.1170	53.4	941
VF02651045	2	65	1.0	45	13	7.38	10.268	404.8	42683.64	5784.18	1699.15	1.2888	49.5	907
VF02800445	2	80	0.4	45	16	4.99	10.403	362.5	39232.62	7856.73	1120.72	0.3113	39.9	1873
VF02800645	2	80	0.6	45	16	6.66	11.543	407.8	54326.71	8158.43	1629.42	0.6678	69.5	1421
VF02800845	2	80	0.8	45	16	7.99	12.217	453.0	67610.66	8460.13	2112.28	0.7376	65.5	1429
VF02801045	2	80	1.0	45	16	9.08	12.638	498.3	79577.69	8761.83	2573.86	0.8511	60.9	1378

V-PLATE n V-stiffeners
edges free out-of-plane
free in-plane

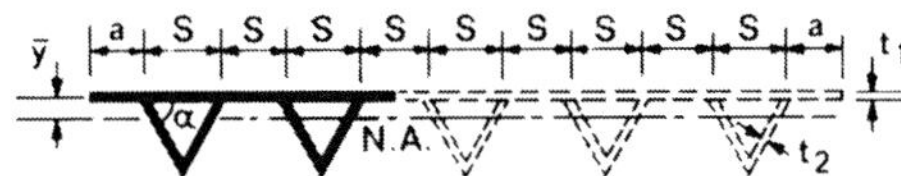

CODE	n	S	t_2	α	a	$\bar{y}$	rad. gyr.	A	I_{NA}	Z_{plate}	$Z_{stiff.}$	σ_{cr}/E	L_1	L_{euler}
To scale multiply by	–	t_1	t_1	–	t_1	t_1	t_1	t_1^2	t_1^4	t_1^3	t_1^3	10^{-3}	t_1	t_1
VF02350460	2	35	0.4	60	7	4.85	8.630	175.0	13034.00	2687.57	511.92	0.8115	24.6	962
VF02350660	2	35	0.6	60	7	6.27	9.349	203.0	17741.38	2829.02	738.00	1.7505	25.6	708
VF02350860	2	35	0.8	60	7	7.35	9.721	231.0	21827.27	2970.47	950.55	2.8462	27.4	576
VF02351060	2	35	1.0	60	7	8.19	9.921	259.0	25493.24	3111.92	1152.56	3.7502	29.4	512
VF02500460	2	50	0.4	60	10	6.93	12.329	250.0	38000.00	5484.83	1044.73	0.3979	35.2	1968
VF02500660	2	50	0.6	60	10	8.96	13.355	290.0	51724.14	5773.50	1506.13	0.8588	36.5	1451
VF02500860	2	50	0.8	60	10	10.50	13.887	330.0	63636.36	6062.18	1939.90	1.3977	39.6	1182
VF02501060	2	50	1.0	60	10	11.70	14.173	370.0	74324.32	6350.85	2352.17	1.8398	42.0	1052
VF02650460	2	65	0.4	60	13	9.01	16.027	325.0	83466.00	9269.36	1765.59	0.2355	45.7	>2000
VF02650660	2	65	0.6	60	13	11.65	17.362	377.0	113637.93	9757.22	2545.36	0.5084	47.4	>2000
VF02650860	2	65	0.8	60	13	13.65	18.053	429.0	139809.09	10245.08	3278.43	0.8275	50.6	>2000
VF02651060	2	65	1.0	60	13	15.21	18.425	481.0	163290.54	10732.94	3975.16	1.0891	54.6	1783
VF02800460	2	80	0.4	60	16	11.09	19.726	400.0	155648.00	14041.16	2674.51	0.1555	56.3	>2000
VF02800660	2	80	0.6	60	16	14.33	21.368	464.0	211862.07	14780.17	3855.70	0.3357	58.4	>2000
VF02800860	2	80	0.8	60	16	16.80	22.219	528.0	260654.55	15519.18	4966.14	0.5465	62.4	>2000
VF02801060	2	80	1.0	60	16	18.72	22.677	592.0	304432.43	16258.18	6021.55	0.7192	67.3	>2000
VF02350470	2	35	0.4	70	7	9.80	14.767	200.9	43801.82	4470.45	1144.17	0.3796	35.9	>2000
VF02350670	2	35	0.6	70	7	12.21	15.566	241.8	58585.00	4798.47	1633.18	0.8219	36.8	1735
VF02350870	2	35	0.8	70	7	13.92	15.888	282.7	71370.98	5126.49	2089.39	1.3690	38.2	1373
VF02351070	2	35	1.0	70	7	15.20	16.006	323.7	82917.51	5454.50	2521.88	1.9716	39.9	1153
VF02500470	2	50	0.4	70	10	14.00	21.096	287.0	127702.11	9123.37	2335.03	0.1861	51.3	>2000
VF02500670	2	50	0.6	70	10	17.44	22.237	345.4	170801.75	9792.79	3333.02	0.4033	52.6	>2000
VF02500870	2	50	0.8	70	10	19.89	22.697	403.9	208078.66	10462.21	4264.05	0.6724	54.5	>2000
VF02501070	2	50	1.0	70	10	21.72	22.865	462.4	241742.00	11131.64	5146.70	0.9694	56.9	>2000
VF02650470	2	65	0.4	70	13	18.20	27.424	373.0	280561.54	15418.49	3946.20	0.1101	67.0	>2000
VF02650670	2	65	0.6	70	13	22.67	28.908	449.1	375251.45	16549.82	5632.80	0.2388	68.3	>2000
VF02650870	2	65	0.8	70	13	25.86	29.507	525.1	457148.82	17681.14	7206.25	0.3982	70.8	>2000
VF02651070	2	65	1.0	70	13	28.23	29.725	601.1	531107.17	18812.47	8697.93	0.5743	73.9	>2000
VF02800470	2	80	0.4	70	16	22.40	33.753	459.1	523067.85	23355.82	5977.68	0.0727	82.2	>2000
VF02800670	2	80	0.6	70	16	27.91	35.578	552.7	699603.99	25069.54	8532.53	0.1577	84.0	>2000
VF02800870	2	80	0.8	70	16	31.82	36.316	646.2	852290.21	26783.27	10915.97	0.2630	88.1	>2000
VF02801070	2	80	1.0	70	16	34.75	36.584	739.8	990175.23	28497.00	13175.56	0.3794	90.9	>2000
VF03350445	3	35	0.4	45	7	2.09	4.476	248.4	4976.02	2378.24	322.96	1.6253	17.4	346
VF03350645	3	35	0.6	45	7	2.80	4.985	278.1	6909.75	2464.86	470.16	3.3695	30.8	265
VF03350845	3	35	0.8	45	7	3.38	5.291	307.8	8616.58	2551.48	610.11	3.7314	29.0	268
VF03351045	3	35	1.0	45	7	3.85	5.486	337.5	10156.41	2638.10	744.05	4.2937	27.0	260
VF03500445	3	50	0.4	45	10	2.99	6.394	354.9	14507.34	4853.55	659.10	0.7970	24.9	715
VF03500645	3	50	0.6	45	10	4.00	7.121	397.3	20145.04	5030.33	959.50	1.6530	44.0	550
VF03500845	3	50	0.8	45	10	4.82	7.559	439.7	25121.22	5207.11	1245.13	1.8303	41.4	556
VF03501045	3	50	1.0	45	10	5.50	7.837	482.1	29610.51	5383.88	1518.48	2.1060	38.5	538
VF03650445	3	65	0.4	45	13	3.89	8.312	461.3	31872.63	8202.51	1113.87	0.4717	32.4	1213
VF03650645	3	65	0.6	45	13	5.21	9.257	516.5	44258.65	8501.26	1621.56	0.9785	57.2	935
VF03650845	3	65	0.8	45	13	6.27	9.826	571.6	55191.32	8800.01	2104.27	1.0834	53.8	945
VF03651045	3	65	1.0	45	13	7.15	10.188	626.8	65054.30	9098.76	2566.22	1.2466	50.1	915
VF03800445	3	80	0.4	45	16	4.78	10.230	567.8	59422.07	12425.10	1687.28	0.3115	39.9	1841
VF03800645	3	80	0.6	45	16	6.41	11.393	635.6	82514.09	12877.65	2456.33	0.6461	70.3	1421
VF03800845	3	80	0.8	45	16	7.72	12.094	703.5	102896.51	13330.19	3187.53	0.7154	66.5	1436
VF03801045	3	80	1.0	45	16	8.80	12.539	771.4	121284.66	13782.74	3887.30	0.8231	61.7	1390
VF03350460	3	35	0.4	60	7	4.66	8.514	273.0	19788.46	4243.52	771.55	0.8116	24.6	949
VF03350660	3	35	0.6	60	7	6.06	9.260	315.0	27011.25	4455.70	1113.93	1.7520	25.5	701
VF03350860	3	35	0.8	60	7	7.13	9.657	357.0	33291.18	4667.88	1436.27	2.8391	27.4	573
VF03351060	3	35	1.0	60	7	7.98	9.877	399.0	38925.99	4880.05	1742.88	3.7132	29.6	512
VF03500460	3	50	0.4	60	10	6.66	12.163	390.0	57692.31	8660.25	1574.59	0.3979	35.2	1941
VF03500660	3	50	0.6	60	10	8.66	13.229	450.0	78750.00	9093.27	2273.32	0.8596	36.5	1436
VF03500860	3	50	0.8	60	10	10.19	13.795	510.0	97058.82	9526.28	2931.16	1.3937	39.2	1176
VF03501060	3	50	1.0	60	10	11.40	14.110	570.0	113486.84	9959.29	3556.89	1.8208	42.3	1052
VF03650460	3	65	0.4	60	13	8.66	15.611	507.0	126750.00	14635.83	2661.06	0.2355	45.7	>2000
VF03650660	3	65	0.6	60	13	11.26	17.197	585.0	173013.75	15367.62	3841.91	0.5089	47.4	>2000
VF03650860	3	65	0.8	60	13	13.25	17.934	663.0	213238.24	16099.41	4953.67	0.8253	51.0	1991
VF03651060	3	65	1.0	60	13	14.81	18.343	741.0	249330.59	16831.20	6011.14	1.0777	54.9	1783
VF03800460	3	80	0.4	60	16	10.66	19.460	624.0	236307.69	22170.25	4030.95	0.1555	56.3	>2000
VF03800660	3	80	0.6	60	16	13.86	21.166	720.0	322560.00	23278.76	5819.69	0.3360	58.4	>2000
VF03800860	3	80	0.8	60	16	16.30	22.073	816.0	397552.94	24387.28	7503.78	0.5450	62.7	>2000
VF03801060	3	80	1.0	60	16	18.23	22.576	912.0	464842.11	25495.79	9105.64	0.7115	67.7	>2000

V-PLATE n V-stiffeners edges free out-of-plane free in-plane

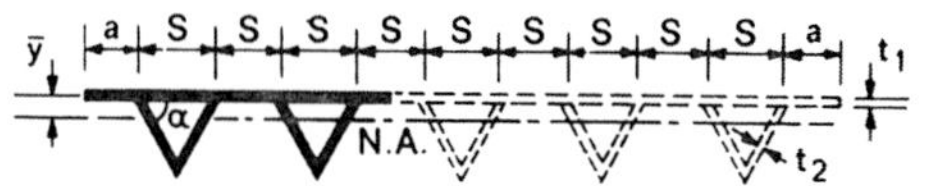

CODE	DIMENSIONS					SECTION PROPERTIES						BUCKLING PROPERTIES		
	n	S	t_2	α	a	$\bar{y}$	rad. gyr.	A	I_{NA}	Z_{plate}	$Z_{stiff.}$	σ_{cr}/E	L_1	L_{euler}
To scale multiply by	$-$	t_1	t_1	$-$	t_1	t_1	t_1	t_1^2	t_1^4	t_1^3	t_1^3	10^{-3}	t_1	t_1
VF03350470	3	35	0.4	70	7	9.47	14.623	311.8	66676.82	7042.24	1726.81	0.3797	35.9	>2000
VF03350670	3	35	0.6	70	7	11.87	15.477	373.2	89398.62	7534.27	2468.54	0.8227	36.8	1723
VF03350870	3	35	0.8	70	7	13.59	15.840	434.6	109042.44	8026.29	3161.09	1.3713	38.2	1367
VF03351070	3	35	1.0	70	7	14.88	15.986	496.0	126751.34	8518.32	3817.70	1.9739	39.8	1150
VF03500470	3	50	0.4	70	10	13.53	20.891	445.4	194393.05	14371.92	3524.10	0.1862	51.3	>2000
VF03500670	3	50	0.6	70	10	16.95	22.110	533.1	260637.38	15376.06	5037.83	0.4037	52.6	>2000
VF03500870	3	50	0.8	70	10	19.41	22.628	620.9	317907.98	16380.19	6451.20	0.6735	54.5	>2000
VF03501070	3	50	1.0	70	10	21.26	22.837	708.6	369537.44	17384.33	7791.22	0.9704	56.8	>2000
VF03650470	3	65	0.4	70	13	17.58	27.158	579.1	427081.53	24288.54	5955.73	0.1102	66.9	>2000
VF03650670	3	65	0.6	70	13	22.04	28.744	693.1	572620.32	25985.53	8513.93	0.2390	68.4	>2000
VF03650870	3	65	0.8	70	13	25.23	29.417	807.1	698443.84	27682.52	10902.53	0.3999	70.7	>2000
VF03651070	3	65	1.0	70	13	27.63	29.688	921.1	811873.76	29379.51	13167.16	0.5750	73.7	>2000
VF03800470	3	80	0.4	70	16	21.64	33.425	712.7	796233.93	36792.12	9021.70	0.0727	82.2	>2000
VF03800670	3	80	0.6	70	16	27.12	35.377	853.0	1067570.70	39362.70	12896.84	0.1578	83.9	>2000
VF03800870	3	80	0.8	70	16	31.05	36.206	993.4	1302151.10	41933.29	16515.08	0.2635	88.1	>2000
VF03801070	3	80	1.0	70	16	34.01	36.539	1133.7	1513625.37	44503.88	19945.51	0.3798	90.7	>2000
VF05350445	5	35	0.4	45	7	2.02	4.418	428.0	8352.65	4127.07	539.71	1.6255	17.4	342
VF05350645	5	35	0.6	45	7	2.72	4.934	477.5	11623.04	4271.44	786.46	3.3167	31.1	264
VF05350845	5	35	0.8	45	7	3.29	5.248	527.0	14516.39	4415.80	1021.37	3.6668	29.2	269
VF05351045	5	35	1.0	45	7	3.76	5.451	576.5	17129.81	4560.17	1246.38	4.2152	27.2	261
VF05500445	5	50	0.4	45	10	2.89	6.311	611.4	24351.74	8422.59	1101.45	0.7971	24.9	705
VF05500645	5	50	0.6	45	10	3.89	7.048	682.1	33886.42	8717.22	1605.03	1.6262	44.3	549
VF05500845	5	50	0.8	45	10	4.70	7.498	752.8	42321.83	9011.84	2084.43	1.7977	41.8	557
VF05501045	5	50	1.0	45	10	5.37	7.787	823.6	49941.13	9306.47	2543.64	2.0666	38.8	540
VF05650445	5	65	0.4	45	13	3.76	8.204	794.8	53500.77	14234.18	1861.45	0.4718	32.4	1197
VF05650645	5	65	0.6	45	13	5.05	9.163	886.8	74448.46	14732.10	2712.49	0.9624	57.7	933
VF05650845	5	65	0.8	45	13	6.11	9.747	978.7	92981.06	15230.02	3522.69	1.0640	54.2	946
VF05651045	5	65	1.0	45	13	6.98	10.123	1070.6	109720.66	15727.94	4298.75	1.2231	50.4	918
VF05800445	5	80	0.4	45	16	4.63	10.098	978.3	99744.73	21561.83	2819.72	0.3115	39.9	1816
VF05800645	5	80	0.6	45	16	6.22	11.277	1091.4	138798.77	22316.08	4108.86	0.6354	70.9	1419
VF05800845	5	80	0.8	45	16	7.51	11.996	1204.5	173350.22	23070.32	5336.15	0.7025	67.0	1438
VF05801045	5	80	1.0	45	16	8.59	12.460	1317.7	204558.86	23824.57	6511.72	0.8075	62.1	1394
VF05350460	5	35	0.4	60	7	4.52	8.423	469.0	33276.12	7355.44	1290.43	0.8116	24.6	938
VF05350660	5	35	0.6	60	7	5.90	9.190	539.0	45519.89	7709.07	1865.10	1.7525	25.5	695
VF05350860	5	35	0.8	60	7	6.97	9.605	609.0	56181.03	8062.70	2406.78	2.8340	27.5	570
VF05351060	5	35	1.0	60	7	7.81	9.840	679.0	65749.03	8416.32	2922.33	3.6875	29.7	512
VF05500460	5	50	0.4	60	10	6.46	12.033	670.0	97014.93	15011.11	2633.53	0.3979	35.2	1920
VF05500660	5	50	0.6	60	10	8.44	13.128	770.0	132711.04	15732.79	3806.32	0.8598	36.4	1425
VF05500860	5	50	0.8	60	10	9.95	13.721	870.0	163793.10	16454.48	4911.79	1.3912	39.3	1170
VF05501060	5	50	1.0	60	10	11.16	14.058	970.0	191688.14	17176.17	5963.95	1.8076	42.5	1052
VF05650460	5	65	0.4	60	13	8.40	15.643	871.0	213141.79	25368.77	4450.66	0.2355	45.7	>2000
VF05650660	5	65	0.6	60	13	10.97	17.067	1001.0	291566.15	26588.42	6432.68	0.5090	47.3	>2000
VF05650860	5	65	0.8	60	13	12.94	17.837	1131.0	359853.45	27808.08	8300.92	0.8238	51.1	1982
VF05651060	5	65	1.0	60	13	14.51	18.275	1261.0	421138.85	29027.73	10079.07	1.0697	55.2	1783
VF05800460	5	80	0.4	60	16	10.34	19.253	1072.0	397373.13	38428.43	6741.83	0.1555	56.3	>2000
VF05800660	5	80	0.6	60	16	13.50	21.005	1232.0	543584.42	40275.95	9744.18	0.3361	58.3	>2000
VF05800860	5	80	0.8	60	16	15.93	21.954	1392.0	670896.55	42123.48	12574.17	0.5440	62.8	>2000
VF05801060	5	80	1.0	60	16	17.86	22.492	1552.0	785154.64	43971.00	15267.71	0.7063	68.0	>2000
VF05350470	5	35	0.4	70	7	9.22	14.509	533.7	112350.14	12185.82	2891.07	0.3797	35.9	>2000
VF05350670	5	35	0.6	70	7	11.60	15.405	636.0	150925.41	13005.87	4137.61	0.8230	36.8	1714
VF05350870	5	35	0.8	70	7	13.33	15.798	738.3	184272.38	13825.91	5302.37	1.3725	38.1	1362
VF05351070	5	35	1.0	70	7	14.63	15.966	840.7	214300.36	14645.96	6406.82	1.9756	39.8	1148
VF05500470	5	50	0.4	70	10	13.17	20.728	762.4	327551.42	24869.03	5900.14	0.1862	51.3	>2000
VF05500670	5	50	0.6	70	10	16.58	22.007	908.6	440015.77	26542.59	8444.11	0.4038	52.5	>2000
VF05500870	5	50	0.8	70	10	19.04	22.569	1054.8	537237.26	28216.15	10821.17	0.6741	54.4	>2000
VF05501070	5	50	1.0	70	10	20.90	22.809	1201.0	624782.38	29889.70	13075.14	0.9712	56.7	>2000
VF05650470	5	65	0.4	70	13	17.12	26.946	991.1	719630.47	42028.65	9971.24	0.1102	66.9	>2000
VF05650670	5	65	0.6	70	13	21.55	28.609	1181.1	966714.64	44856.97	14270.54	0.2391	68.3	>2000
VF05650870	5	65	0.8	70	13	24.75	29.339	1371.2	1180310.26	47685.29	18287.78	0.3992	70.7	>2000
VF05651070	5	65	1.0	70	13	27.17	29.651	1561.2	1372646.89	50513.60	22096.98	0.5754	73.7	>2000
VF05800470	5	80	0.4	70	16	21.07	33.165	1219.8	1341650.62	63664.71	15104.36	0.0727	82.2	>2000
VF05800670	5	80	0.6	70	16	26.52	35.211	1453.7	1802304.58	67949.02	21616.92	0.1579	83.9	>2000
VF05800870	5	80	0.8	70	16	30.46	36.110	1687.6	2200523.81	72233.33	27702.20	0.2637	88.0	>2000
VF05801070	5	80	1.0	70	16	33.44	36.494	1921.5	2559108.63	76517.64	33472.35	0.3801	90.6	>2000

| V-PLATE | n V-stiffeners edges pinned free in-plane |

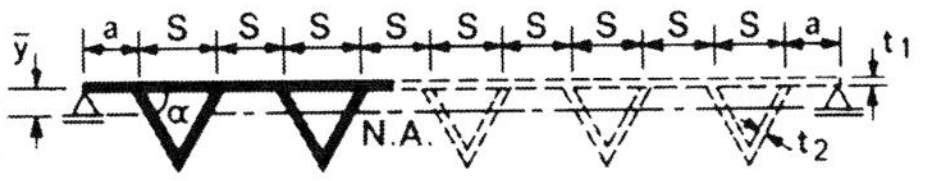

a S S S S S S S S a t_1 $\bar{y}$ α N.A. t_2

| CODE | n | DIMENSIONS | | | | SECTION PROPERTIES | | | | | | BUCKLING PROPERTIES | | |

CODE	n	S	t_2	α	a	$\bar{y}$	rad. gyr.	A	I_{NA}	Z_{plate}	$Z_{stiff.}$	σ_{cr}/E	L_1	L_{euler}
To scale multiply by	–	t_1	t_1	–	t_1	t_1	t_1	t_1^2	t_1^4	t_1^3	t_1^3	10^{-3}	t_1	t_1
VP01350445	1	35	0.4	45	35	1.39	3.777	124.8	1780.66	1282.75	110.52	2.3442	414.5	1826
VP01350645	1	35	0.6	45	35	1.93	4.334	134.7	2530.39	1311.62	162.51	2.7054	435.4	1644
VP01350845	1	35	0.8	45	35	2.40	4.713	144.6	3212.06	1340.49	212.67	3.0249	442.2	1507
VP01351045	1	35	1.0	45	35	2.80	4.985	154.5	3838.75	1369.37	261.20	3.2638	446.6	1410
VP01500445	1	50	0.4	45	50	1.98	5.396	178.3	5191.43	2617.85	225.55	0.7992	24.9	>2000
VP01500645	1	50	0.6	45	50	2.76	6.192	192.4	7377.24	2676.78	331.65	1.5841	45.6	>2000
VP01500845	1	50	0.8	45	50	3.42	6.733	206.6	9364.61	2735.70	434.01	1.6933	43.8	>2000
VP01501045	1	50	1.0	45	50	4.00	7.121	220.7	11191.69	2794.63	533.06	1.8435	41.8	>2000
VP01650445	1	65	0.4	45	65	2.58	7.015	231.8	11405.57	4424.17	381.18	0.4731	32.3	>2000
VP01650645	1	65	0.6	45	65	3.58	8.049	250.2	16207.79	4523.75	560.49	0.9374	59.3	>2000
VP01650845	1	65	0.8	45	65	4.45	8.753	268.5	20574.05	4623.34	733.48	1.0020	56.9	>2000
VP01651045	1	65	1.0	45	65	5.21	9.257	286.9	24588.14	4722.92	900.87	1.0910	54.3	>2000
VP01800445	1	80	0.4	45	80	3.17	8.634	285.3	21264.09	6701.70	577.40	0.3124	39.8	>2000
VP01800645	1	80	0.6	45	80	4.41	9.907	307.9	30217.16	6852.55	849.03	0.6190	74.2	>2000
VP01800845	1	80	0.8	45	80	5.48	10.773	330.5	38357.44	7003.40	1111.07	0.6615	70.0	>2000
VP01801045	1	80	1.0	45	80	6.41	11.393	353.1	45841.16	7154.25	1364.63	0.7203	67.0	>2000
VP01350460	1	35	0.4	60	35	3.19	7.368	133.0	7221.05	2263.21	266.26	0.8137	24.6	>2000
VP01350660	1	35	0.6	60	35	4.33	8.292	147.0	10106.25	2333.94	388.99	1.7645	25.4	1967
VP01350860	1	35	0.8	60	35	5.27	8.873	161.0	12676.09	2404.66	506.25	2.8568	27.7	1481
VP01351060	1	35	1.0	60	35	6.06	9.260	175.0	15006.25	2475.39	618.85	3.4990	30.8	1290
VP01500460	1	50	0.4	60	50	4.56	10.526	190.0	21052.63	4618.80	543.39	0.3990	35.1	>2000
VP01500660	1	50	0.6	60	50	6.19	11.845	210.0	29464.29	4763.14	793.86	0.8657	36.3	>2000
VP01500860	1	50	0.8	60	50	7.53	12.676	230.0	36956.52	4907.48	1033.15	1.4023	39.5	>2000
VP01501060	1	50	1.0	60	50	8.66	13.229	250.0	43750.00	5051.81	1262.95	1.7146	44.0	>2000
VP01650460	1	65	0.4	60	65	5.93	13.684	247.0	46252.63	7805.78	918.33	0.2361	45.6	>2000
VP01650660	1	65	0.6	60	65	8.04	15.399	273.0	64733.04	8049.71	1341.62	0.5125	47.1	>2000
VP01650860	1	65	0.8	60	65	9.79	16.479	299.0	81193.48	8293.64	1746.03	0.8303	51.3	>2000
VP01651060	1	65	1.0	60	65	11.26	17.197	325.0	96118.75	8537.57	2134.39	1.0146	57.1	>2000
VP01800460	1	80	0.4	60	80	7.29	16.842	304.0	86231.58	11824.13	1391.07	0.1559	56.2	>2000
VP01800660	1	80	0.6	60	80	9.90	18.952	336.0	120685.71	12193.64	2032.27	0.3384	58.0	>2000
VP01800860	1	80	0.8	60	80	12.05	20.282	368.0	151373.91	12563.14	2644.87	0.5483	63.1	>2000
VP01801060	1	80	1.0	60	80	13.86	21.166	400.0	179200.00	12932.65	3233.16	0.6698	70.3	>2000
VP01350470	1	35	0.4	70	35	6.74	13.064	145.9	24907.10	3693.68	602.53	0.3807	35.9	>2000
VP01350670	1	35	0.6	70	35	8.87	14.341	166.4	34220.29	3857.69	872.74	0.8296	36.5	>2000
VP01350870	1	35	0.8	70	35	10.53	15.056	186.9	42357.10	4021.70	1128.06	1.3913	37.8	1999
VP01351070	1	35	1.0	70	35	11.87	15.477	207.3	49665.90	4185.70	1371.41	2.0052	39.4	1593
VP01500470	1	50	0.4	70	50	9.63	18.663	208.5	72615.46	7538.12	1229.65	0.1867	51.2	>2000
VP01500670	1	50	0.6	70	50	12.67	20.486	237.7	99767.61	7872.83	1781.10	0.4071	52.2	>2000
VP01500870	1	50	0.8	70	50	15.05	21.508	267.0	123490.08	8207.54	2302.16	0.6833	53.9	>2000
VP01501070	1	50	1.0	70	50	16.95	22.110	296.2	144798.54	8542.25	2798.79	0.9857	56.1	>2000
VP01650470	1	65	0.4	70	65	12.52	24.262	271.0	159536.17	12739.42	2078.11	0.1105	66.8	>2000
VP01650670	1	65	0.6	70	65	16.47	26.632	309.0	219189.43	13305.08	3010.06	0.2410	68.0	>2000
VP01650870	1	65	0.8	70	65	19.56	27.960	347.0	271307.70	13870.74	3890.65	0.4047	70.0	>2000
VP01651070	1	65	1.0	70	65	22.04	28.744	385.0	318122.40	14436.41	4729.96	0.5842	74.1	>2000
VP01800470	1	80	0.4	70	80	15.41	29.861	333.6	297432.93	19297.58	3147.90	0.0729	82.0	>2000
VP01800670	1	80	0.6	70	80	20.28	32.778	380.3	408648.12	20154.44	4559.62	0.1591	83.6	>2000
VP01800870	1	80	0.8	70	80	24.07	34.413	427.1	505815.36	21011.31	5893.52	0.2672	86.2	>2000
VP01801070	1	80	1.0	70	80	27.12	35.377	473.9	593094.83	21868.17	7164.91	0.3857	89.7	>2000
VP02350445	2	35	0.4	45	35	1.61	4.029	214.6	3482.88	2157.16	219.25	1.5635	542.2	>2000
VP02350645	2	35	0.6	45	35	2.22	4.577	234.4	4911.07	2214.91	321.35	1.7759	566.5	>2000
VP02350845	2	35	0.8	45	35	2.73	4.937	254.2	6195.49	2272.65	419.35	1.9346	578.0	>2000
VP02351045	2	35	1.0	45	35	3.16	5.185	274.0	7367.32	2330.40	513.81	2.0446	586.6	>2000
VP02500445	2	50	0.4	45	50	2.31	5.755	306.6	10154.16	4402.37	447.45	0.7986	24.9	>2000
VP02500645	2	50	0.6	45	50	3.17	6.539	334.9	14317.98	4520.22	655.81	1.2820	934.1	>2000
VP02500845	2	50	0.8	45	50	3.89	7.053	363.1	18062.66	4638.07	855.82	1.3935	956.1	>2000
VP02501045	2	50	1.0	45	50	4.52	7.408	391.4	21479.06	4755.92	1048.59	1.4712	970.5	>2000
VP02650445	2	65	0.4	45	65	3.00	7.482	398.5	22308.68	7440.00	756.19	0.4727	32.3	>2000
VP02650645	2	65	0.6	45	65	4.12	8.501	435.3	31456.61	7639.17	1108.32	0.9433	60.4	>2000
VP02650845	2	65	0.8	45	65	5.06	9.169	472.1	39683.67	7838.34	1446.34	1.0180	56.1	>2000
VP02651045	2	65	1.0	45	65	5.87	9.630	508.8	47189.49	8037.51	1772.12	1.1177	53.4	>2000
VP02800445	2	80	0.4	45	80	3.69	9.208	490.5	41591.42	11270.06	1145.47	0.3121	39.8	>2000
VP02800645	2	80	0.6	45	80	5.07	10.462	535.8	58646.46	11571.76	1678.88	0.6222	72.4	>2000
VP02800845	2	80	0.8	45	80	6.23	11.284	581.0	73984.67	11873.46	2190.91	0.6720	69.1	>2000
VP02801045	2	80	1.0	45	80	7.23	11.852	626.3	87978.22	12175.16	2684.39	0.7380	65.8	>2000

V-PLATE   n V-stiffeners
edges pinned
free in-plane

CODE	n	S	t_2	α	a	$\bar{y}$	rad. gyr.	A	I_{NA}	Z_{plate}	$Z_{stiff.}$	σ_{cr}/E	L_1	L_{euler}
To scale multiply by	$-$	t_1	t_1	$-$	t_1	t_1	t_1	t_1^2	t_1^4	t_1^3	t_1^3	10^{-3}	t_1	t_1
VP02350460	2	35	0.4	60	35	3.67	7.794	231.0	14031.82	3819.17	526.78	0.8136	24.6	>2000
VP02350660	2	35	0.6	60	35	4.92	8.670	259.0	19467.57	3960.62	766.57	1.7618	25.5	>2000
VP02350860	2	35	0.8	60	35	5.91	9.194	287.0	24260.98	4102.07	994.44	2.8423	27.7	>2000
VP02351060	2	35	1.0	60	35	6.74	9.526	315.0	28583.33	4243.52	1212.44	3.3713	819.3	>2000
VP02500460	2	50	0.4	60	50	5.25	11.134	330.0	40909.09	7794.23	1075.07	0.4011	37.6	>2000
VP02500660	2	50	0.6	60	50	7.02	12.385	370.0	56756.76	8082.90	1564.43	0.8644	36.5	>2000
VP02500860	2	50	0.8	60	50	8.45	13.135	410.0	70731.71	8371.58	2029.47	1.3952	39.4	>2000
VP02501060	2	50	1.0	60	50	9.62	13.608	450.0	83333.33	8660.25	2474.36	1.7368	43.6	>2000
VP02650460	2	65	0.4	60	65	6.82	14.474	429.0	89877.27	13172.25	1816.86	0.2361	45.6	>2000
VP02650660	2	65	0.6	60	65	9.13	16.101	481.0	124694.59	13660.11	2643.89	0.5117	47.4	>2000
VP02650860	2	65	0.8	60	65	10.98	17.075	533.0	155397.56	14147.97	3429.81	0.8261	51.2	>2000
VP02651060	2	65	1.0	60	65	12.51	17.691	585.0	183083.33	14635.83	4181.67	1.0278	56.6	>2000
VP02800460	2	80	0.4	60	80	8.40	17.814	528.0	167563.64	19953.23	2752.17	0.1559	56.2	>2000
VP02800660	2	80	0.6	60	80	11.23	19.817	592.0	232475.68	20692.23	4004.95	0.3379	58.3	>2000
VP02800860	2	80	0.8	60	80	13.52	21.015	656.0	289717.07	21431.24	5195.45	0.5455	63.0	>2000
VP02801060	2	80	1.0	60	80	15.40	21.773	720.0	341333.33	22170.25	6334.36	0.6785	69.7	>2000
VP02350470	2	35	0.4	70	35	7.66	13.671	256.9	48005.90	6265.47	1187.71	0.3806	35.9	>2000
VP02350670	2	35	0.6	70	35	9.91	14.815	297.8	65362.78	6593.49	1712.52	0.8281	36.6	>2000
VP02350870	2	35	0.8	70	35	11.62	15.409	338.7	80430.62	6921.50	2205.97	1.3848	37.9	>2000
VP02351070	2	35	1.0	70	35	12.96	15.731	379.7	93949.79	7249.52	2675.00	1.9916	39.5	>2000
VP02500470	2	50	0.4	70	50	10.95	19.530	367.0	139958.88	12786.67	2423.90	0.1866	51.2	>2000
VP02500670	2	50	0.6	70	50	14.16	21.164	425.4	190562.04	13456.09	3494.94	0.4063	52.4	>2000
VP02500870	2	50	0.8	70	50	16.60	22.013	483.9	234491.61	14125.52	4501.98	0.6800	54.1	>2000
VP02501070	2	50	1.0	70	50	18.51	22.472	542.4	273906.10	14794.94	5459.19	0.9790	56.4	>2000
VP02650470	2	65	0.4	70	65	14.23	25.389	477.0	307489.66	21609.47	4096.39	0.1104	66.9	>2000
VP02650670	2	65	0.6	70	65	18.41	27.514	553.1	418664.80	22740.90	5906.44	0.2405	68.1	>2000
VP02650870	2	65	0.8	70	65	21.58	28.617	629.1	515178.06	23872.13	7608.34	0.4027	70.3	>2000
VP02651070	2	65	1.0	70	65	24.07	29.214	705.1	601771.71	25003.45	9226.02	0.5802	74.4	>2000
VP02800470	2	80	0.4	70	80	17.51	31.248	587.1	573271.57	32733.88	6205.18	0.0729	82.1	>2000
VP02800670	2	80	0.6	70	80	22.66	33.863	680.7	780542.11	34447.60	8947.04	0.1588	83.6	>2000
VP02800870	2	80	0.8	70	80	26.56	35.221	774.2	960477.63	36161.33	11525.06	0.2660	86.5	>2000
VP02801070	2	80	1.0	70	80	29.62	35.956	867.8	1121919.40	37875.05	13975.51	0.3831	90.0	>2000
VP03350445	3	35	0.4	45	35	1.71	4.124	304.4	5176.07	3031.57	327.75	1.1243	696.3	>2000
VP03350645	3	35	0.6	45	35	2.33	4.667	334.1	7276.06	3118.20	479.74	1.2842	716.5	>2000
VP03350845	3	35	0.8	45	35	2.86	5.017	363.8	9156.93	3204.82	625.36	1.3932	728.4	>2000
VP03351045	3	35	1.0	45	35	3.30	5.255	393.5	10868.30	3291.44	765.48	1.4650	738.1	>2000
VP03500445	3	50	0.4	45	50	2.44	5.891	434.9	15090.59	6186.89	668.88	0.8507	1083.5	>2000
VP03500645	3	50	0.6	45	50	3.33	6.667	477.3	21213.01	6363.66	979.07	0.9631	1124.1	>2000
VP03500845	3	50	0.8	45	50	4.08	7.167	519.7	26696.59	6540.44	1276.24	1.0406	1149.8	>2000
VP03501045	3	50	1.0	45	50	4.72	7.508	562.1	31685.99	6717.22	1562.20	1.0923	1171.7	>2000
VP03650445	3	65	0.4	45	65	3.17	7.658	565.3	33154.02	10455.84	1130.41	0.4726	32.3	>2000
VP03650645	3	65	0.6	45	65	4.33	8.667	620.5	46604.97	10754.59	1654.62	0.7689	1596.6	>2000
VP03650845	3	65	0.8	45	65	5.31	9.317	675.6	58652.41	11053.34	2156.84	0.8289	1636.2	>2000
VP03651045	3	65	1.0	45	65	6.13	9.760	730.8	69614.12	11352.10	2640.13	0.8693	1667.2	>2000
VP03800445	3	80	0.4	45	80	3.90	9.425	695.8	61811.05	15838.43	1712.34	0.3120	39.8	>2000
VP03800645	3	80	0.6	45	80	5.33	10.667	763.6	86888.47	16290.98	2506.41	0.6236	72.2	>2000
VP03800845	3	80	0.8	45	80	6.53	11.468	831.5	109349.24	16743.53	3267.16	0.6761	68.7	>2000
VP03801045	3	80	1.0	45	80	7.55	12.013	899.4	129785.82	17196.08	3999.24	0.7428	65.5	1933
VP03350460	3	35	0.4	60	35	3.87	7.951	329.0	20798.94	5375.13	786.60	0.8135	24.5	>2000
VP03350660	3	35	0.6	60	35	5.15	8.804	371.0	28758.61	5587.31	1142.86	1.7607	25.6	>2000
VP03350860	3	35	0.8	60	35	6.16	9.304	413.0	35753.39	5799.48	1480.72	2.4145	990.0	>2000
VP03351060	3	35	1.0	60	35	6.99	9.613	455.0	42050.48	6011.66	1803.50	2.4957	1002.7	>2000
VP03500460	3	50	0.4	60	50	5.53	11.359	470.0	60638.30	10969.66	1605.32	0.4011	37.6	>2000
VP03500660	3	50	0.6	60	50	7.35	12.578	530.0	83844.34	11402.67	2332.36	0.8638	36.5	>2000
VP03500860	3	50	0.8	60	50	8.81	13.292	590.0	104237.29	11835.68	3021.88	1.3927	39.4	>2000
VP03501060	3	50	1.0	60	50	9.99	13.734	650.0	122596.15	12268.69	3680.61	1.8357	1605.8	>2000
VP03650460	3	65	0.4	60	65	7.19	14.766	611.0	133222.34	18538.72	2712.98	0.2361	45.6	>2000
VP03650660	3	65	0.6	60	65	9.56	16.351	689.0	184206.01	19270.51	3941.69	0.5114	47.4	>2000
VP03650860	3	65	0.8	60	65	11.45	17.279	767.0	229009.32	20002.30	5106.97	0.8246	51.2	>2000
VP03651060	3	65	1.0	60	65	12.99	17.854	845.0	269343.75	20734.09	6220.23	1.0315	56.5	>2000
VP03800460	3	80	0.4	60	80	8.84	18.174	752.0	248374.47	28082.32	4109.61	0.1559	56.1	>2000
VP03800660	3	80	0.6	60	80	11.76	20.124	848.0	343426.42	29190.83	5970.85	0.3377	58.4	>2000
VP03800860	3	80	0.8	60	80	14.09	21.267	944.0	426955.93	30299.34	7736.00	0.5446	58.4	>2000
VP03801060	3	80	1.0	60	80	15.99	21.974	1040.0	502153.85	31407.85	9422.36	0.6810	69.5	>2000

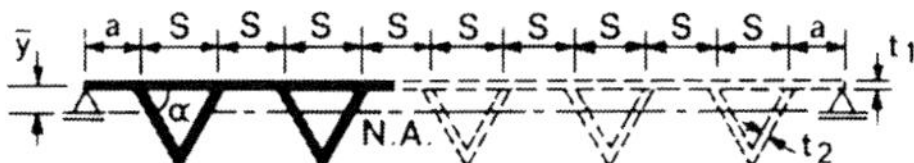

CODE	n	\$S\$	\$t_2\$	\$\alpha\$	\$a\$	\$\bar{y}\$	rad. gyr.	\$A\$	\$I_{NA}\$	\$Z_{plate}\$	\$Z_{stiff.}\$	\$\sigma_{cr}/E\$	\$L_1\$	\$L_{euler}\$
To scale multiply by	–	\$t_1\$	\$t_1\$	–	\$t_1\$	\$t_1\$	\$t_1\$	\$t_1^2\$	\$t_1^4\$	\$t_1^3\$	\$t_1^3\$	\$10^{-3}\$	\$t_1\$	\$t_1\$
VP03350470	3	35	0.4	70	35	8.03	13.887	367.8	70932.61	8837.26	1770.91	0.3805	35.9	>2000
VP03350670	3	35	0.6	70	35	10.32	14.975	429.2	96254.28	9329.29	2548.88	0.8275	36.7	>2000
VP03350870	3	35	0.8	70	35	12.03	15.522	490.6	118198.55	9821.31	3279.11	1.3823	38.0	>2000
VP03351070	3	35	1.0	70	35	13.37	15.805	552.0	137892.44	10313.34	3972.64	1.9867	39.6	>2000
VP03500470	3	50	0.4	70	50	11.47	19.839	525.4	206800.60	18035.22	3614.10	0.1866	51.2	>2000
VP03500670	3	50	0.6	70	50	14.74	21.394	613.1	280624.73	19039.36	5201.79	0.4060	52.4	>2000
VP03500870	3	50	0.8	70	50	17.19	22.174	700.9	344602.19	20043.49	6692.06	0.6788	54.2	>2000
VP03501070	3	50	1.0	70	50	19.10	22.579	788.6	402018.76	21047.63	8107.42	0.9765	56.4	>2000
VP03650470	3	65	0.4	70	65	14.91	25.791	683.1	454340.93	30479.53	6107.83	0.1104	66.9	>2000
VP03650670	3	65	0.6	70	65	19.16	27.812	797.1	616532.54	32176.52	8791.02	0.2404	68.2	>2000
VP03650870	3	65	0.8	70	65	22.35	28.826	911.1	757091.01	33873.51	11309.58	0.4020	70.4	>2000
VP03651070	3	65	1.0	70	65	24.83	29.353	1025.1	883235.22	35570.50	13701.54	0.5785	73.3	>2000
VP03800470	3	80	0.4	70	80	18.35	31.742	840.7	847055.27	46170.17	9252.10	0.0729	82.1	>2000
VP03800670	3	80	0.6	70	80	23.58	34.230	981.0	1149438.91	48740.76	13316.58	0.1587	83.7	>2000
VP03800870	3	80	0.8	70	80	27.51	35.478	1121.4	1411490.57	51311.35	17131.66	0.2655	86.6	>2000
VP03801070	3	80	1.0	70	80	30.56	36.126	1261.7	1646668.84	53881.93	20755.00	0.3822	90.1	>2000
VP05350445	5	35	0.4	45	35	1.79	4.204	484.0	8555.49	4780.40	544.58	0.6315	1188.0	>2000
VP05350645	5	35	0.6	45	35	2.44	4.742	533.5	11994.16	4924.77	796.19	0.7386	1184.6	>2000
VP05350845	5	35	0.8	45	35	2.97	5.083	583.0	15063.44	5069.14	1036.83	0.8058	1184.8	>2000
VP05351045	5	35	1.0	45	35	3.42	5.312	632.5	17850.03	5213.50	1268.10	0.8471	1190.2	>2000
VP05500445	5	50	0.4	45	50	2.56	6.006	691.4	24943.11	9755.92	1111.38	0.5259	1627.7	>2000
VP05500645	5	50	0.6	45	50	3.48	6.774	762.1	34968.41	10050.55	1624.87	0.6054	1629.9	>2000
VP05500845	5	50	0.8	45	50	4.25	7.262	832.8	43916.73	10345.18	2115.97	0.6551	1645.6	>2000
VP05501045	5	50	1.0	45	50	4.89	7.589	903.6	52040.91	10639.81	2587.96	0.6863	1668.6	>2000
VP05650445	5	65	0.4	45	65	3.32	7.808	898.8	54800.01	16487.51	1878.24	0.4725	32.3	1615
VP05650645	5	65	0.6	45	65	4.52	8.806	990.8	76825.60	16985.43	2746.03	0.9466	58.5	992
VP05650845	5	65	0.8	45	65	5.52	9.440	1082.7	96485.05	17483.35	3575.99	1.0286	55.7	1015
VP05651045	5	65	1.0	45	65	6.36	9.866	1174.6	114333.88	17981.27	4373.65	1.1278	53.4	1005
VP05800445	5	80	0.4	45	80	4.09	9.610	1106.3	102166.97	24975.16	2845.14	0.3120	39.8	>2000
VP05800645	5	80	0.6	45	80	5.57	10.838	1219.4	145230.61	25729.41	4159.67	0.6249	72.0	1573
VP05800845	5	80	0.8	45	80	6.79	11.619	1332.5	179882.92	26483.66	5416.89	0.6791	68.5	1607
VP05801045	5	80	1.0	45	80	7.83	12.143	1445.7	213159.58	27237.90	6625.18	0.7446	65.5	1583
VP05350460	5	35	0.4	60	35	4.04	8.083	525.0	34300.00	8487.05	1305.70	0.8135	24.5	>2000
VP05350660	5	35	0.6	60	35	5.35	8.915	595.0	47288.60	8840.68	1894.43	1.3872	1521.8	>2000
VP05350860	5	35	0.8	60	35	6.38	9.393	665.0	58671.05	9194.30	2451.81	1.4766	1498.9	>2000
VP05351060	5	35	1.0	60	35	7.22	9.682	735.0	68906.25	9547.93	2983.73	1.5256	1490.6	>2000
VP05500460	5	50	0.4	60	50	5.77	11.547	750.0	100000.00	17320.51	2664.69	0.3988	35.1	>2000
VP05500660	5	50	0.6	60	50	7.64	12.736	850.0	137867.65	18042.20	3866.18	0.8634	36.6	>2000
VP05500860	5	50	0.8	60	50	9.12	13.418	950.0	171052.63	18763.88	5003.70	1.3908	39.4	1422
VP05501060	5	50	1.0	60	50	10.31	13.832	1050.0	200892.86	19485.57	6089.24	1.7456	43.5	1199
VP05650460	5	65	0.4	60	65	7.51	15.011	975.0	219700.00	29271.66	4503.33	0.2361	45.6	>2000
VP05650660	5	65	0.6	60	65	9.93	16.556	1105.0	302895.22	30491.31	6533.85	0.5111	47.5	>2000
VP05650860	5	65	0.8	60	65	11.85	17.444	1235.0	375802.63	31710.96	8456.26	0.8235	51.2	>2000
VP05651060	5	65	1.0	60	65	13.40	17.982	1365.0	441361.61	32930.62	10290.82	1.0331	56.4	>2000
VP05800460	5	80	0.4	60	80	9.24	18.475	1200.0	409600.00	44340.50	6821.62	0.1559	56.1	>2000
VP05800660	5	80	0.6	60	80	12.23	20.377	1360.0	564705.88	46188.02	9897.43	0.3375	58.5	>2000
VP05800860	5	80	0.8	60	80	14.59	21.470	1520.0	700631.58	48035.54	12809.48	0.5438	62.9	>2000
VP05801060	5	80	1.0	60	80	16.50	22.131	1680.0	822857.14	49883.06	15588.46	0.6820	69.4	>2000
VP05350470	5	35	0.4	70	35	8.34	14.065	589.7	116658.27	13980.84	2935.78	0.3805	35.9	>2000
VP05350670	5	35	0.6	70	35	10.67	15.104	692.0	157656.24	14800.89	4219.00	0.8270	36.7	>2000
VP05350870	5	35	0.8	70	35	12.39	15.608	794.3	193518.74	15620.93	5421.84	1.3804	38.0	>2000
VP05351070	5	35	1.0	70	35	13.72	15.860	896.7	225541.00	16440.97	6563.55	1.9830	39.6	>2000
VP05500470	5	50	0.4	70	50	11.92	20.094	842.4	340111.57	28532.33	5991.39	0.1865	51.2	>2000
VP05500670	5	50	0.6	70	50	15.24	21.576	988.6	460222.26	30205.89	8610.21	0.4058	52.4	>2000
VP05500870	5	50	0.8	70	50	17.70	22.298	1134.8	564194.57	31879.45	11064.99	0.6779	54.2	>2000
VP05501070	5	50	1.0	70	50	19.60	22.657	1281.0	657553.94	33553.01	13395.01	0.9747	56.5	>2000
VP05650470	5	65	0.4	70	65	15.50	26.122	1095.1	747225.13	48219.64	10125.45	0.1104	66.9	>2000
VP05650670	5	65	0.6	70	65	19.81	28.049	1285.1	1011108.31	51047.95	14551.26	0.2402	68.2	>2000
VP05650870	5	65	0.8	70	65	23.01	28.987	1475.2	1239535.47	53876.27	18699.83	0.4015	70.5	>2000
VP05651070	5	65	1.0	70	65	25.48	29.454	1665.2	1444646.00	56704.58	22637.57	0.5774	73.4	>2000
VP05800470	5	80	0.4	70	80	19.07	32.150	1347.8	1393097.01	73042.76	15337.96	0.0729	82.1	>2000
VP05800670	5	80	0.6	70	80	24.38	34.522	1581.7	1885070.38	77327.08	22042.14	0.1586	83.8	>2000
VP05800870	5	80	0.8	70	80	28.32	35.677	1815.6	2310940.95	81611.39	28326.36	0.2651	86.7	>2000
VP05801070	5	80	1.0	70	80	31.36	36.251	2049.5	2693340.92	85895.70	34291.23	0.3814	90.2	>2000

CODE	DIMENSIONS					SECTION PROPERTIES						BUCKLING PROPERTIES		
	n	S	t_2	α	a	$\bar{y}$	rad. gyr.	A	I_{NA}	Z_{plate}	$Z_{stiff.}$	$\sigma_{cr/E}$	L_1	L_{euler}
To scale multiply by	$-$	t_1	t_1	$-$	t_1	t_1	t_1	t_1^2	t_1^4	t_1^3	t_1^3	10^{-3}	t_1	t_1
VQ02350445	2	35	0.4	45	7	2.18	4.551	158.6	3285.35	1503.83	214.51	1.6275	17.4	>2000
VQ02350645	2	35	0.6	45	7	2.91	5.050	178.4	4549.33	1561.57	311.88	3.5205	18.5	1848
VQ02350845	2	35	0.8	45	7	3.50	5.345	198.2	5661.73	1619.32	404.30	4.0303	27.2	1730
VQ02351045	2	35	1.0	45	7	3.97	5.529	216.0	6663.85	1677.07	492.65	4.6005	25.9	1621
VQ02500445	2	50	0.4	45	10	3.12	6.502	226.6	9578.28	3069.04	437.78	0.7983	24.8	>2000
VQ02500645	2	50	0.6	45	10	4.15	7.214	254.9	13263.36	3186.89	636.49	1.7287	26.5	>2000
VQ02500845	2	50	0.8	45	10	4.99	7.635	283.1	16506.51	3304.74	825.11	1.9830	38.8	>2000
VQ02501045	2	50	1.0	45	10	5.68	7.898	311.4	19428.15	3422.59	1005.41	2.2629	36.9	>2000
VQ02650445	2	65	0.4	45	13	4.06	8.453	294.5	21043.47	5186.67	739.85	0.4725	32.3	>2000
VQ02650645	2	65	0.6	45	13	5.41	9.378	331.3	29139.60	5385.84	1075.68	1.0238	34.5	>2000
VQ02650845	2	65	0.8	45	13	6.49	9.926	368.1	36264.80	5585.01	1394.44	1.1752	50.3	>2000
VQ02651045	2	65	1.0	45	13	7.38	10.268	404.8	42683.64	5784.18	1699.15	1.3410	47.9	>2000
VQ02800445	2	80	0.4	45	16	4.99	10.403	362.5	39232.62	7856.73	1120.72	0.3120	39.7	>2000
VQ02800645	2	80	0.6	45	16	6.66	11.543	407.8	54326.71	8158.43	1629.42	0.6761	42.4	>2000
VQ02800845	2	80	0.8	45	16	7.99	12.217	453.0	67610.66	8460.13	2112.28	0.7769	63.3	>2000
VQ02801045	2	80	1.0	45	16	9.08	12.638	498.3	79577.69	8761.83	2573.86	0.8866	60.5	>2000
VQ02350460	2	35	0.4	60	7	4.85	8.630	175.0	13034.00	2687.57	511.92	0.8144	24.5	>2000
VQ02350660	2	35	0.6	60	7	6.27	9.349	203.0	17741.38	2829.02	738.00	1.7730	25.2	>2000
VQ02350860	2	35	0.8	60	7	7.35	9.721	231.0	21827.27	2970.47	950.55	2.8874	27.5	>2000
VQ02351060	2	35	1.0	60	7	8.19	9.921	259.0	25493.24	3111.92	1152.56	3.7610	29.3	1800
VQ02500460	2	50	0.4	60	10	6.93	12.329	250.0	38000.00	5484.83	1044.73	0.3994	35.0	>2000
VQ02500660	2	50	0.6	60	10	8.96	13.355	290.0	51724.14	5773.50	1506.13	0.8702	36.0	>2000
VQ02500860	2	50	0.8	60	10	10.50	13.887	330.0	63636.36	6062.18	1939.90	1.4176	39.3	>2000
VQ02501060	2	50	1.0	60	10	11.70	14.173	370.0	74324.32	6350.85	2352.17	1.8456	41.9	>2000
VQ02650460	2	65	0.4	60	13	9.01	16.027	325.0	83486.00	9269.36	1765.59	0.2364	45.5	>2000
VQ02650660	2	65	0.6	60	13	11.65	17.362	377.0	113637.93	9757.22	2545.36	0.5152	46.8	>2000
VQ02650860	2	65	0.8	60	13	13.65	18.053	429.0	139809.09	10245.08	3278.43	0.8395	51.1	>2000
VQ02651060	2	65	1.0	60	13	15.21	18.425	481.0	163290.54	10732.94	3975.16	1.0927	54.4	>2000
VQ02800460	2	80	0.4	60	16	11.09	19.726	400.0	155648.00	14041.16	2674.51	0.1561	56.0	>2000
VQ02800660	2	80	0.6	60	16	14.33	21.368	464.0	211862.07	14780.17	3855.70	0.3402	57.7	>2000
VQ02800860	2	80	0.8	60	16	16.80	22.219	528.0	260654.55	15519.18	4966.14	0.5544	62.8	>2000
VQ02801060	2	80	1.0	60	16	18.72	22.677	592.0	304432.43	16258.18	6021.55	0.7216	67.1	>2000
VQ02350470	2	35	0.4	70	7	9.80	14.767	200.9	43801.82	4470.45	1144.17	0.3814	35.8	>2000
VQ02350670	2	35	0.6	70	7	12.21	15.566	241.8	58585.00	4798.47	1633.18	0.8342	36.4	>2000
VQ02350870	2	35	0.8	70	7	13.92	15.888	282.7	71370.98	5126.49	2089.39	1.4067	37.5	>2000
VQ02351070	2	35	1.0	70	7	15.20	16.006	323.7	82917.51	5454.50	2521.88	2.0394	39.5	>2000
VQ02500470	2	50	0.4	70	10	14.00	21.096	287.0	127702.11	9123.37	2335.03	0.1870	51.1	>2000
VQ02500670	2	50	0.6	70	10	17.44	22.237	345.4	170801.75	9792.79	3333.02	0.4094	52.0	>2000
VQ02500870	2	50	0.8	70	10	19.89	22.697	403.9	208078.66	10462.21	4264.05	0.6911	53.5	>2000
VQ02501070	2	50	1.0	70	10	21.72	22.865	462.4	241742.00	11131.64	5146.70	1.0026	55.5	>2000
VQ02650470	2	65	0.4	70	13	18.20	27.424	373.0	280561.54	15418.49	3946.20	0.1107	66.7	>2000
VQ02650670	2	65	0.6	70	13	22.67	28.908	449.1	375251.45	16549.82	5632.80	0.2424	67.7	>2000
VQ02650870	2	65	0.8	70	13	25.86	29.507	525.1	457148.82	17681.14	7206.25	0.4033	69.5	>2000
VQ02651070	2	65	1.0	70	13	28.23	29.725	601.1	531107.17	18812.47	8697.93	0.5941	72.1	>2000
VQ02800470	2	80	0.4	70	16	22.40	33.753	459.1	523067.85	23355.82	5977.68	0.0731	81.9	>2000
VQ02800670	2	80	0.6	70	16	27.91	35.578	552.7	699603.99	25069.54	8532.53	0.1601	83.2	>2000
VQ02800870	2	80	0.8	70	16	31.82	36.316	646.2	852290.21	26783.27	10915.97	0.2704	85.5	>2000
VQ02801070	2	80	1.0	70	16	34.75	36.584	739.8	990175.23	28497.00	13175.56	0.3925	88.7	>2000
VQ03350445	3	35	0.4	45	7	2.09	4.476	248.4	4976.02	2378.24	322.96	1.6275	17.4	>2000
VQ03350645	3	35	0.6	45	7	2.80	4.985	278.1	6909.75	2464.86	470.16	3.4371	29.7	>2000
VQ03350845	3	35	0.8	45	7	3.38	5.291	307.8	8616.58	2551.48	610.11	3.7968	28.4	>2000
VQ03351045	3	35	1.0	45	7	3.85	5.486	337.5	10156.41	2638.10	744.05	4.4147	461.5	>2000
VQ03500445	3	50	0.4	45	10	2.99	6.394	354.9	14507.34	4853.55	659.10	0.7982	24.8	>2000
VQ03500645	3	50	0.6	45	10	4.00	7.121	397.3	20145.04	5030.33	959.50	1.6876	42.6	>2000
VQ03500845	3	50	0.8	45	10	4.82	7.559	439.7	25121.22	5207.11	1245.13	1.8639	40.6	>2000
VQ03501045	3	50	1.0	45	10	5.50	7.837	482.1	29610.51	5383.88	1518.48	2.1339	38.1	>2000
VQ03650445	3	65	0.4	45	13	3.89	8.312	461.3	31872.63	8202.51	1113.87	0.4725	32.3	>2000
VQ03650645	3	65	0.6	45	13	5.21	9.257	516.5	44258.65	8501.26	1621.56	0.9993	55.1	>2000
VQ03650845	3	65	0.8	45	13	6.27	9.826	571.6	55191.32	8800.01	2104.27	1.1058	53.1	>2000
VQ03651045	3	65	1.0	45	13	7.15	10.188	626.8	65054.30	9098.76	2566.22	1.2635	49.5	>2000
VQ03800445	3	80	0.4	45	16	4.78	10.230	567.8	59422.07	12425.10	1687.28	0.3120	39.8	>2000
VQ03800645	3	80	0.6	45	16	6.41	11.393	635.6	82514.09	12877.65	2456.33	0.6599	67.6	>2000
VQ03800845	3	80	0.8	45	16	7.72	12.094	703.5	102896.51	13330.19	3187.53	0.7289	64.9	>2000
VQ03801045	3	80	1.0	45	16	8.80	12.539	771.4	121284.66	13782.74	3887.30	0.8344	60.9	>2000

V-PLATE

n V-stiffeners
edges pinned
free in-plane

CODE	n	S	t_2	α	a	$\bar{y}$	rad. gyr.	A	I_{NA}	Z_{plate}	$Z_{stiff.}$	σ_{cr}/E	L_1	L_{euler}
To scale multiply by	−	t_1	t_1	−	t_1	t_1	t_1	t_1^2	t_1^4	t_1^3	t_1^3	10^{-3}	t_1	t_1
VG03350460	3	35	0.4	60	7	4.66	8.514	273.0	19788.46	4243.52	771.55	0.8137	24.6	>2000
VG03350660	3	35	0.6	60	7	6.06	9.260	315.0	27011.25	4455.70	1113.93	1.7635	25.5	>2000
VG03350860	3	35	0.8	60	7	7.13	9.657	357.0	33291.18	4667.88	1436.27	2.8522	27.5	>2000
VG03351060	3	35	1.0	60	7	7.98	9.877	399.0	38925.99	4880.05	1742.88	3.7203	29.5	>2000
VG03500460	3	50	0.4	60	10	6.66	12.163	390.0	57692.31	8660.25	1574.59	0.3989	35.1	>2000
VG03500660	3	50	0.6	60	10	8.66	13.229	450.0	78750.00	9093.27	2273.32	0.8653	36.4	>2000
VG03500860	3	50	0.8	60	10	10.19	13.795	510.0	97058.82	9526.28	2931.16	1.4001	39.3	>2000
VG03501060	3	50	1.0	60	10	11.40	14.110	570.0	113486.84	9959.29	3556.89	1.8246	42.2	>2000
VG03650460	3	65	0.4	60	13	8.66	15.811	507.0	126750.00	14635.83	2661.06	0.2361	45.6	>2000
VG03650660	3	65	0.6	60	13	11.26	17.197	585.0	173013.75	15367.62	3841.91	0.5123	47.2	>2000
VG03650860	3	65	0.8	60	13	13.25	17.934	663.0	213238.24	16099.41	4953.67	0.8291	51.1	>2000
VG03651060	3	65	1.0	60	13	14.81	18.343	741.0	249330.59	16831.20	6011.14	1.0800	54.8	>2000
VG03800460	3	80	0.4	60	16	10.66	19.460	624.0	236307.69	22170.25	4030.95	0.1559	56.2	>2000
VG03800660	3	80	0.6	60	16	13.86	21.166	720.0	322560.00	23278.76	5819.69	0.3383	58.2	>2000
VG03800860	3	80	0.8	60	16	16.30	22.073	816.0	397552.94	24387.28	7503.78	0.5475	62.9	>2000
VG03801060	3	80	1.0	60	16	18.23	22.576	912.0	464842.11	25495.79	9105.64	0.7131	67.6	>2000
VG03350470	3	35	0.4	70	7	9.47	14.623	311.8	66676.82	7042.24	1726.81	0.3807	35.9	>2000
VG03350670	3	35	0.6	70	7	11.87	15.477	373.2	89358.62	7534.27	2468.54	0.8291	36.6	>2000
VG03350870	3	35	0.8	70	7	13.59	15.840	434.6	109042.44	8026.29	3161.09	1.3887	37.8	>2000
VG03351070	3	35	1.0	70	7	14.88	15.986	496.0	126751.34	8518.32	3817.70	2.0014	39.4	>2000
VG03500470	3	50	0.4	70	10	13.53	20.891	445.4	194393.05	14371.92	3524.10	0.1867	51.2	>2000
VG03500670	3	50	0.6	70	10	16.95	22.110	533.1	260637.38	15376.06	5037.83	0.4068	52.3	>2000
VG03500870	3	50	0.8	70	10	19.41	22.628	620.9	317907.98	16380.19	6451.20	0.6820	54.0	>2000
VG03501070	3	50	1.0	70	10	21.26	22.837	708.6	369537.44	17384.33	7791.22	0.9939	56.2	>2000
VG03650470	3	65	0.4	70	13	17.58	27.158	579.1	427081.53	24288.54	5955.73	0.1105	66.8	>2000
VG03650670	3	65	0.6	70	13	22.04	28.744	693.1	572620.32	25985.53	8513.93	0.2408	68.0	>2000
VG03650870	3	65	0.8	70	13	25.23	29.417	807.1	698443.84	27682.52	10902.53	0.4039	70.1	>2000
VG03651070	3	65	1.0	70	13	27.63	29.688	921.1	811873.76	29379.51	13167.16	0.5831	74.2	>2000
VG03800470	3	80	0.4	70	16	21.64	33.425	712.7	796233.93	36792.12	9021.70	0.0729	82.0	>2000
VG03800670	3	80	0.6	70	16	27.12	35.377	853.0	1067570.70	39362.70	12896.84	0.1590	83.7	>2000
VG03800870	3	80	0.8	70	16	31.06	36.206	993.4	1302151.10	41933.29	16515.08	0.2668	86.3	>2000
VG03801070	3	80	1.0	70	16	34.01	36.539	1133.7	1513625.37	44503.88	19945.51	0.3850	89.7	>2000
VG05350445	5	35	0.4	45	7	2.02	4.418	428.0	8352.65	4127.07	539.71	1.1766	1036.7	>2000
VG05350645	5	35	0.6	45	7	2.72	4.934	477.5	11623.04	4271.44	786.46	1.4013	1032.4	>2000
VG05350845	5	35	0.8	45	7	3.29	5.248	527.0	14516.39	4415.80	1021.37	1.5464	1037.6	>2000
VG05351045	5	35	1.0	45	7	3.76	5.451	576.5	17129.81	4560.17	1246.38	1.6389	1056.2	>2000
VG05500445	5	50	0.4	45	10	2.89	6.311	611.4	24351.74	8422.59	1101.45	0.7981	24.8	>2000
VG05500645	5	50	0.6	45	10	3.89	7.048	682.1	33886.42	8717.22	1605.03	1.2125	1294.5	>2000
VG05500845	5	50	0.8	45	10	4.70	7.498	752.8	42321.83	9011.84	2084.43	1.3315	1302.5	>2000
VG05501045	5	50	1.0	45	10	5.37	7.787	823.6	49941.13	9306.47	2543.64	1.4126	1330.8	>2000
VG05650445	5	65	0.4	45	13	3.76	8.204	794.8	53500.77	14234.18	1861.45	0.4734	32.3	>2000
VG05650645	5	65	0.6	45	13	5.05	9.163	886.8	74448.46	14732.10	2712.49	0.9681	57.1	>2000
VG05650845	5	65	0.8	45	13	6.11	9.747	978.7	92981.06	15230.02	3522.69	1.0689	53.9	>2000
VG05651045	5	65	1.0	45	13	6.98	10.123	1070.6	109720.66	15727.94	4298.75	1.2160	1664.3	>2000
VG05800445	5	80	0.4	45	16	4.63	10.098	978.3	99744.73	21561.83	2819.72	0.3119	39.8	>2000
VG05800645	5	80	0.6	45	16	6.22	11.277	1091.4	138798.77	22316.08	4108.86	0.6392	70.2	>2000
VG05800845	5	80	0.8	45	16	7.51	11.996	1204.5	173350.22	23070.32	5336.15	0.7057	66.6	>2000
VG05801045	5	80	1.0	45	16	8.59	12.460	1317.7	204558.86	23824.57	6511.72	0.8106	63.4	>2000
VG05350460	5	35	0.4	60	7	4.52	8.423	469.0	33276.12	7355.44	1290.43	0.8135	24.5	>2000
VG05350660	5	35	0.6	60	7	5.90	9.190	539.0	45519.89	7709.07	1865.10	1.7605	25.6	>2000
VG05350860	5	35	0.8	60	7	6.97	9.605	609.0	56161.03	8062.70	2406.78	2.9196	1266.5	>2000
VG05351060	5	35	1.0	60	7	7.81	9.840	679.0	65749.03	8416.32	2922.33	3.0571	1261.2	>2000
VG05500460	5	50	0.4	60	10	6.46	12.033	670.0	97014.93	15011.11	2633.53	0.3989	35.1	>2000
VG05500660	5	50	0.6	60	10	8.44	13.128	770.0	132711.04	15732.79	3806.32	0.8637	36.5	>2000
VG05500860	5	50	0.8	60	10	9.95	13.721	870.0	163793.10	16454.48	4911.79	1.3927	39.3	>2000
VG05501060	5	50	1.0	60	10	11.16	14.058	970.0	191688.14	17176.17	5963.95	1.8038	42.4	>2000
VG05650460	5	65	0.4	60	13	8.40	15.643	871.0	213141.79	25368.77	4450.66	0.2361	45.6	>2000
VG05650660	5	65	0.6	60	13	10.97	17.067	1001.0	291566.15	26588.42	6432.68	0.5113	47.5	>2000
VG05650860	5	65	0.8	60	13	12.94	17.837	1131.0	359853.45	27808.08	8300.92	0.8247	51.1	>2000
VG05651060	5	65	1.0	60	13	14.51	18.275	1261.0	421138.85	29027.73	10079.07	1.0705	55.1	>2000
VG05800460	5	80	0.4	60	16	10.34	19.253	1072.0	397373.13	38428.43	6741.83	0.1559	56.1	>2000
VG05800660	5	80	0.6	60	16	13.50	21.005	1232.0	543584.42	40275.95	9744.18	0.3376	58.5	>2000
VG05800860	5	80	0.8	60	16	15.93	21.954	1392.0	670896.55	42123.48	12574.17	0.5446	62.9	>2000
VG05801060	5	80	1.0	60	16	17.86	22.492	1552.0	785154.64	43971.00	15267.71	0.7068	68.0	>2000

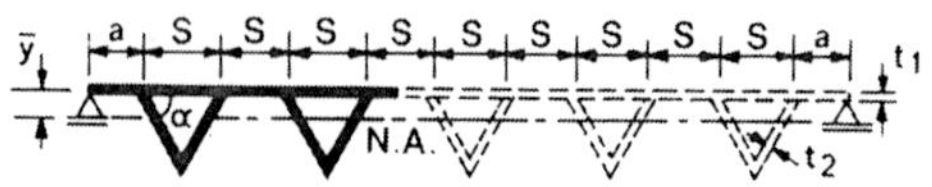

V-PLATE   n V-stiffeners
edges pinned
free in-plane

CODE	n	S	t_2	α	a	$\bar{y}$	rad. gyr.	A	I_{NA}	Z_{plate}	$Z_{stiff.}$	σ_{cr}/E	L_1	L_{euler}
To scale multiply by	−	t_1	t_1	−	t_1	t_1	t_1	t_1^2	t_1^4	t_1^3	t_1^3	10^{-3}	t_1	t_1
VQ05350470	5	35	0.4	70	7	9.22	14.509	533.7	112350.14	12185.82	2891.07	0.3805	35.9	>2000
VQ05350670	5	35	0.6	70	7	11.60	15.405	636.0	150925.41	13005.87	4137.61	0.8274	36.7	>2000
VQ05350870	5	35	0.8	70	7	13.33	15.798	738.3	184272.38	13825.91	5302.37	1.3820	38.0	>2000
VQ05351070	5	35	1.0	70	7	14.63	15.966	840.7	214300.36	14645.96	6406.82	1.9866	39.6	>2000
VQ05500470	5	50	0.4	70	10	13.17	20.728	762.4	327551.42	24869.03	5900.14	0.1866	51.2	>2000
VQ05500670	5	50	0.6	70	10	16.58	22.007	908.6	440015.77	26542.59	8444.11	0.4060	52.4	>2000
VQ05500870	5	50	0.8	70	10	19.04	22.569	1054.8	537237.26	28216.15	10821.17	0.6787	54.2	>2000
VQ05501070	5	50	1.0	70	10	20.90	22.809	1201.0	624782.38	29889.70	13075.14	0.9765	56.4	>2000
VQ05650470	5	65	0.4	70	13	17.12	26.946	991.1	719630.47	42028.65	9971.24	0.1104	66.9	>2000
VQ05650670	5	65	0.6	70	13	21.55	28.609	1181.1	966714.64	44856.97	14270.54	0.2403	68.2	>2000
VQ05650870	5	65	0.8	70	13	24.75	29.339	1371.2	1180310.26	47685.29	18287.78	0.4019	70.4	>2000
VQ05651070	5	65	1.0	70	13	27.17	29.651	1561.2	1372646.89	50513.60	22096.98	0.5785	73.3	>2000
VQ05800470	5	80	0.4	70	16	21.07	33.165	1219.8	1341650.62	63664.71	15104.36	0.0729	82.1	>2000
VQ05800670	5	80	0.6	70	16	26.52	35.211	1453.7	1802304.58	67949.02	21616.92	0.1587	83.7	>2000
VQ05800870	5	80	0.8	70	16	30.46	36.110	1687.6	2200523.81	72233.33	27702.20	0.2655	86.6	>2000
VQ05801070	5	80	1.0	70	16	33.44	36.494	1921.5	2559108.63	76517.64	33472.35	0.3822	90.1	>2000

10. References

1. Desai C.S. and Abel J.F., "Introduction to the finite element method" van Nostrand, 1972.

2. Marguerre K., "Zur Theorie der gekrümmten Platte großer Formänderung" Proc.Fifth Int. Congress Appl.Mechanics, Cambridge 1938.

3. Bulson P.S., "The stability of flat plates" Chatto and Windus, 1970.

4. Timoshenko S.P. and Gere J.M. "Theory of elastic stability" McGraw Hill, 1961.

5. Klöppel E.K. and Scheer I. "Beulwerte ausgesteifter Rechteckplatten Heft II" W.Ernst (Berlin),1968.

6. "Inquiry into the basis of design and method of erection of steel box girder bridges" Appendix I Her Majesty's Stationery Office, London 1973

7. DIN 4114,Teil 1 "Stabilitätsfälle im Stahlbau - Knicken von Stäben und Stabwerken" Entwurf Oktober 1978

8. Cheung Y.K."Finite strip method of structural analysis" Pergamon Press 1976.

9. Wilkinson J.H. and Reinsch C. "Handbook for automatic computation" Vol II Linear Algebra, Part 2 , Springer (Berlin) 1971.

10. DASt - Richlinie 012 "Beulsicherheitsnachweise für Platten" Deutscher Ausschuß für Stahlbau, Oktober 1978.

Michael Lawo und Georg Thierauf
Stabtragwerke

Matrizenmethoden der Statik und Dynamik, Teil 1: Statik. Mit 171 Abb. 1980. 475 S. DIN A 5. Kart.

Inhalt: Vorwort — Einführung — Stabwerke — System und Elementeinteilung — Stabelemente als starre Körper — Gleichgewicht von Stabwerken — Flexibilitätsmatrizen und Verformungen statisch bestimmter Stabwerke — Das Weggrößenverfahren — Sonderfälle der Belastung — Die Berechnung von Einflußlinien nach dem Weggrößenverfahren — Das Kraftgrößenverfahren — Das Verfahren der Übertragungsmatrizen — Iterative Verfahren der linearen Statik — Elastische Stabilität ebener Stabtragwerke — Nichtlineare Verformungen ebener Stabwerke im elastischen Bereich — Aspekte der Programmierung — Anhang — Literaturverzeichnis — Lehrbuchverzeichnis — Namen- und Sachregister.

Die Stoffauswahl des Buches orientiert sich an den üblichen Lehrinhalten der Baustatik: Ermittlung von Zustands-, Biege- und Einflußlinien, Weg- und Kraftgrößenverfahren, Drehwinkel- und Übertragungsverfahren, Verfahren von Kani, Einführungen in die Stabilitätsberechnung und die geometrisch nichtlineare Berechnung. Auf der Grundlage der Matrizenrechnung wird ein Einblick in die computerorientieren Verfahren gegeben. Alle Rechenverfahren werden an Beispielen erläutert. Das Buch ist vor allem für Studierende des Bauingenieurwesens an Hoch- und Fachschulen gedacht, eignet sich aber auch zum Selbststudium.

Christian Petersen
Statik und Stabilität der Baukonstruktion

Elasto- und plasto-statische Berechnungsverfahren druckbeanspruchter Tragwerke: Nachweisformen gegen Knicken, Kippen, Beulen. Mit 932 Abb. und 189 Tabellen. 1980. XVI, 960 S. 17,5 X 24,5 cm. Gbd.

<u>Inhalt</u>: Grundlagen der elasto-statischen Stabilitätstheorie — Grundlagen der plasto-statischen Stabilitätstheorie — Elasto-statische Berechnung von Stab- und Rahmentragwerken — Plasto-statische Berechnung von Stab- und Rahmentragwerken — Ausgewählte Stab- und Rahmensysteme — Ausgewählte Baukonstruktionen — Biegedrillknicken und Kippen — Kippbiegung Theorie II. Ordnung — Beulen und Beulbiegung Theorie II. Ordnung ebener und gekrümmter Flächentragwerke — Anhang — Literaturverzeichnis.

Nach Darlegung der Grundlagen und Nachweisverfahren der elasto- und plastostatischen Verzweigungs- und Verformungstheorie II. Ordnung (Knicken, Kippen, Beulen) werden die Berechnungsverfahren für den Stabilitätsnachweis der wichtigsten Tragsysteme des Konstruktiven Ingenieurbaues — Türme, Pfeiler, Rahmen, Fachwerke, Verbände, Gerüste, verspannte Masten, Bogen, Träger, Platten und Schalen — zusammengestellt und durch Diagramme für die Baupraxis aufbereitet. Das Werk ist Lehr- und Handbuch zugleich.